Servicing TV/VCR Combo Units

Other Books by Homer Davidson and PROMPT® Publications

SMD Electronics Projects

Troubleshooting Consumer Electronics Audio Circuits

Servicing TV/VCR Combo Units

By

Homer Davidson

PROMPT© Publications is an imprint of Sams Technical Publishing, 5436 W. 78th St., Indianapolis, IN 46268.

International Standard Book Number: 0-61224-0

Library of Congress Catalog Card Number: 2001088013

Acquisitions Editor: Alice J. Tripp
Editor: Cricket A. Franklin
Assistant Editor: Kim Heusel
Typesetting: Cricket A. Franklin
Cover Design: Christy Pierce
Graphics Conversion: Christy Pierce
Illustrations: Courtesy the author

PRINTED IN THE UNITED STATES OF AMERICA

9 8 7 6 5 4 3 2 1

Contents

Part II: Reparing VCR Circuits and Mechanical Movements 165

Chapter 8: Servicing VCR Speed Problems 167

Introduction

There are millions and millions of TVs and VCR units now in service within American households. The combination of the TV and VCR as one unit has been readily available for only the last few years. The early TV/VCR combinations started out with a nine-inch TV screen and now come in sizes up to 27 inches. *Servicing TV/VCR Combo Units* contains 15 chapters. Part I covers troubleshooting TV circuits and includes different service problems within its seven chapters. Part II, which is comprised of eight chapters, covers repair of VCR circuits and mechanical movements.

The different electronic and mechanical problems found in *Servicing TV/VCR Combo Units* consist of actual service problems. And, several different case histories of the TV/VCR are found throughout each chapter.

There are many different TV/VCR service problems and every page in *Servicing TV/VCR Combo Units* will help you to solve them. This book is well illustrated with many photos, schematics, and drawings to help you locate the areas where servicing problems are found in TV/VCR combinations.

I would like to point out that I give you part numbers where I am able, but I ***strongly urge*** that you double-check the part number before you order by it. Part numbers have a way of changing over a period of time and there is always a possibility that the number listed in the book is wrong. So do yourself a favor and always double-check all part numbers.

Part I:

Troubleshooting the TV Circuits

Chapter 1: Troubleshooting Low-Voltage Power Supply Circuits

The TV/VCR combination chassis was first manufactured in a 9-inch TV chassis. It expanded to include 13- and 19-inch receivers, and can now be found in units as large as 27 inches. Today, a VCR mechanism can be found in a Panasonic 27-inch color stereo TV with a built-in DVD/video CD/CD player. The TV/VCR combo has followed yesterday's trail of combining the AM/FM radio chassis and phonograph into the TV console.

The hookup problem of connecting several cables to the existing VCR and TV has been solved. VCR and TV internal switching has eliminated many of the outside cables. You can now play your favorite VCR tapes or record directly from the TV program with ease. Besides easy operation, you can now record or play back tapes by pushing a few front buttons that might consist of:

- ✔ Power
- ✔ Stop
- ✔ Eject
- ✔ Rewind (REW)
- ✔ Play
- ✔ Fast Forward (FF)
- ✔ Record
- ✔ Volume up and down
- ✔ Channel up and down

Combining the TV and VCR into one unit has increased sales, and the TV/VCR combo is here to stay!

Servicing the TV/VCR combo should be a snap for electronics technicians who already have mastered the repair of the individual units. In this book, troubleshooting the TV chassis is covered in Part I, while the VCR mechanism is found in Part II. Regular service problems that can be found in TVs are also covered in the TV/VCR combo. Likewise, breakdown problems found in VCRs can be found in the TV/VCR combo.

Required Test Equipment

Most test equipment already on the electronics technician's service bench can be used to successfully service most TV/VCR combo units. Required test equipment includes:

- ✔ NTSC pattern generator
- ✔ AC milli-voltmeter (RMS)
- ✔ Alignment tape
- ✔ Blank tape
- ✔ DC voltmeter
- ✔ Oscilloscope
- ✔ Frequency counter
- ✔ DC power supply
- ✔ Variable isolation transformer
- ✔ ESR meter *(Fig. 1-1)*

Fig. 1-1. Checking for broken trace or PC wiring with an ESR meter.

In-circuit tests with the ESR (equivalent series resistance) meter can be used to quickly determine if a capacitor is defective. This meter can spot a defective capacitor that is open, leaky, or experiencing a loss of capacitance. Most capacitor testers can locate a defective electrolytic, but the ESR meter can also point out an ESR problem.

In addition to checking capacitors, the ESR meter can locate defective SMD chip components. A poorly soldered connection or broken foil or trace can easily be located with the ESR meter.

Obtaining schematic diagrams and exact service literature is a must in servicing a TV/VCR combination. Extra deck-extension cables are required when servicing one or two warranty products. For instance, when repairing the VCR deck of a Sears model 934.44727790, there are three types of U15 deck-extension cables (part #N1090XA) to tie the VCR deck to the main CBA TV chassis (*Fig. 1-2*). These deck-extension cables must be obtained from the manufacturer.

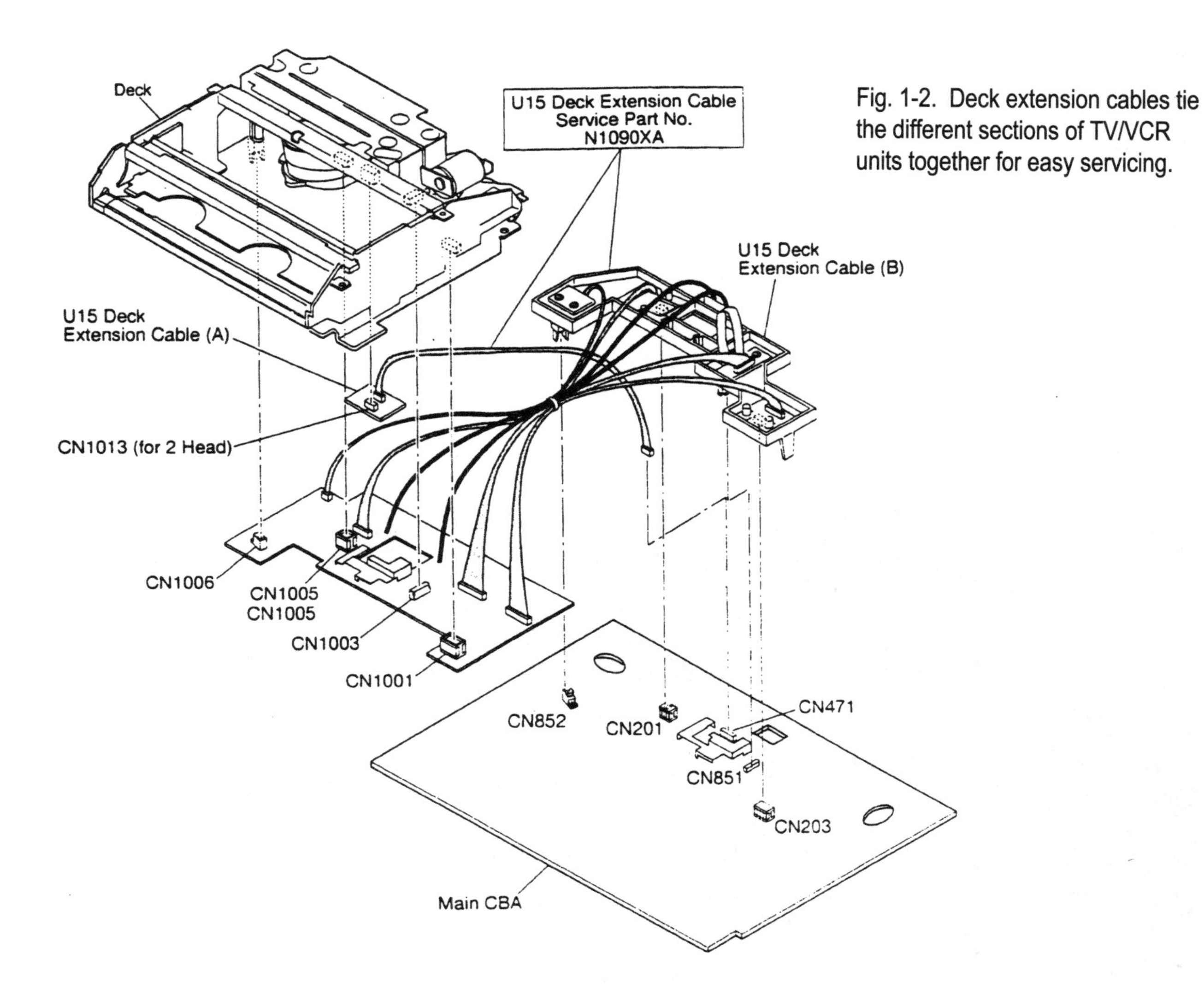

Fig. 1-2. Deck extension cables tie the different sections of TV/VCR units together for easy servicing.

Chip Components

Rectangular chip resistors, capacitors, and transistors are found in low-voltage and signal circuits of the TV/VCR chassis. These chip components are very tiny parts that may appear black or brown and are found on the foil or PC wiring side. Two or more SMD parts might be found in one component or package. A tie-in or solid conductor chip might look like an SMD resistor or capacitor. Once the chip parts are removed, they must not be used again; always new components.

Surface-mounted components such as resistors, capacitors, and transistors have numbers and letters stamped on the top of the part. White numbers on a black SMD resistor component indicate the chip's vaule. For instance, a chip resistor with numbers 472 equals a 4.7K-ohm resistor. The first two numbers indicate the resistance value and the last number indicates the total of zeros *(Fig. 1-3)*. Usually chip resistors have 1/10 or lower wattage ratings.

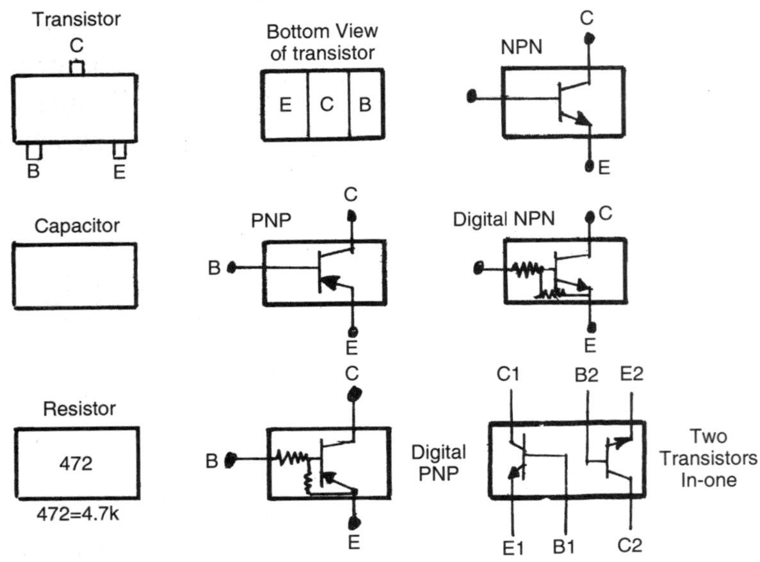

Fig. 1-3. SMD chip resistors, capacitors, and transistors are found in TV/VCR combos.

The miniature chip transistor might have three different legs or end-soldered terminals. A digital transistor has an internal resistor in series with the base terminal and a resistor that connects internally to the base and emitter terminals. Remember to check for a digital transistor when a test reveals a higher resistance measurement between elements. The digital transistor might be identified in the part lists as a resistor built-in transistor.

Removing Chip Parts

Grasp the leadless SMD component body with a pair of tweezers and alternately apply heat from a 30-watt or less soldering iron. A battery-powered or heat-controlled soldering iron is ideal. When the solder on both ends has melted, remove the chip with a twisting motion. Usually, the original leadless component has been glued to the PC board before the original board went through the solder bath *(Fig. 1-4)*.

Be extremely careful not to lift the chip off the board until the part is completely disconnected from the foil pattern. Inspect the trace or foil for areas that might be pulled up by the soldering process. If a trace is broken in the process, patch and repair with a bare piece of solid hookup wire.

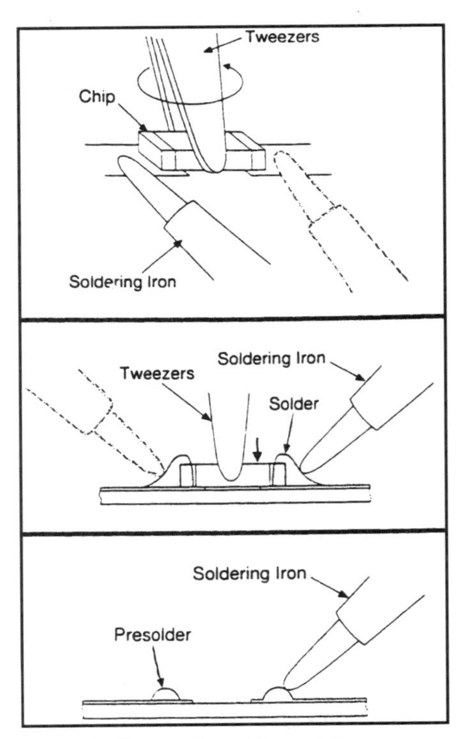

Fig. 1-4. Removing chip resistors or capacitors from PC board wiring.

To install a new chip, presolder or place a dab of solder on the tab or pads where the leadless chip was removed. Hold the component in position as solder is applied to each end or terminal. Make sure the SMD chip is lying flat. Do not use the iron tip on the chip to push it down; the heat can ruin the SMD component. Do not apply glue when replacing the new component on the PC board.

Be very careful when removing a flat-pack IC or microprocessor from the PCB. The original IC or flat-pack component is glued to the PC board. Place masking tape around the flat IC to protect other parts from heat damage when using the hot-air desoldering machine. Mark Terminal 1 on the PCB if a dot is not found here. Apply hot air to remove the flat-pack IC for about 5 or 7 seconds. Remove the flat-pack IC with tweezers while applying hot air on the IC terminals. Be careful not to pull up foil or traces when removing the IC from the PCB.

The IC terminals can be removed with desoldering braid and a soldering iron. Place solder flux on all pins and lay the mesh braid over the contacts and apply heat from the soldering iron. Go down each side and remove most of the excess solder (*Fig 1-5*). Apply the iron tip to each pin and lift the pin up with a sharp tool or tweezers. Be careful not to pull up a trace or foil from the PCB.

Replace the new IC component by removing excess solder over the foil or trace pin connections. The IC should lie flat on the PCB. Line up Terminal 1 with the dot that is found on the IC. Make sure Terminal 1 is in the right spot. Place a dab of solder on each corner of the IC terminals. Solder up all terminals. Recheck to make sure no pins are tied together by solder bridges or excess solder.

Before handling any critical semiconductors or microprocessors, make sure you are grounded. Electrostatic breakdown of the semiconductors may occur due to electrostatic charges. Be sure to wear an arm-grounding band to ground the workbench.

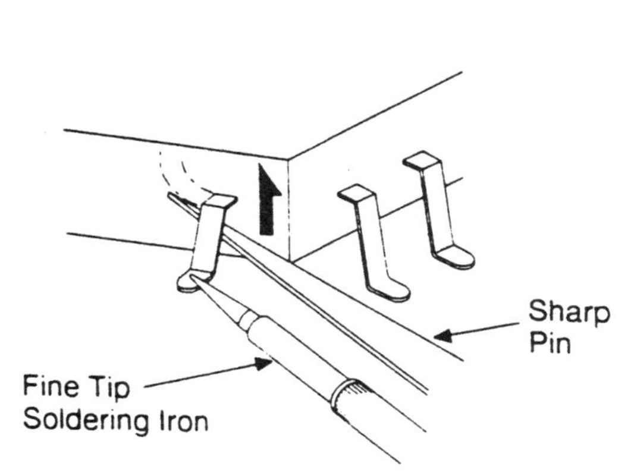

Fig. 1-5. Removing SMD flat pack IC pin terminals with a soldering iron and a sharp tool.

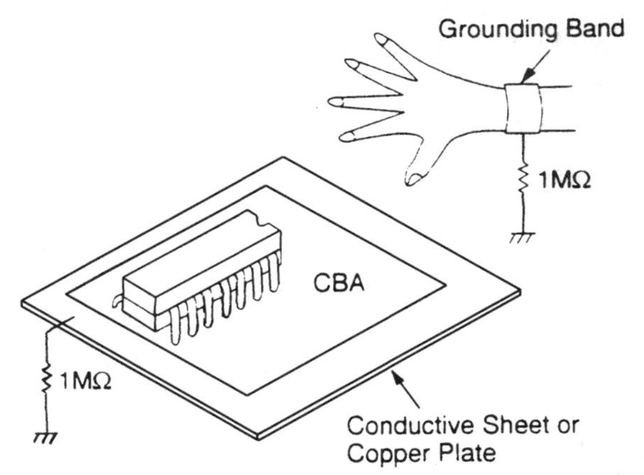

Fig. 1-6. Use a conductive mat and hand grounding band to prevent damage to SMD ICs and microprocessors.

Be sure to place a conductive sheet or copper plate with proper grounding (1 megohm) on the workbench or copper plate for grounding. The TV/VCR chassis rests on the conductive sheet and is grounded at all times. Because the static electricity charge in clothing will not escape through the body-grounding band, keep clothing away from contacting critical ICs and microprocessors *(Fig. 1-6)*.

Removing the Chassis

Follow the manufacturer's disassembly flow chart, if available.

Remove the back cover by unscrewing the six to eight screws holding the plastic rear cover to the portable TV.

Discharge the picture tube to the CRT metal band or CRT ground wire.

Remove the CRT HV cable.

If the HV CBA unit is mounted off to one side, remove screws holding the HV unit.

Pull the tray chassis backward. The main chassis board and tape deck are fastened to the plastic tray chassis assembly.

Usually, the tape deck is mounted on top of the main chassis board and must be removed to service the main chassis. Remove the four or six end screws that hold the tape deck to the main chassis. Unplug cables for each section. The tray chassis must be removed to get at the tape deck, main chassis, and CRT.

To replace a picture tube, remove the back cover, tray chassis, and main chassis to get at the picture tube. You must remove the HV-flyback assembly—if mounted on one side—before the tray can be removed. *Figure 1-7* is an example of how to remove the various chassis components (as found in a Zenith TVSA1320 color TV/VCR combination).

Remove the four screws holding the CRT to the front plastic cabinet and pull the CRT backward. Lay the front cabinet down on a blanket or soft pad so as not to damage the front picture tube area. Remove the four screws and pull the CRT up and away from the front cabinet. Replace the new tube and reverse the order of removal. Be very careful not to use large screwdrivers or tools near the CRT that could possibly crack the neck of the picture tube.

Low–Voltage Circuits

The raw AC power supplies voltage to the switching transformer and transistor. Most raw low-voltage circuits are quite common with one another, including a 4- to 6-amp fuse, bridge rectifiers, line-filter degaussing coil, a thermistor or posistor, isolation resistor, and main electrolytic capacitor. You might find a surge absorber across the AC

Fig. 1-7. Rermoving the various boards and chassis' in a TV/VCR combination.

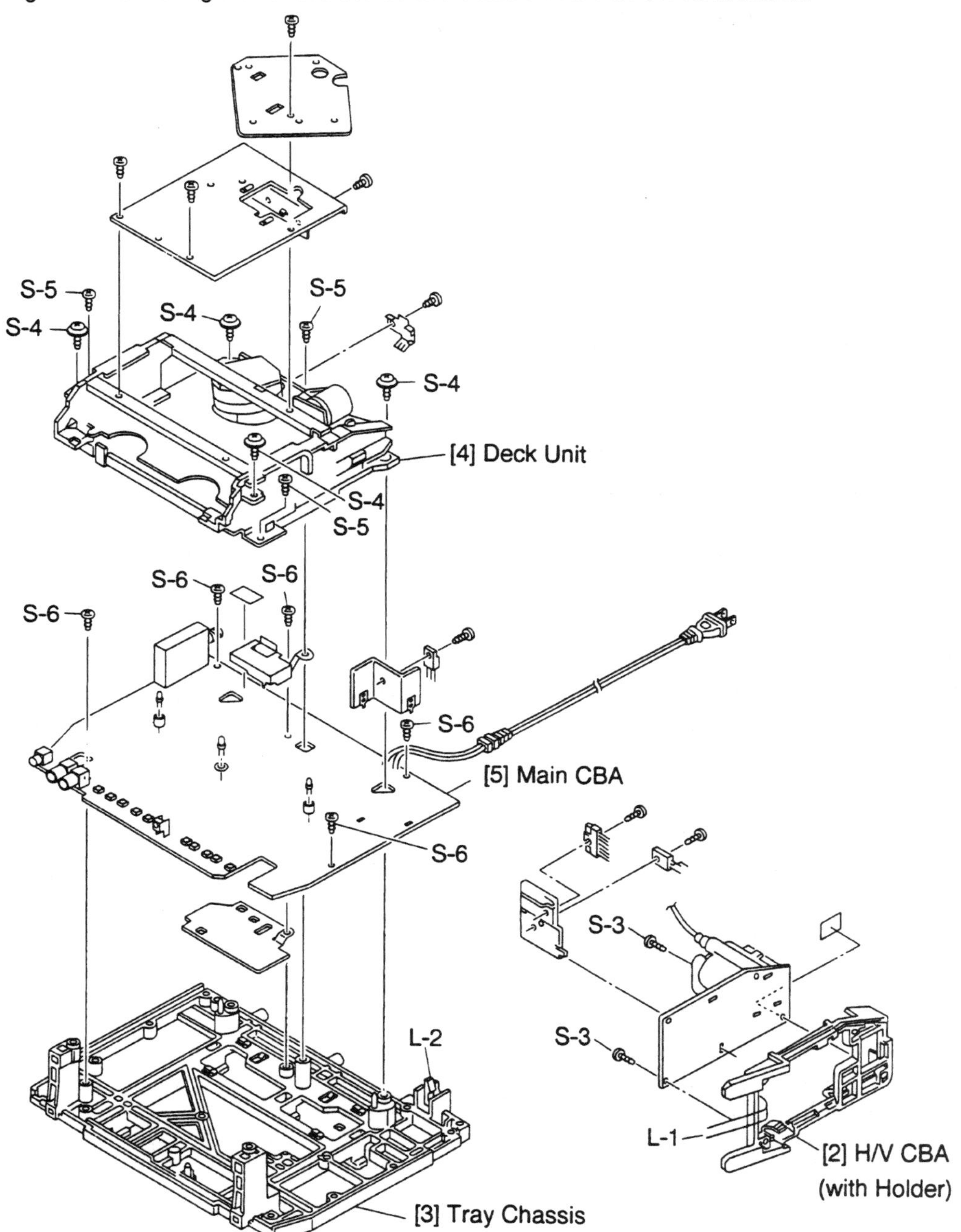

line after the main power fuse to prevent lightning and line-voltage surges from entering the low-voltage circuits.

A main fuse protects the low-voltage power-supply circuits. If the main fuse is blown, check to make sure that all components in the power-supply circuit are not defective before connecting the AC (DC) plug to the AC (DC) power supply. You might cause some components in the power supply to fail.

The line-filter component prevents any power-line noise from entering the chassis. The 117V - 120V AC power-line voltage is fed to the degaussing coil and the bridge rectifier circuits *(Fig. 1-8)*. A thermistor or posistor is found in series, with one line applied to the degaussing coil. When AC is applied to the TV, the line voltage is found across the degaussing coil, creating a magnetic field to degauss the front of the CRT each time the set is turned on.

Fig. 1-8. Block diagram of the raw DC power supply found in TV/VCR combinations.

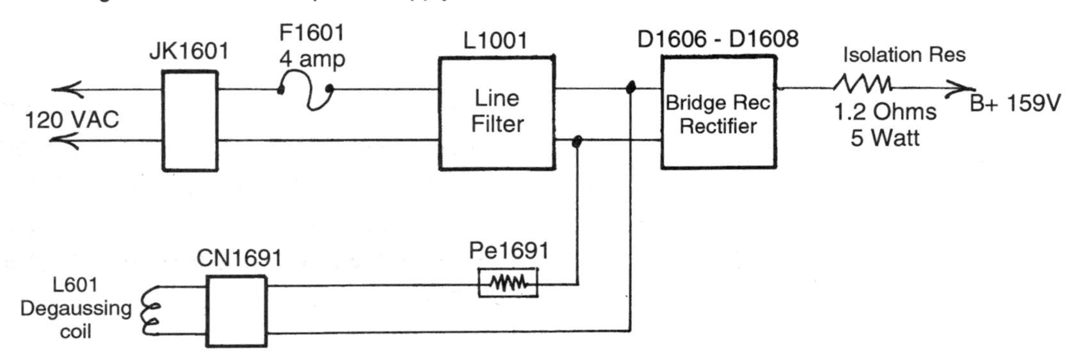

A thermistor is a temperature-sensitive resistor that changes value in response to temperature change. The thermistor has a low resistance when cold. When the thermistor gets warm, resistance increases, cutting off voltage to the degaussing coil. In some TV receivers, a relay might control the degaussing action of the degaussing coil and is controlled by the main computer or control IC. Remember, some TV/VCR combos have a separate power supply for the TV and the VCR circuits.

Degaussing Problems

Each time the TV is turned on or plugged in, the degaussing circuits should begin to function and demagnetize the picture tube. When the degaussing circuits are not functioning, different colored areas can be found on the raster or picture. A strong PM speaker placed near or on top of the TV can magnetize the front of the picture tube. Sometimes if a vacuum sweeper is run close by or turned off in front of the TV, color spots may appear on the screen. And sometimes when the TV's position in a room is changed, the magnetic north pole can magnetize the picture tube.

Some degaussing problems that can occur and their causes include:

- ✔ No degaussing action as a result of a badly soldered joint on the thermistor or posistor.
- ✔ Poor color with a halo effect can be the result of a badly soldered joint on a thermistor.
- ✔ A halo of colors throughout the picture might be caused by a bad posistor or thermistor.
- ✔ A color bull's-eye pattern over a normal black-and-white picture can be the result of a badly soldered connection on the posistor or thermistor.

Double-check the posistor or thermistor when the TV is hit by lightning or a power surge. Lightning or a power-line surge can also destroy the surge protector found across the AC line in some TV/VCRs. An ohmmeter continuity test of the degaussing coil and posistor can determine if the degaussing component is defective.

Rectifier Circuits

Four silicon diodes are found in most bridge-rectifier circuits or in one component (*Fig. 1-9*). The input AC circuits are marked with a cycle symbol and the rectifier output DC voltage is marked with positive and negative markings. When a bridge rectifier component is not available, four different silicon diodes can be wired in a bridge circuit. Sometimes only one or two silicon diodes are defective when there is an overloaded condition. The bridge rectifier part must be replaced when any of the diodes are shorted or leaky.

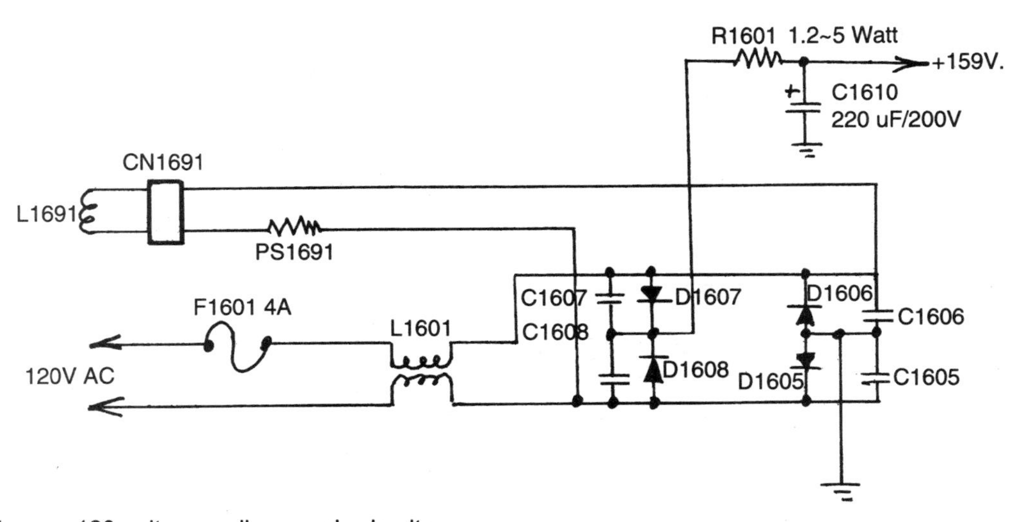

Fig. 1-9. The raw 120-volt power line supply circuits.

The DC-output raw voltage is filtered with a fairly large electrolytic capacitor. Here, a 220 µF - 200V electrolytic is found to filter out the ripple of the low-voltage circuits before it is applied to the switching transformer. A shorted main filter capacitor can damage silicon rectifiers and blow the main power fuse. The leaky filter capacitor can damage one or more silicon diodes. The open or loss of capacity of the main electrolytic can produce hum bars in the picture, hum in the speaker, and low output voltage.

Troubleshooting the Low-Voltage Power Supply

To determine if the main fuse is open when the TV/VCR chassis is dead, check for a B+ voltage across the main filter capacitor *(Fig. 1-10)*. Suspect shorted silicon diodes or a shorted isolation resistor or filter capacitor if no voltage is found at the main filter capacitor. Remember, the raw supply voltage is quite high compared to the B+ voltage applied to the horizontal output transistor (150V - 165V).

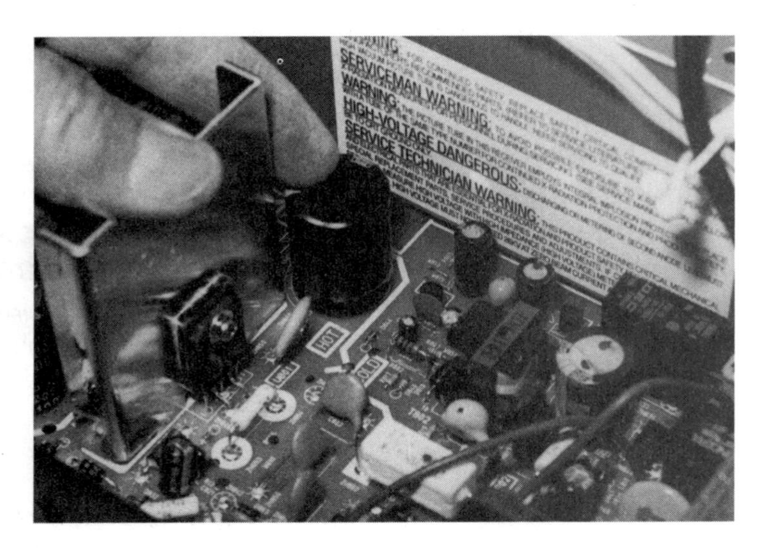

Fig. 1-10. The main electrolytic filter capacitor found in the main board of a TV/VCR.

Check each low-voltage silicon diode in the bridge circuit with the diode test of the DMM (digital multimeter). Replace low-voltage silicon diodes with 2.5-amp diodes. Often, a leaky or shorted diode opens the line fuse. An intermittently dead system may be the result of a defective low-voltage diode and damaged isolation resistor. Also, several silicon diodes can be destroyed when lightning strikes the TVs power source.

Some other problems to look for when troubleshooting the low-voltage power supply include:

- ✔ If the chassis is dead, but the line fuse is okay, the electrolytic capacitor may be bad.
- ✔ A dried-up or open electrolytic capacitor may result in low B+ voltage.
- ✔ If the chassis is dead but there is a humming sound, suspect the main filter capacitors.
- ✔ Intermittent start-up with a motorboat engine-type sound can be caused by a defective main electrolytic capacitor.

Filter Capacitor Problems

A bad filter capacitor can cause the following symptoms:

- ✔ Floating hum bars in the raster and hum in the speaker *(Fig. 1-11)*.
- ✔ Severe weaving in the picture.
- ✔ A shrunken picture that is pulled in on all four sides
- ✔ A severe hourglass pattern moving from bottom to the top of the screen.
- ✔ An intermittent TV operation with hum bars moving up the screen or raster.
- ✔ A picture in which both sides are pulled in, there are fast-moving wavy edges, the raster sides pull in 4 to 6 inches several minutes after, and severe hum bars from top to bottom of the screen.
- ✔ Very light horizontal and narrow lines moving up the screen.
- ✔ A loud popping noise in the audio with the top blown off or bulging sides.

Fig. 1-11. A dried-up filter capacitor in the low voltage supply might cause black hum bars in the raster.

Repairing Filter Capacitor Problems

Determine if the low-voltage filter capacitor is defective by shunting a new electrolytic across the one in the chassis. Choose a filter capacitor with the same capacity and working voltage or higher capacity and higher working voltage. Be sure to shut the set down. Discharge the original capacitor with a screwdriver or clip-wire across both terminals.

Observe correct polarity and clip a new filter across the old filter. Make sure the positive terminal is on the positive lead, as the new capacitor might explode.

Do not clip in the new capacitor while the set is operating, or you can damage other semiconductors in the TV.

Check the main filter capacitor with the ESR meter. Sometimes a defective capacitor will have the same capacity but still have internal resistance problems. The ESR meter can check for correct capacity and the equivalent series resistance of the suspected electrolytic with in-circuit tests. Replace the suspected electrolytic when it has lost capacitance and has poor ESR measurement.

Blown Line Fuse - Zenith TVSA1320

The 4-amp line fuse (F601) was blown in a 13-inch Zenith TV/VCR unit. After replacing the fuse, the switching transistor (Q601) was checked for leakage or shorted conditions. Q601 checked normal in the switching circuits. Since isolation resistor (R626, 1.2 ohms) was found normal, the low-voltage silicon diodes were tested. D609 and D612 were found to be leaky and were replaced with 2.5-amp rectifiers *(Fig. 1-12)*.

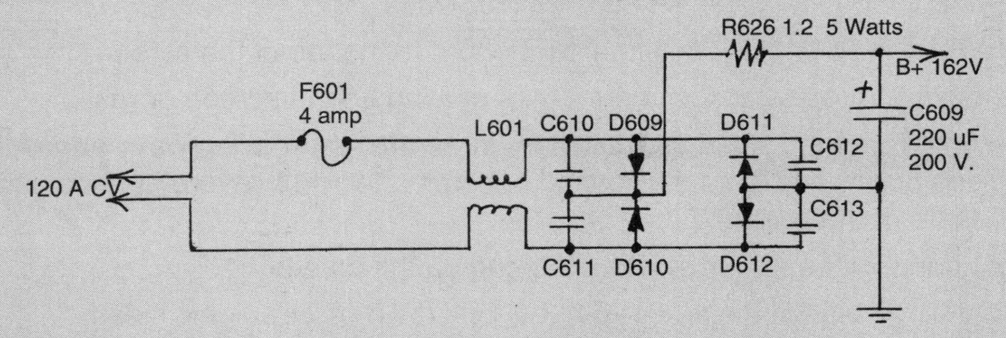

Fig. 1-12. The AC power line supply in the hot ground circuits of a Zenith TVSA1320 with open F601 and leaky silicon diodes.

Switching Power Supply Circuits

In early RCA TV chassis', the chopper power supply was introduced in the CTC124, CTC168, and CTC169 chassis. A switch-mode power supply (SMPS) was found in the Sylvania C8 and C5 TV chassis. You might find a similar switch-mode power supply or switching supply in the present-day TV/VCR combo. Although the SMPS circuits have been used in earlier TV chassis' for a number of years, servicing these switching circuits is a little more difficult.

Sometimes when making the various tests with test instruments, a slip of a test probe might damage other components within the switching power supply. Knowing where to place the ground probe when taking tests in the hot or cold ground circuits takes a little more service time. Always use a power-line isolation transformer to protect the TV chassis and test instruments.

Notice that in some low-voltage switching circuits no AC on/off switch is found. The switching power supply is on all the time, and switching voltage sources is accomplished with switching transistors and diodes. The switching power-supply circuits have a hot and cold ground system *(Fig. 1-13)*. The oscillator or switching transistor and primary winding of the switching transformer are above ground or on the "hot" ground side. All components on the secondary side of the switching transformer are in the "cold" power-supply circuit.

When taking waveform or voltage measurements on the hot side, make sure the test instrument is grounded to the hot side. Always plug the TV set into a variable isolation

Fig. 1-13. Switching transformer found in the TV/VCR power supply circuits.

transformer. Clip the test instrument to the negative terminal of the main filter capacitor for taking measurements in the primary circuits of the switching transformer. Likewise, when taking test measurements in the secondary (cold) side of the switching transformer, clip the test ground lead to the filter capacitor on the secondary side or common transformer ground terminal.

Some important items to take note of when working with switching-power supply follow.

- ✔ Extreme care must be exercised when servicing the switching power-supply circuits.
- ✔ Not adhering to the hot and cold systems can damage components in other circuits.
- ✔ Voltage measurements will not be the same as found on the circuit diagram.
- ✔ A slip of a test probe or clip can damage parts in either section .

Switching Transformer Circuits

The switched-mode transformer (T1601) is driven by AC power derived from the raw low-voltage source. Power-line isolation for the chassis is provided by the switching transformer and the optoisolator (IC1601). The raw DC voltage is applied to the collector terminal of switching transistor Q1601.

AC voltage is developed in the secondary circuits of the switching transformer (T1601). The secondary AC voltage is rectified by silicon diodes D1651, D1652, D1653, and D1654. DC voltage is filtered with electrolytic capacitors C1655, C1656, C1657, and C1658 *(Fig. 1-14)*.

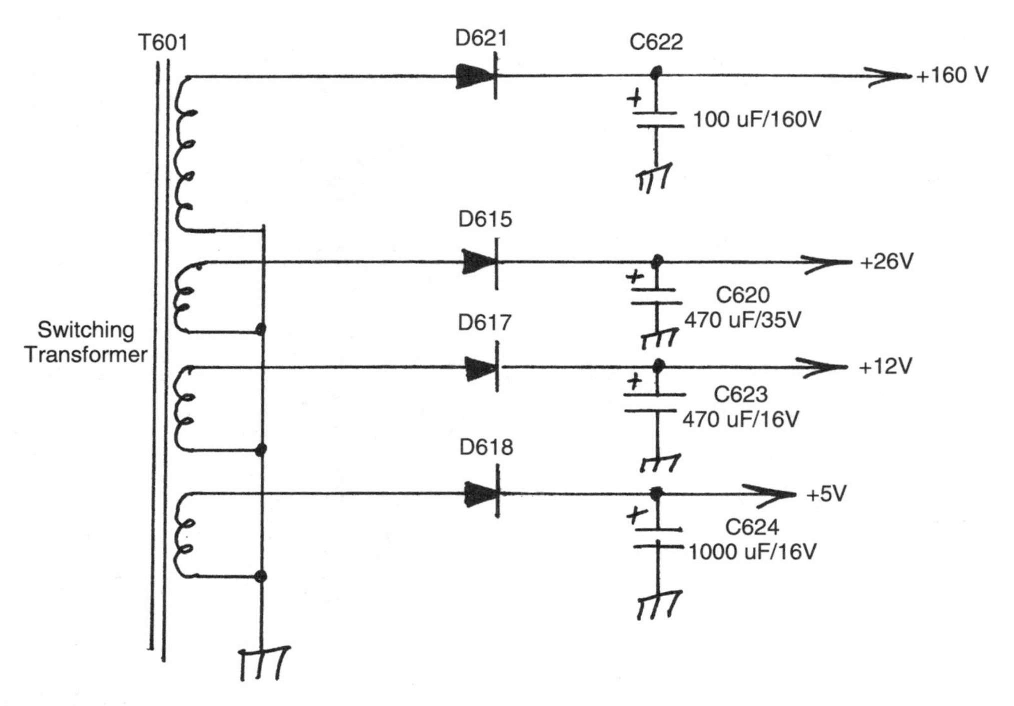

Fig. 1-14. Block diagram of secondary circuits in the low-voltage power supply of the cold ground circuits.

Troubleshooting the Switching Power Supply

Most service problems found in the switching low-voltage power supply are a leaky or shorted switching transistor, blown line fuse, defective IC regulator, transistors, and diodes. Since the switching transformer does most of the work – such as the horizontal output transistor in the horizontal TV circuits – check the power transistor for leakage or shorted conditions. A higher than normal collector voltage on the switching transistor might indicate an open switching transistor *(Fig. 1-16)*.

When the collector voltage of the switching transistor is high, check the voltage on the base terminal and waveform. Make a quick test on limiter Q1602, protection transistor Q1603, and error-voltage detector IC1601. Also do a quick test on all zener and switching diodes within the primary switching circuits with the diode test of the DMM.

The error-voltage detector optoisolator also can be tested with the diode test of the DMM. Place the red lead of the DMM to Pin 4 and black probe to Pin 3. An infinite measurement indicates the part is normal. If the resistance measurement is low, the optoisolator is probably leaky.

Reverse the test leads. If the same measurement or an even lower measurement results, suspect a leaky component *(Fig. 1-17)*. The TV might not start up with low B+ voltage and there may be hum in the audio circuits caused by an optoisolator.

In a Sears 934.44727790 9-inch TV/VCR power supply:

- ✔ The +117 output voltage feeds the horizontal, CRT, and TV circuits.
- ✔ A +33V source supplies voltage to the tuner and servo system control circuit.
- ✔ The 5V source supplies the audio, tuner, system control, and TV micon circuits.
- ✔ The VCR servo/system control circuits are supplied with the +6V source.
- ✔ A 12V source feeds the audio input IC and system control circuits.
- ✔ The 9V source feeds the tuner, video, and chroma circuits *(Fig. 1-15)*.
- ✔ The +6V source is fed to a +5V regulator IC671.
- ✔ Regulator IC670 (+6V) is fed from the AL+12V source.
- ✔ Q671, Q672, Q673, Q674, and Q670 provide switching of the various voltage circuits.

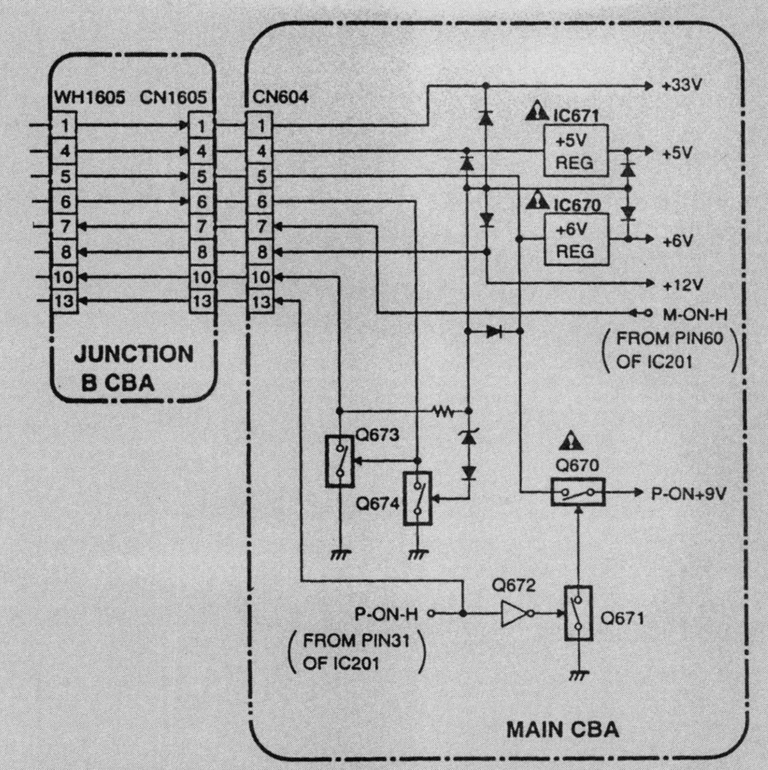

Fig. 1-15. The voltage regulator transistors found in a Sears 9-inch color TV/VCR.

Check the emitting diode across Pins 1 and 2. After placing the positive terminal at Pin 1 and the black probe at Pin 2, a regular diode measurement should result. Reverse the DMM test leads. Normal infinite measurement should be registered. If the measurement between Pins 1 and 2 have a low ohm measurement, then the diode portion is leaky or shorted. Replace IC1601 when any two terminals show a really low ohm measurement.

Badly soldered joints on the switching transformer (T1601) can cause the set to intermittently shut off or turn on by itself. An intermittent dead condition can result from bad soldered connections of any secondary silicon diode. A shorted silicon diode at either diode source will not blow the line face, but the picture might shut off. Check these soldered joints with the ESR meter. When the chassis shuts off intermittently, check R1511 and R1613 (1.2 ohms) for poorly soldered joints.

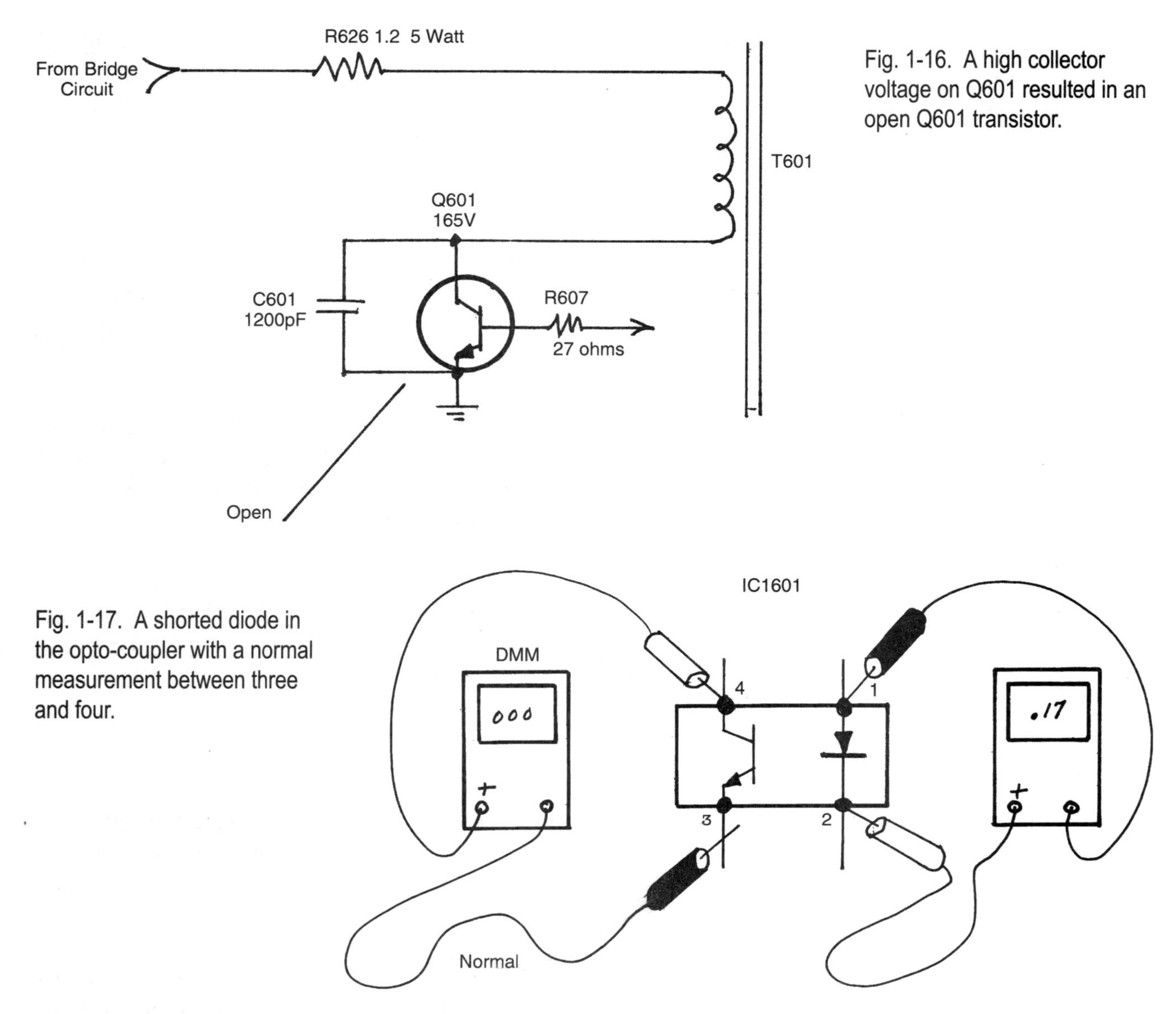

Fig. 1-16. A high collector voltage on Q601 resulted in an open Q601 transistor.

Fig. 1-17. A shorted diode in the opto-coupler with a normal measurement between three and four.

When the line fuse is open and the isolation resistor (R1601 - 1.2 ohms, 5W) is damaged in a dead TV chassis, suspect a shorted switching transistor (Q1601). When the limiter transistor (Q1602) and protection transistor (Q1603) become leaky or shorted, check for possible damaged resistors or diodes in the same circuits. A quick in-circuit diode or transistor test using the diode test of the DMM can quickly determine if a component is defective. If the chassis shuts down in one or two hours, a defective error-voltage detector IC may be to blame. Check the switching transistor for poorly soldered contacts and a loose mounting screw.

Do not overlook a shorted or leaky component in the secondary voltage circuits, which lowers the voltage source and shuts down the chassis. For instance, a leaky vertical output IC might lower or destroy components in scan-derived flyback circuits, thereby shutting down the chassis. Likewise, a leaky audio output IC can load down the +12V source and also cause the chassis to shut down. A defective tuner can load down the power supply and produce a dead/will-not-power-up symptom.

If the switching power supply is operating and one of the voltage sources is missing, go directly to the silicon diode that supplies the same source. Check the silicon diode in the secondary circuits with the diode test of the DMM. Discharge each filter capacitor to be tested. Connect the ESR meter across each filter capacitor and electrolytic in that voltage source.

A shorted or leaking horizontal output transistor may be the cause of a TV shutdown when the main line fuse is okay. The +117V source feeds the horizontal output transformer, horizontal output transistor, and horizontal drive transistor circuits. A defective component in the H.V block can shut the chassis down in the low-voltage power supply.

Badly soldered connections of components, and broken foil or traces within the primary or hot ground and in the cold ground section can also cause the chassis to go dead or shut down. Also check the solder connection on isolation resistor (R1602, 2.2 ohms, 25W) if the chassis goes dead and the line fuse checks out okay.

The intermittently dead problem also can be caused by a badly soldered joint on the primary winding of T1601 on Pins 4 and 6. If the chassis is dead but fuse is okay it might be due to a defective main filter capacitor (C1610, 220 μF, 200V). Locate broken connections and open traces or foil with the ESR meter.

Low–Voltage Regulators

Defective regulators within low-voltage sources can cause many different problems. The transistor regulator can go open with no voltage output or become shorted or leaky, thus producing high or low output voltage. The transistor regulator should be checked for possible burned, leaky, or shorted diodes.

A defective low-voltage regulator also can cause the chassis to go dead even though the line fuse is okay. The TV might contain either transistors or IC components as voltage regulators.

Check the solder contact of the regulator transistor's emitter terminal when the line fuse is okay but the chassis is dead. The TV might shut off at once because of an intermittent regulator transistor or poor terminal connection, while a defective voltage regulator might cause the TV to operate for a few minutes and then shut off, but not be in shutdown mode. An open regulator transistor might produce a no-raster, no-audio, and normal high-voltage problem, while a no-raster, relay-clicked-with-no-sound problem can be caused by a low-voltage regulator. Check for a damaged diode in the same regulator circuit *(Fig. 1-18)*.

Don't overlook transistor voltage regulators within the secondary flyback circuits. Usually, a defective regulator in the secondary circuits will not open the line fuse. An open secondary voltage regulator can cause the chassis to go dead, but the line fuse still

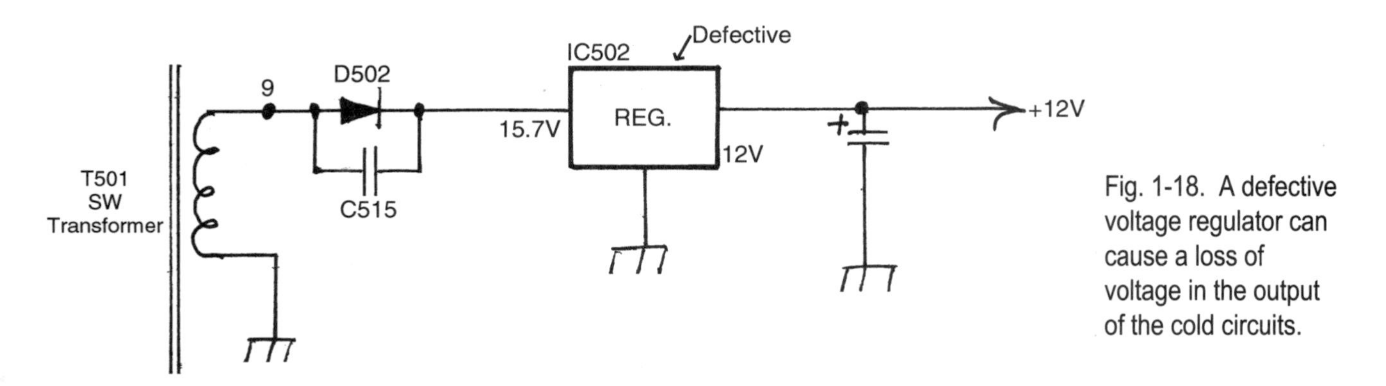

Fig. 1-18. A defective voltage regulator can cause a loss of voltage in the output of the cold circuits.

checks out okay. With a defective voltage regulator, the set might momentarily come on, but then turn off and appear dead. When the suspected regulator is removed from the circuit, the transistor tests normal. Replace the suspected intermittent transistor regulator, even if it tests good.

Be aware that low-voltage regulators can be damaged when lightning strikes and by high line-voltage surges.

Line–Voltage Regulators

You might find that line-voltage regulator IC components after the bridge rectifier in the AC-DC chassis are not found in switching power supplies. The line-voltage regulator might have a 115V - 135V DC output source. Notice that the last three digits in the regulator part number show the correct output-regulated voltage. For instance, the IC line regulator with a part number STR30130 has a regulated output voltage of 130V.

Keep in mind that the line-voltage regulator is susceptible to damage from storms or power surges. Also, line-voltage IC regulators can cause many different service

Intermittently Dead - Emerson VT1920

An Emerson 19-inch TV/VCR came in with an intermittent and dead operation. Sometimes the set would come on and other times not. A badly soldered contact on the collector terminal was found, causing the intermittent problem *(Fig. 1-19)*. A bad solder joint was found at R501, and also the 220 μF, 160V electrolytic was replaced. Problems on the electrolytic were revealed by the ESR meter.

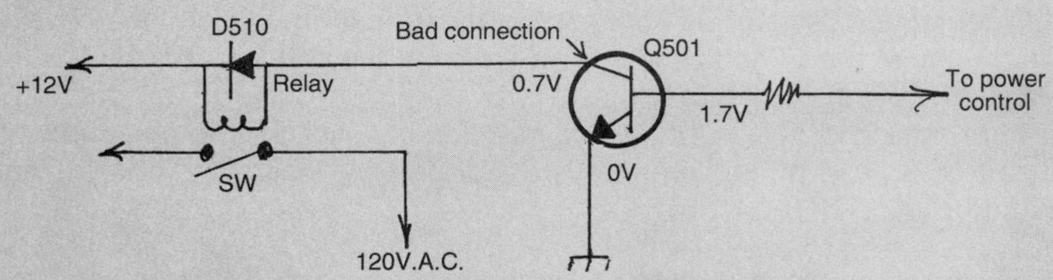

Fig. 1-19. A poorly soldered collector terminal connection of Q501 caused intermittent operation in an Emerson VT1920 chassis.

symptoms. With a defective line regulator, the TV might start up, turn off, and come right back on. Suspect the HV regulator when the set comes on with a loud hum and then shuts down.

A defective or bad line-voltage regulator may also be responsible when:

- ✔ The sides pull in and the vertical size of picture varies.
- ✔ Foldover occurs at the top of the picture.
- ✔ There is shrinkage of the picture on both sides when cold.
- ✔ After the chassis warms up, the picture pulsates.
- ✔ The TV might intermittently turn on, might not come on, and then stay on.
- ✔ There is no raster, the sound is weak, or the sound doesn't change with the channels.
- ✔ The chassis shuts down with a line voltage greater than 95V AC.
- ✔ The relay might click off and the chassis appears intermittently dead.
- ✔ The chassis is dead with an open fuse
- ✔ The TV does not turn off

Relay Problems

In some TV/VCR units, a relay might turn the set off and on or be found in the degaussing circuits. Inspect the foil or trace of PC wiring around the relay circuits after storm damage. Check open trace wiring with the ESR meter. See Figure 1-20 for a sample of relay circuitry.

Relay Does Not Click

Possible Causes:

- ✔ Open solenoid
- ✔ Bad solenoid connection
- ✔ Defective transistor driver
- ✔ Bad AC connection
- ✔ Open relay-driver transistor

Relay Does Not Click, Dead Chassis

Possible Causes:

- ✔ Open relay coil
- ✔ Electrolytic capacitor in relay circuits
- ✔ Damaged relay (from storms and high-voltage breakdown)

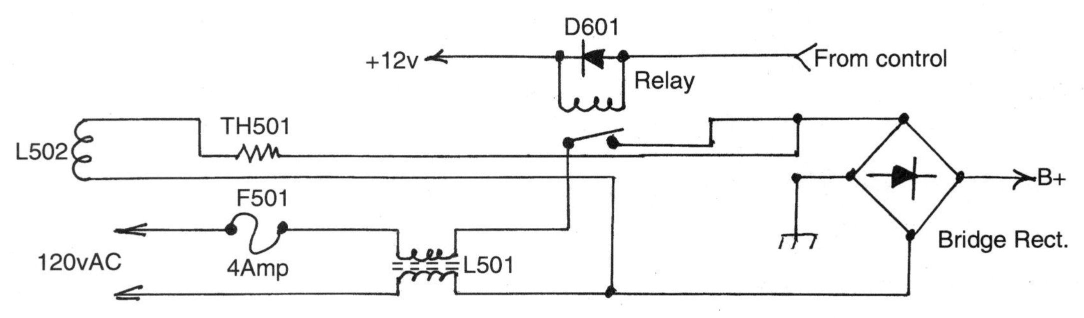

Fig. 1-20. Relay on and off circuits found in some TV/VCR chassis'.

Relay Does Not Click, No Shut Off

Possible Causes:

- ✔ Defective relay (may be stuck in one position)
- ✔ Defective relay driver transistor
- ✔ Shorted diode in the relay circuit

Intermittent Shut Off

Possible Causes:

- ✔ Poor relay contact
- ✔ Relay driver transistor (make sure it is not a shutdown problem first)
- ✔ Bad solder joint on AC side of relay

No Raster Shut Off, Audio Shut Off

Possible Cause:

- ✔ Bad relay

Relay Chatters

Possible Causes:

- ✔ Electrolytic capacitors in relay circuits
- ✔ Main filter capacitor (accompanied by low DC voltage)
- ✔ Badly soldered terminal on small resistors in relay circuit

Ticking Noise

Possible Cause:

- ✔ Large filter capacitor in the relay driver circuit

DC Power Adjustment

In several TV/VCR combo units, electrical adjustments are required after replacing circuit components and certain mechanical parts. It is important to perform these adjustments only after all repairs and replacements have been completed.

The purpose of DC voltage adjustment is to obtain correct operation. A sign of incorrect DC voltage might be a dark picture and the unit not operating correctly. Simply locate the positive (+) and ground connectors for accurate adjustments.

For instance, in a Sears TV/VCR combo model number 934.44727790, look for test points J660 (+117V) and J657 (GND) connection. Connect the DC voltmeter to these test points. Adjust VR911 so that the voltage on J660 becomes +117V, plus or minus 0.5V DC. J660 is located at the junction of L1651 (47 μH) and electrolytic C1691 (470 μF - 160V).

DC Power Circuits

Some small-screen TV/VCR units can operate within an automobile or with a portable battery supply. A separate DC input jack is provided to power the DC circuit. The DC power supply is switched into the TV/VCR circuits with switching transistors.

A 6-amp fuse (F901) protects the DC power-supply circuits. The 12V source is applied to the primary winding of T901 and switching DC-on transistor Q901 and switching control IC IC901 *(Fig. 1-21)*. D901, D902, and D903 provide proper DC polarity and with a reference voltage applied to T901. If at any time the battery polarity is switched and applied to JK901, D901 and D902 protect the positive applied voltage, and the 6-amp fuse will open. Electrolytic capacitors C900 (1000 μF - 25V) and C901 (3300 μF - 25V) help filter the DC-applied voltage.

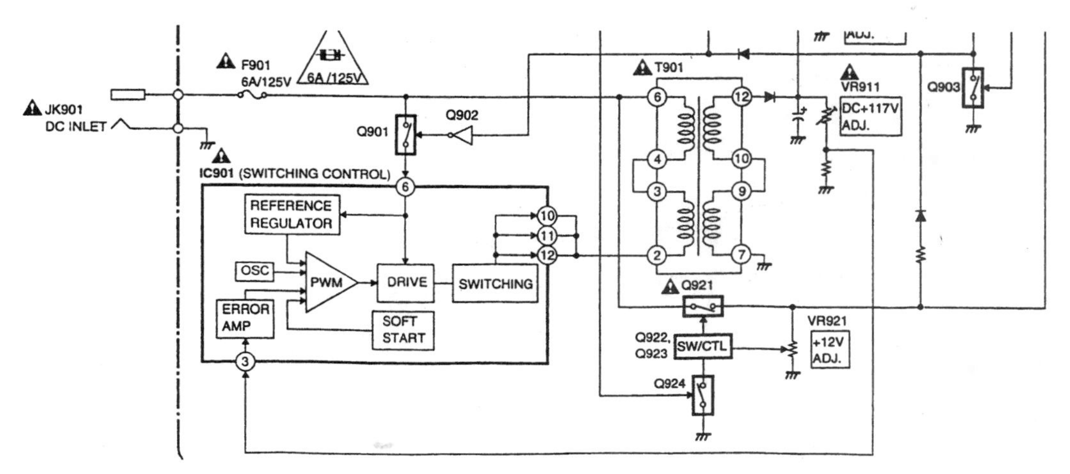

Fig. 1-21. Block diagram of the 12-volt DC power supply in a Sears TV/VCR.

The switching control IC901 provides an AC-output voltage on the secondary winding of switching transformer T901. The secondary voltage is rectified by silicon diode D911. C913 and C1691 provide filter action on the +117V source. VR911 (10K) adjusts the DC +117V. VR921 (500 ohms) adjusts the 12V DC source.

After making repairs in the DC power supply or within the TV/VCR unit, make the DC 117V and DC 12V adjustments. Supply a 13.2V DC source from the DC jack and connect the DC voltmeter to J660 and J657 (GND) in a Sears model number 934.44727790. Adjust VR911 so that the voltage of J660 becomes +117V plus or minus 0.5V DC.

To make the DC 12V adjustment (DC power), clip the DC voltmeter leads to J648 (+12V) and J645 (GND). *Adjust VR921 to read +12V plus or minus 0.5V DC on jack J648.* Jack J648 is located at the +12V source after D923. Both the + DC 117V and DC 12V sources should be adjusted after making repairs in the TV/VCR chassis.

Troubleshooting the DC Power Supply

Connect a 13.2V source to the input jack of a TV/VCR chassis from an external battery or power supply. If the 6-amp fuse is blown, check the components in the primary winding of switching transformer T901. Improper external battery polarity can cause F901 to open. A leaky or shorted D901, D902, and D903 can open the 6-amp fuse. A leaky or shorted switching control IC901, or switching DC-on transistor Q901 and Q902, can cause the fuse to open.

Most components that cause a dead DC power supply are an open 6-amp fuse, polarity diodes, switching control IC, 117V (D911) rectifier, and a shorted or open filter capacitor (C913). Check each diode with the diode test of the DMM. Defective transistors can be checked in circuit with the diode test of the DMM or a transistor checker. Test all filter and electrolytic capacitors in the secondary circuits with the ESR meter.

Continually Blows Fuses

- ✔ A shorted or leaky component in hot side of switching power supply *(Fig. 1-22)* (check each silicon diode in the bridge rectifier circuits with the diode test of the DMM then test the switching power transistor for shorted conditions between emitter and collector terminals)
- ✔ The shorted raw B+ main filter capacitor can cause the line fuse to open.
- ✔ A leaky line-voltage regulator can blow the fuse.
- ✔ When the main fuse is blown by lightning or an excessive power-line surge, look for a burned surge protector.
- ✔ Sometimes a shorted horizontal output transistor can blow the main fuse.

Fig. 1-22. The location of the power line fuse and DC power supply.

Intermittent Operation

Just about any component in the hot or cold chassis side of the power supply can cause intermittent operation. Causes include:

- ✔ Poor on/off switch contact.
- ✔ A switching transistor.
- ✔ Poorly soldered connections on the primary winding of the switching transformer.
- ✔ The intermittent limiter and protection transistor.
- ✔ A poor connection on the isolator resistor (1.2 - 2.5 ohms).
- ✔ Poor relay contacts.
- ✔ The line-voltage regulator. Poorly soldered connections on the regulator can produce the same results.
- ✔ A badly soldered joint on the thermistor.
- ✔ An open trace between the main filter capacitor and low-voltage source capacitors.
- ✔ Poor connections of the low-voltage diodes
- ✔ Switching diodes.
- ✔ Poor connections on silicon diodes in the secondary winding of the various low-voltage sources.
- ✔ Incorrect capacity in the electrolytic capacitors in the low-voltage sources.
- ✔ Main filter capacitors (check with ESR meter).

Intermittently Dead Symptom – Panasonic PV-M2021

In a Panasonic PV-M2021 TV/VCR, the TV sometimes would appear dead and at other times operate intermittently. The output-DC voltage was monitored at D1006 and at the switching transistor. When the set went dead, no voltage was found at D1006 and high voltage (+168V) was found at the collector terminal of switching transistor Q1001 (*Fig. 1 -23*). The collector voltage should be 135V. No voltage was measured on the emitter terminal of Q1001. Switching transistor Q1001 was replaced with a 25C4130 transistor and solved the dead/intermittent problem.

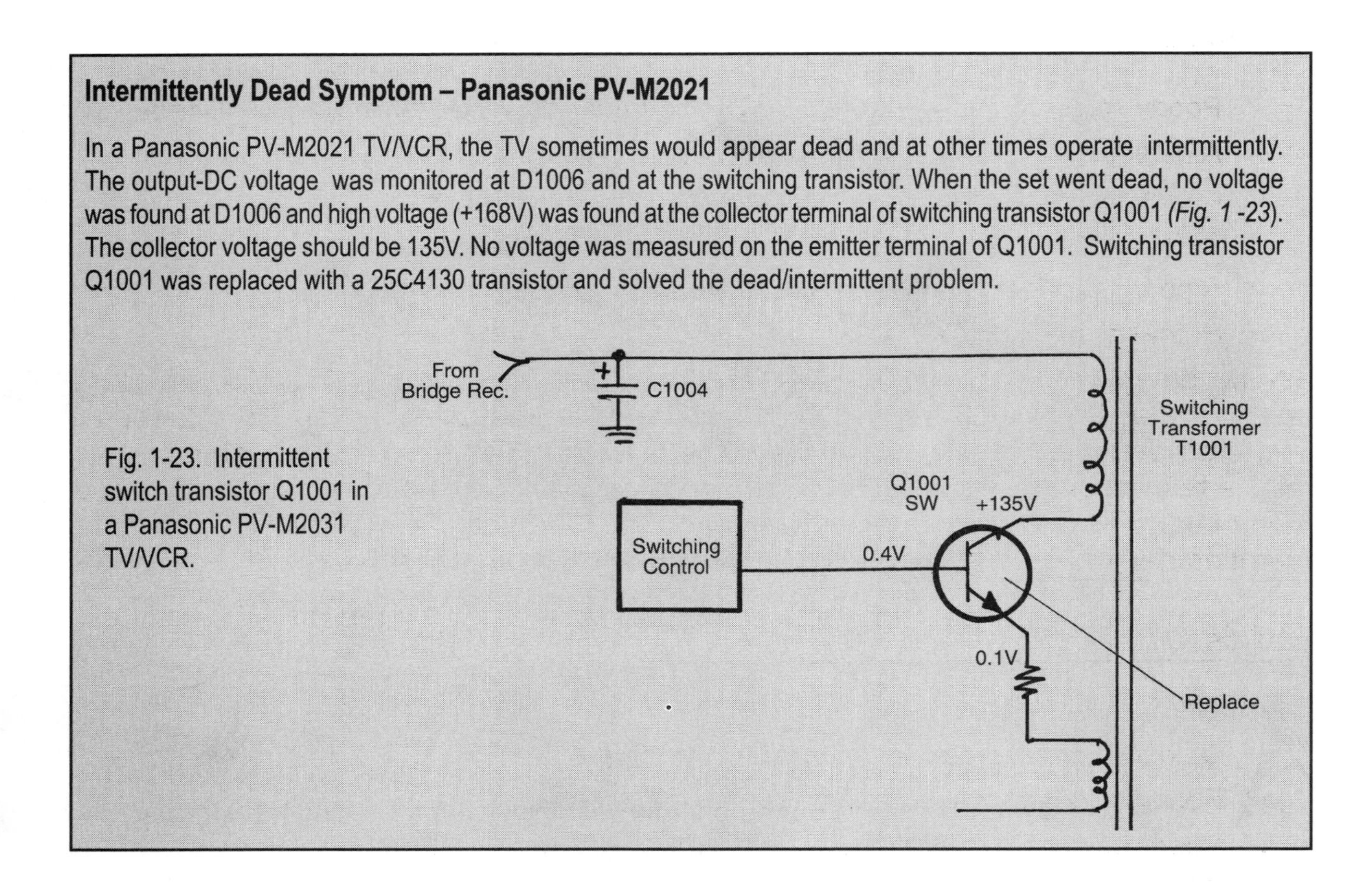

Fig. 1-23. Intermittent switch transistor Q1001 in a Panasonic PV-M2031 TV/VCR.

TV Shutdown

The most common reasons for chassis shutdown in the power-supply circuits are bad relays, diodes, capacitors, and poor contacts. Shutdown can be immediate or after any length of time from a few seconds to a few hours.

Immediate shutdown can be caused by a variety of problems including:

- ✔ Poorly soldered connections on an isolation resistor (2.2 to 4.7 ohms).
- ✔ Zener diodes in low-voltage sources.
- ✔ An intermittent relay-driver transistor.
- ✔ Poor contacts.
- ✔ A defective line-voltage regulator.
- ✔ Poor terminal connections.
- ✔ Bad electrolytic capacitors in low-voltage power supply.

Non-immediate shutdown or intermittent shutdown can be caused by:

- ✔ Poorly soldered contacts on the switching transformer and low-voltage rectifiers.
- ✔ A defective capacitor in the voltage sources (after four or five minutes).
- ✔ A bad error-detector IC optocoupler (after one to two hours)
- ✔ Defective low-voltage regulator transistors (check all voltage sources).
- ✔ A poor jumper connection or SMD solid-tie-in jumper
- ✔ Problems in the horizontal circuits.

Several different symptoms can point at a defective line-volt regulator. If the picture narrows when the channels are changed, if the TV operation becomes intermittent, or if shutdown occurs, these can all be signs of a defective line-voltage regulator or a fusible resistor tied to it. When the set comes on with a loud hum in the speaker and then shuts down, look for a defective line-voltage regulator. When the set comes on momentarily and then goes off, again look for a line-voltage regulator.

Shuts Off - Not Shuts Down

Possible Causes:

- ✔ Defective off/on switch or switching transistors
- ✔ Poorly soldered connections on flyback transformer
- ✔ Poorly soldered connections on switching transformer
- ✔ Defective line-voltage regulator (operates about 30 minutes and then shuts off)
- ✔ Poorly soldered joints on resistors tied to switching transformer

Shutdown - Variac Voltage Adjustment

A defective line-voltage regulator will cause the set to shut down at a line voltage greater than 100V AC. Often, the set will operate with lower line voltage and shut down with regular or greater line voltage. Use a Variac or variable line-voltage transformer for shutdown tests. Improper adjustment of the B+, as well as the 12V adjustment control being set too high causes line-voltage shut own. The chassis might also shut down with the line voltage over 90V because of a defective isolation resistor.

Check for defective silicon and zener diodes in the secondary low-voltage circuits when the line voltage is raised. Suspect secondary low-voltage regulator transistors for chassis shutdown. Line voltages greater than 102V AC may shut down the chassis with the original electrolytic capacitors of 1 μF to 4.7 μF. The capacitors should be checked for ESR problems.

TV Turns On By Itself

Possible Causes

- ✔ Badly soldered joints or traces to capacitors in low-voltage circuits (discharge each electrolytic and then check with the ESR meter)
- ✔ Defective AC relay
- ✔ Defective relay-driver transistor
- ✔ Defective microprocessor or TV Micon
- ✔ Defective switching transistor
- ✔ Poor ground connections
- ✔ Defective EEPROM or memory IC
- ✔ Leaky switching and zener diodes

Start-up Problems

When the TV won't start up, check all components in the hot power supply. Also keep in mind, a no-start up symptom can result from many different components in the power supply being damaged during a heavy lighting storm.

Possible Causes

- ✔ Leaky switching, limiter, and protection transistors
- ✔ Open start-up transistor
- ✔ Defective start-up transistor
- ✔ Defective error-voltage detector IC
- ✔ Poor terminal connections
- ✔ Defective relay
- ✔ Poorly soldered connections
- ✔ Defective line-voltage regulator
- ✔ Small electrolytics (1 µF to 100 µF) in the low-voltage sources (check all electrolytic capacitors in the hot primary circuit of the low-voltage power supply - if one (or more) is found to be defective, replace 1 µF electrolytics with 10 µF capacitors)
- ✔ Bad optocoupler IC (accompanied by low B+ voltage and hum in the speaker)
- ✔ Defective switching and zener diodes in low-voltage power supply (sometimes diodes test normal when checked out of the circuit, but replace them anyway)
- ✔ Defective voltage-regulator transistors and zener diodes in secondary voltage sources
- ✔ Main filter capacitor (accompanied by a "thump-thump" noise in the speaker and if it does start up, sides of picture are pulled in)

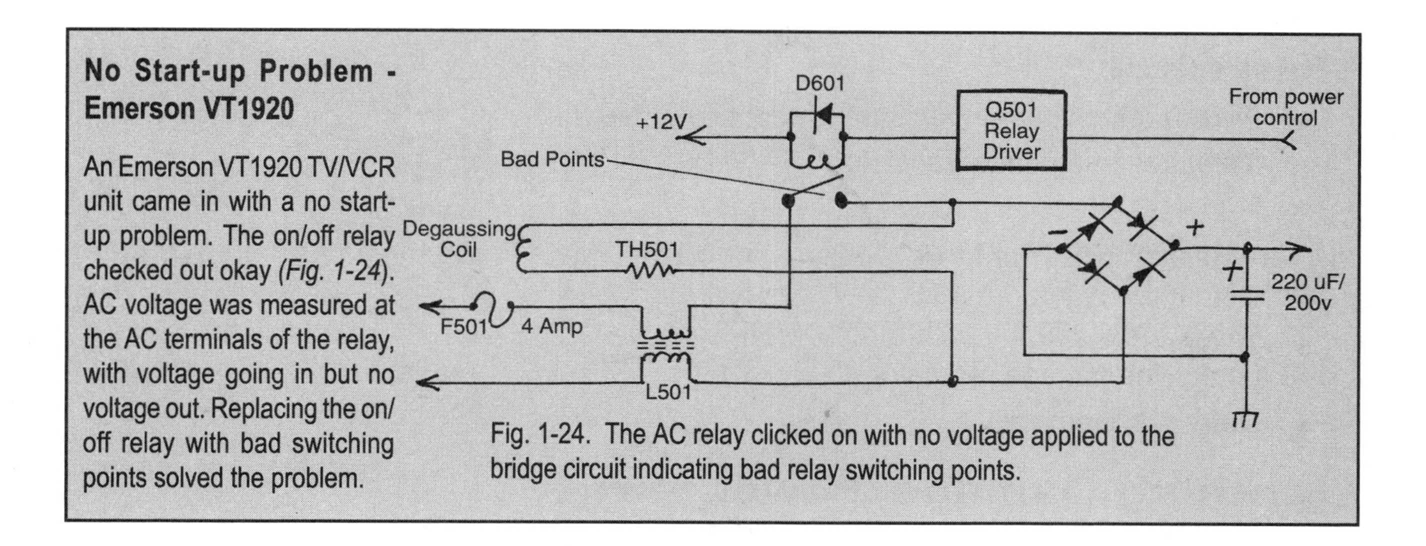

No Start-up Problem - Emerson VT1920

An Emerson VT1920 TV/VCR unit came in with a no start-up problem. The on/off relay checked out okay *(Fig. 1-24)*. AC voltage was measured at the AC terminals of the relay, with voltage going in but no voltage out. Replacing the on/off relay with bad switching points solved the problem.

Fig. 1-24. The AC relay clicked on with no voltage applied to the bridge circuit indicating bad relay switching points.

Poor Raster Width

Often, poor raster width in the TV is caused a variety of malfunctions, including low voltage, improper adjustments, or defective regulators.

Possible Causes

- ✔ Low voltage within horizontal circuits
- ✔ Improperly adjusted B+ control *(Fig. 1-25)*
- ✔ Defective line-voltage regulator
- ✔ Poorly soldered pin contacts
- ✔ Main filter capacitor(severe weaving and sides of picture pull in – check with ESR meter)
- ✔ Defective optoisolator voltage IC (poor focus and shrinkage on both sides of picture)

Fig. 1-25. Poor raster width can be caused by a defective filter capacitor, low voltage, or improper B+ adjustment.

TV Does Not Turn Off

Possible Causes

- ✔ Frozen (stuck in one position) points of AC relay
- ✔ Defective relay drive transistor

In sets with an EEPROM IC this is often a problem; expect to see the no turn off symptom.

- ✔ Leaky switching transistors and diodes
- ✔ Leaky line-voltage regulator IC
- ✔ Defective memory IC

Lightning Damage/Power Surge

Storm damage occurs more often in the spring and summer months than at any other time. Lightning damage can cause considerable harm to TVs. Often, the damage depends on how near the lightning strikes – whether it is directly outside or several blocks away. High line-voltage surges can occur at any time and can destroy the power-supply circuits in the TV. If a surge suppressor is located on the AC power line, the component might be blown apart, protecting other parts. Most commonly, lightning or power surges destroy fuses and silicon diodes. Sometimes, the fuse is not blown open.

Typical lightning damage consists of a blown fuse, one or two blown silicon diodes, and a damaged line-voltage regulator. In other cases, the fuse might be okay, but one or two silicon diodes are shorted. The main filter capacitor also can be damaged by lightning.

Extensive lightning damage might open the line fuse and destroy a couple of silicon diodes, the electrolytic filter capacitor, isolation resistor, and line-voltage regulator. Essentially, a lightning strike can destroy many parts of a TV.

- Switching and voltage-regulator transistors
- ICs
- Micro control processors
- EEPROMS
- Small resistors
- Silicon diodes
- Zener diodes
- Relays
- Relay-driver transistor
- Small electrolytics
- Tuners

Lightning Damage - Zenith TVSA1320

A 13-inch TV/VCR came in with storm or lighting damage. Upon inspecting the chassis, several pieces of foil were stripped from the board with black spots at the AC power cord. The F601 (4-amp) fuse was found open with shorted silicon diodes (D609 - D612). The power-surge protector was blown apart. The switching, limiter, and protection transistors and error-voltage detector were checked in the circuit. Switching transistor Q601 was replaced, and stripped wiring was repaired with regular hookup wire.

A power-line surge might destroy silicon diodes, the filter capacitor, and the isolation resistor. When 220V AC enters the home, expect extensive damage to the main line fuse, several silicon diodes, regulator transistors, switching diodes, horizontal output transistor, and open trace connections in the low-voltage power supply (*Fig. 1-26*). A broken power-line ground wire also can send high voltage into the TV receiver.

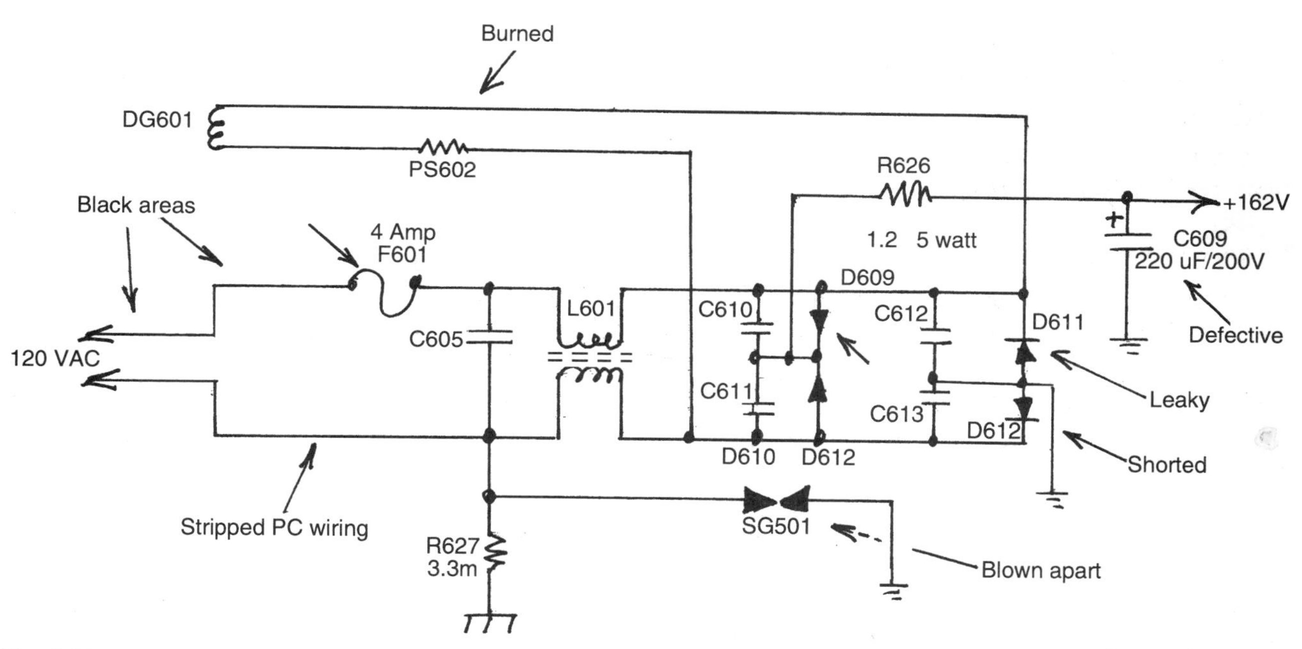

Fig. 1-26. Different components in the low-voltage power supply that can suffer lightning damage.

When a power surge or lightning strike is the cause of a TV problem, check the primary winding for open conditions of a standby or small AC transformer. Inspect the chassis for stripped foil or traces of the PC wiring. Check the PC wiring in the low-voltage power supply with the ESR meter. Repair the PC wiring with hookup wire. If extreme damage has occurred to several different parts and there are stripped traces over the whole TV/VCR chassis, then the chassis might be totaled.

Isolation Transformer

All AC-operated TV/VCR units should be plugged into a variable isolation AC transformer. The isolation transformer protects the TV and any test equipment clipped to the TV chassis *(Fig. 1-27)*. If an isolation transformer isn't used, components in the TV can be damaged, test equipment ruined, and the electronics technician might receive a terrible line-voltage shock. The isolation transformer is ideal in lowering or raising the AC input voltage in order to uncover improper line-voltage operation, intermittent functions, and flushing out tough problems.

Fig. 1-27. Plug the TV/VCR chassis into a variable isolation transfomer when servicing.

Chapter 2: Servicing the TV Tuner IF Circuits

The TV tuner picks up the radio frequency (RF) signal from the outside antenna, cable, or DIRECTV® dish and then amplifies it and selects the correct channel. The intermediate frequency (IF) from the tuner is amplified by transistors or a large IC component. For example, in the Emerson VT1920 TV the IF tuner signal is amplified by a preamp transistor then fed to the sawtooth filter and into the IF/video circuits. The tuner IF signal in a Zenith TVSA1320 model is fed to a SAW filter network and into the VIF amp/ video detector IC *(Fig. 2-1)*.

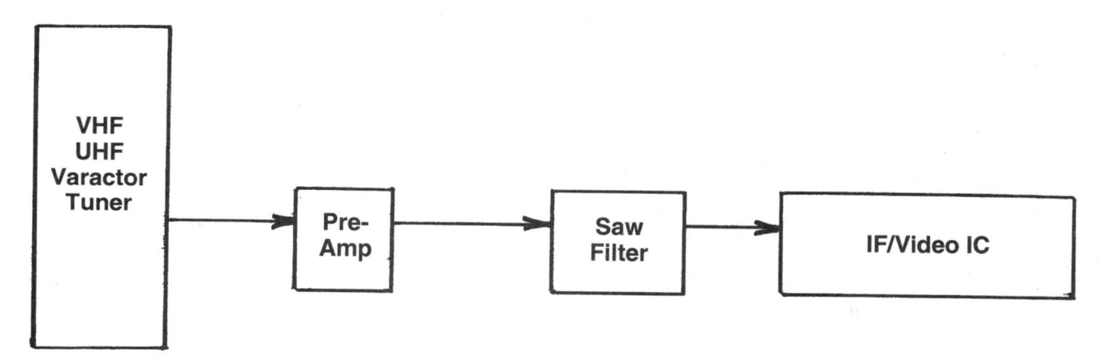

Fig. 2-1. Block diagram of the tuner, IF, and video amp IC circuits.

Most tuners found in TV/VCRs are tuned by varactor diodes. A varactor diode has a special type PN junction that has certain internal capacitance. By varying the reverse-bias voltage, the diode acts as a voltage-variable capacitor. The varactor diode might be referred to as a tuning diode. Varactor tuning is a method of changing the frequency of the RF, oscillator, and mixer circuits in the TV tuner.

Early varactor tuning was found in separate VHF and UHF tuners. Today, both tuners are included in one component. By changing the voltage on the varactor tuner, the entire VHF and UHF bands can be selected. You might find both tuners with components mounted directly on the PC board. For instance, Panasonic model PV-M2021 has a VHF/UHF tuner/TV demodulator circuit constructed on a single PC board, while in a Magnavox TV, the varactor tuner is found on the IF video and sync module *(Fig. 2-2)*.

It's much easier to service the UHF/VHF varactor tuner when it's mounted on the main PC chassis as compared to when the tuner components are spread out on the PCB. A separate defective tuner can easily be removed, sent in for repair, or interchanged.

Most manufacturers request that the tuner be replaced. Tuner components under a metal shield must be repaired in the field. A defective tuner with components mounted separately on a module or PC board creates many problems for the electronics technician. The tuner circuits might also include an SMD IC inside the tuner unit.

Fig. 2-2. The varactor tuner is mounted on a tuner / video module PCB.

The Varactor Tuner

The combination varactor tuner has separate UHF and VHF sections. Both have separate antenna inputs with several operating voltages. A typical varactor-tuner connection might contain a separate operating and tuning voltage. The tuning voltage will change as each channel is selected. As different voltages are applied to the varactor diode, the diode changes capacity, thereby tuning the various coils.

A typical varactor tuner has RF, oscillator, and mixer stages. Each circuit has its own varactor diode for selective tuning *(Fig. 2-3)*. Clock data, a frequency synthesizer, and BT control select the different tuning voltages. The terminal connections on a Zenith varactor tuner are AL + 33V, AL + 5V, IF, PLL-Data, PLL-CLK, PLL-ENA, AGC-IN, P-on + 9V, and lock.

In a Panasonic PV-M2021 TV, the tuner board has a frequency synthesizer (IC7601) that selects the various voltages for the different bands. B1 band selection is a 12-volt source that selects channels 2CH-6CH. When channels 2 through 6 are selected, terminals B2, B3, and BU are at zero volts.

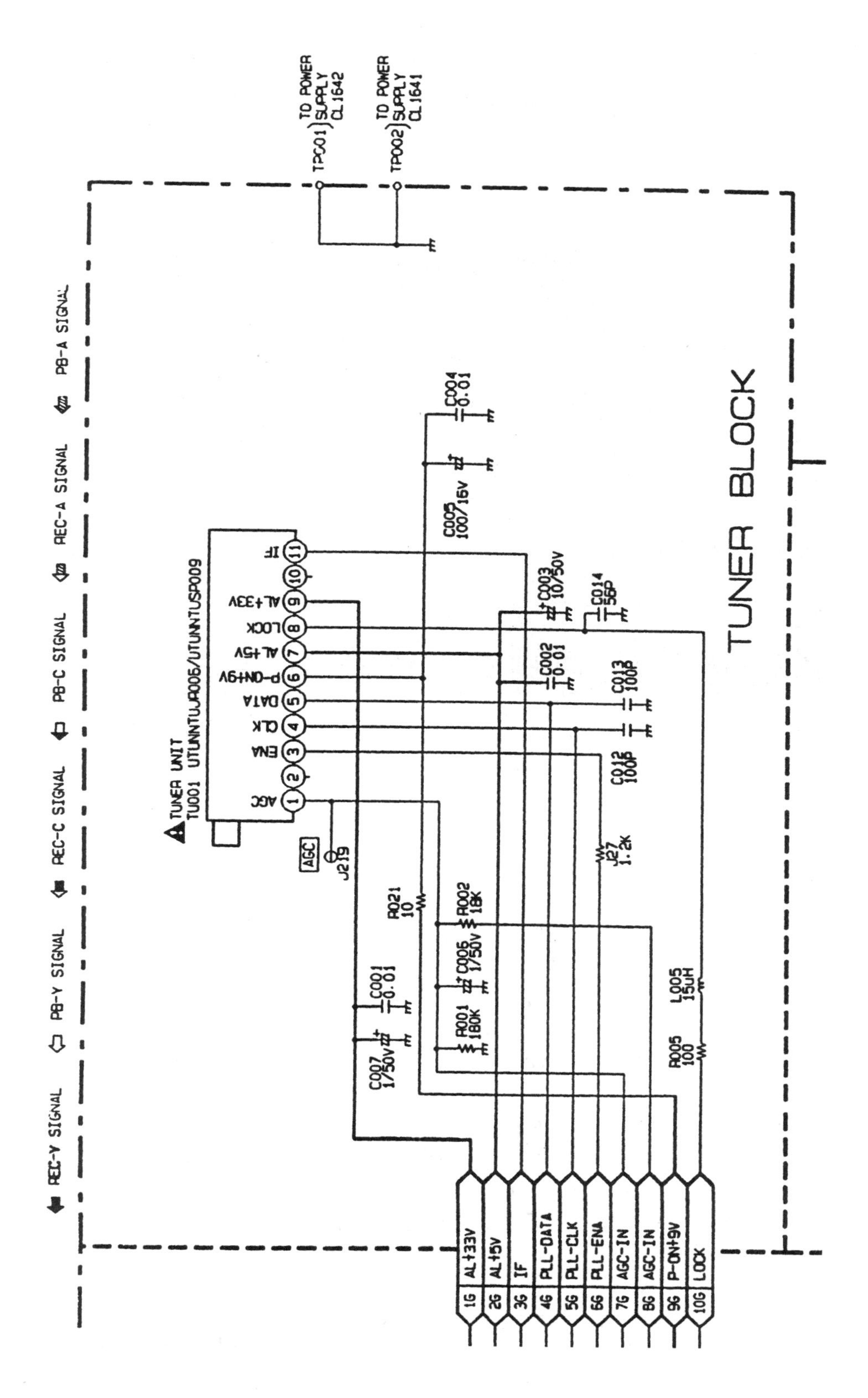

Fig. 2-3. Tuner block diagram of a Zenith 13-inch TV/VCR combination.

On Channels 7CH-13CH, B1 has 6 volts, B2 has 12 volts, B3 has 0 volts, and BU has 0 volts. The BU terminal represents the UHF bands 14CH-69CH. BU has an applied 12-volt source, while B1, B2, and B3 are at zero volts *(Fig. 2-4)*. Notice that each band has a changing voltage for each channel. By measuring the changing voltage applied to the tuning voltage terminals, you can determine if the tuner or synthesizer IC is defective. Check the tuning IC when the tuning voltage is missing or no change of voltage is found on the tuner.

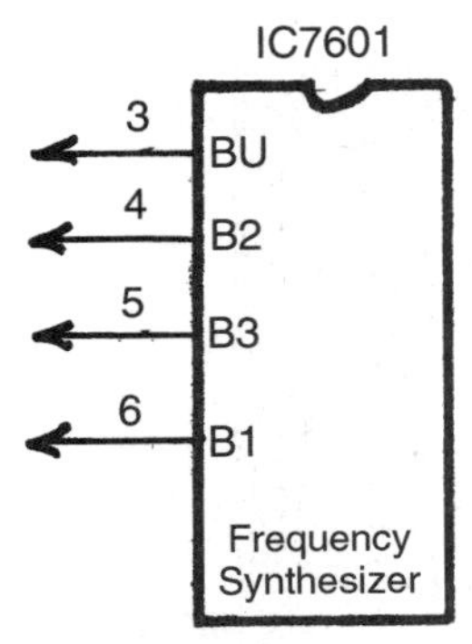

B1	B2	B3	BU	Channels
12v	0v	0v	0v	2 CH--6 CH
6v	12v	0v	0v	7 CH---13 CH
0v	0v	0v	12v	14 CH--69 CH

Fig. 2-4. The band selector voltage chart of a Panasonic PVM2021 frequency synthesizer IC applied to the VHF/UHF tuner.

Tuner Problems

There are many set conditions that can be caused by a tuner that does not function properly. For example, if the tuner is dead or has no action, there is no audio or video. An intermittently functioning tuner can cause sound problems and an intermittent picture. A defective tuner can cause a set to have the any of the following symptoms.

- ✔ Station drift
- ✔ Snowy picture
- ✔ Intermittent loss of video and audio after several minutes of operation *(Fig. 2-5)*
- ✔ Cycles through all channels without stopping when the channel up/down button is pressed once
- ✔ Cannot tune in VHF stations
- ✔ Loss of a single VHF station when cold and loss of several high VHF channels after warm up.
- ✔ No channels below 48
- ✔ Poor audio or a hissing sound.
- ✔ No up or down tuning
- ✔ Tuner functions are locked, no channel up or down, and no volume control
- ✔ Only one channel can be received

The RF, oscillator, mixer, and IC control circuits are found mounted inside the varactor tuner assembly *(Fig. 2-6)*. Tuners that are constructed on a separate module board or under a shield on the main PCB are also varactor-tuning systems. Poor shield and ground connections have been the cause of many different service problems in the modular or PCB tuners.

Fig. 2-5. A defective varactor tuner can cause many different picture and sound symptoms.

Fig. 2-6. An inside view of various parts in the different shielded areas of a varactor tuner.

SMD Tuner Components

SMD resistors and capacitors are always found in the varactor tuning systems. Likewise, SMT transistors and IC components can be found in today's varactor tuning systems. Metal glaze film (MGF) chips and ceramic (C) chips might be found in the VHF/UHF tuners – even the transistors are SMD components. Regular ICs were found in the early varactor tuners. Now most ICs or microprocessors are the SMD-type components located inside the tuner.

Exercise extreme care when servicing tuner components found on a modular or PCB. The foil or traces can be easily damaged when removing SMD parts from the PC wiring. Leave the IC or microprocessors in the original package until it's time to mount the part on the PC wiring and remember that too much heat from a soldering iron can destroy sensitive semiconductors.

Locating a Defective Tuner

A defective tuner can be located by taking critical voltage measurements at the tuner terminals. The defective varactor tuner can be spotted when correct voltages are measured on the tuner. If a different voltage is found on the tuning voltage when a different channel is selected, but the channel does not change, replace the tuner. Usually the tuning voltage will remain constant at the varactor tuner when the stations drift or become snowy. Substitute another varactor tuner, if one is handy, in order to determine if the tuner or tuning voltage source is defective. The tuning source, control, or synthesizer IC also may be defective if no voltage or a change in voltage is found on the tuner.

If the voltage-control IC or microprocessor does not show a changing tuning voltage, a defective control component may be the problem. Before removing the control component, take critical operating voltages on the IC or microprocessor. First check the supply-voltage terminal (Vcc). Next check the oscillator clock functions. A defective crystal oscillator component can result in channels going off frequency. A 4.5-MHz crystal can cause intermittent tuner action.

Tuner Problems

If the tuner itself is defective, no stations are able to be tuned in, a snowy picture may result, there is no tuner action, and there is no audio. An SMD IC in the tuner can also cause no tuner action. Remember to be *very careful* when replacing SMD parts so as not to disturb or push other components out of line.

Will Not Tune Channels

If the TV will not tune channels, first determine if the tuning voltage is changing at the tuner. Sometimes resoldering all stakes and grounds under the tuner shield can bring the tuner back to life. If there has been a lightning strike, it can result in a tuner that won't change channels. Liquid spilled onto or into the tuner can also result in the inability to change channels.

Possible Causes:

- ✔ Defective IC or microprocessor in tuner
- ✔ Defective EEPROM component
- ✔ Defective voltage regulator or decoupling electrolytic providing voltage to the control or synthesizer *(Fig. 2-7)*
- ✔ Defective SMD or regulator IC in the tuner
- ✔ Defective bypass and zener diodes in tuner voltage sources can result from lightning striking the tuner.

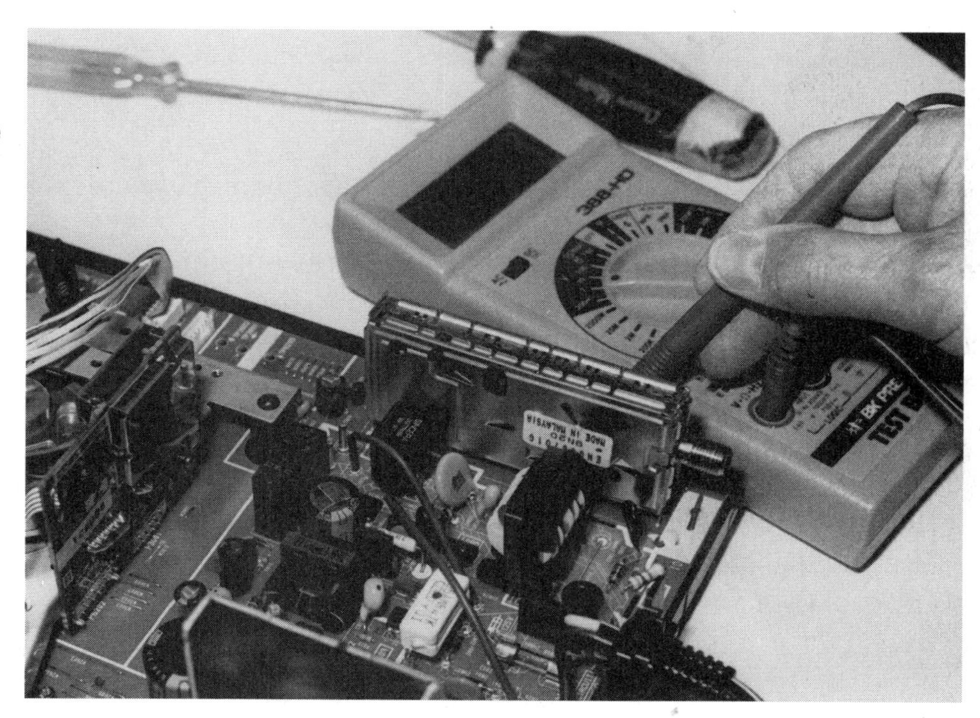

Fig. 2-7. Taking critical voltage measurements on the varactor tuner with a DMM.

Intermittent Tuner Action

Possible Causes:

- ✔ Broken trace or foil at the output pin
- ✔ Defective zener diode and isolation resistor to the tuner voltage source
- ✔ Poorly soldered stakes or grounds (also snowy picture with poor audio)
- ✔ Microscopic cracks of pin connectors inside tuner (also loss of audio and video)
- ✔ Defective tuner (audio and video loss after 15 to 30 minutes of operation)
- ✔ Defective regulator providing voltage to the control or microprocessor

No Tuner Action – Panasonic PV-M2021A

In a Panasonic PV-M 2021A TV, there was no tuner action. The voltage was checked at the tuner-modulator board. No BT control voltage was being fed to the RF and oscillator transistors. When the BT control transistors were replaced (Q7601 and Q7602), the tuner was restored to normal operation *(Fig. 2-8)*.

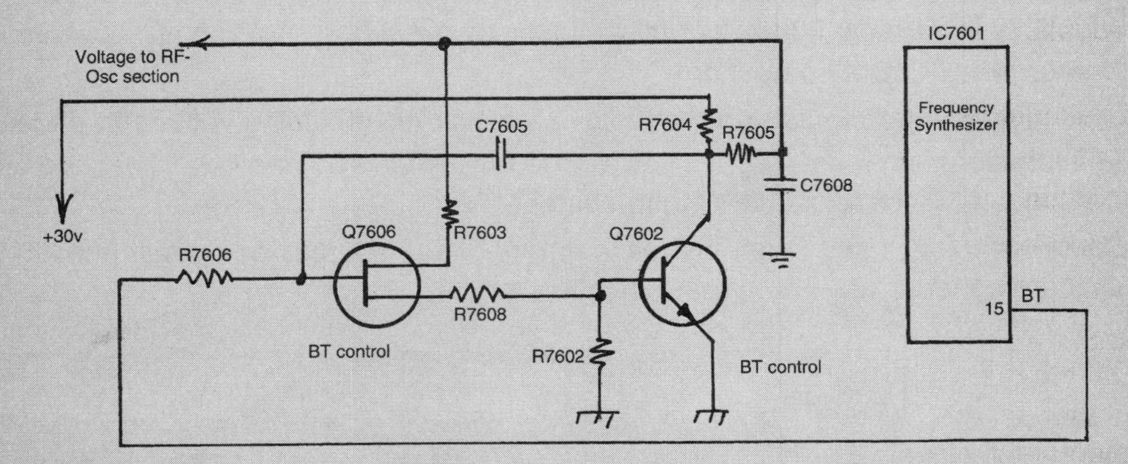

Fig. 2-8. Both Q7601 and Q7602 (BT) transistors were replaced to restore voltage in a Panasonic PVM2021A varactor tuner.

Loss of Channels

Possible Causes:

- ✔ If there is a loss of low VHF stations when the set is cold and a loss of high VHF stations after the set warms up, then the problem is badly soldered joints near a large coil beside the SMD IC. Resoldering all grounds in the tuner and components around the SMD IC can correct this symptom.
- ✔ A defective RF transistor can cause the TV to tune in UHF only and no VHF.
- ✔ When neither VHF nor UHF channels can be tuned in, replace the varactor tuner.
- ✔ When Channels 7 to 13 cannot be tuned in, a transistor in the tuner module may be defective.
- ✔ The loss of low Channels 2 to 6 can result from poorly soldered joints in grid traces of the tuner.
- ✔ When only Channels 7 to 13 and cable channels cannot be tuned in, resolder all grounds in the tuner, including metal shields.
- ✔ When only one channel can be received, check the SMD IC in the tuner.
- ✔ Replace the frequency synthesizer or microprocessor control when the tuner won't tune channels above Channel 10.
- ✔ If Channels 2 to 9 are okay but, when Channel 9 is tuned in, channels begin to hunt and stop at Channel 2, then a defective microprocessor is the cause.

No Up or Down Action

Possible Causes:

- ✔ If, after warm-up, you cannot turn the channels or sound up and down, the IC microprocessor that controls the voltage fed to the tuner may be the source of trouble. Poor resistor and board connections in the supply voltage of IC control cause the problem.
- ✔ Leakage between up button contacts can cause channels to change and not stop.
- ✔ If all functions are locked up with no channel or volume up and down then the cause may be a defective IC in the tuner.
- ✔ Liquid spilled into the TV chassis will cause the channel up and down functions to fail.
- ✔ Leaky bypass or filter capacitors on the tuner result in no channel up and down operation.
- ✔ If after several hours of operation the channels won't change, check for a defective memory IC.

Dead Channels – Zenith TVSA1320

With a Zenith TVSA1320 TV/VCR, no VHF or UHF stations could be tuned in. A voltage check on the tuner indicated +5V at AL +5V connection and very little voltage on Pin 9. When the feed line to Pin 9 was checked with an ohmmeter, low resistance was found from Pin 9 to the common ground. One end of electrolytic capacitor C007 (1 µF, 50V) was disconnected. Pin 9 voltage returned to +33V. Replacing C007 solved the problem (*Fig 2-9*).

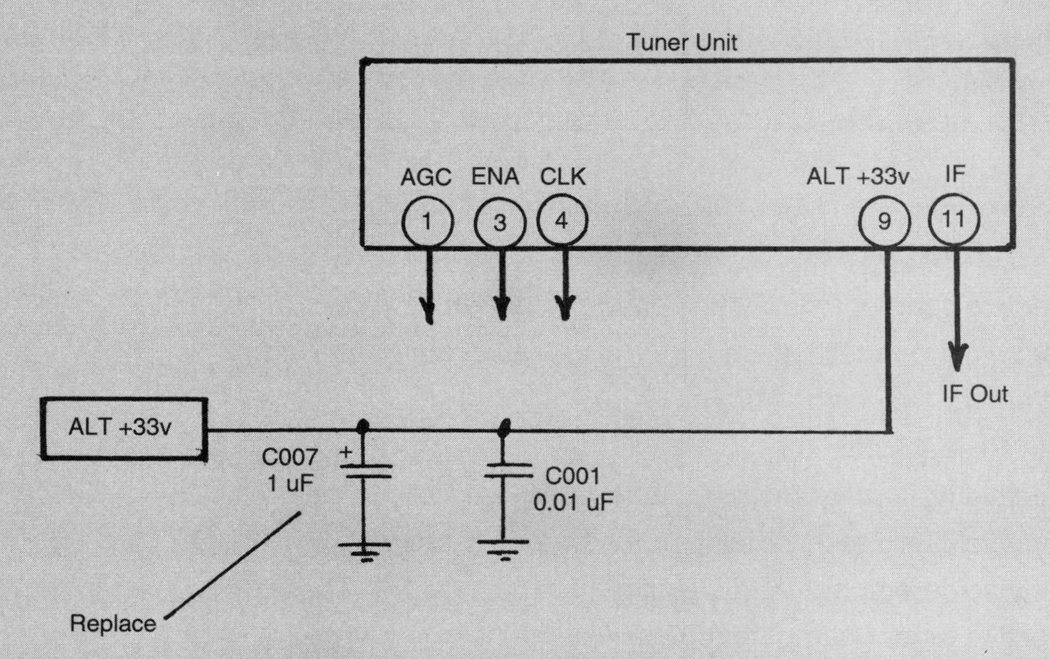

Fig. 2-9. Replacement of C007 (1 µF) electrolytic in a Zenith TVSA1320 TV/VCR restored voltage to the varactor tuner.

Drifting Stations

- ✔ Defective varactor tuner (channel 13 drifts with no video and poor audio)
- ✔ Bad grounds on the bottom of the tuner
- ✔ Bad board resistor connection to the control IC (check for voltage change)
- ✔ Crystal tied to the frequency-synthesizer IC
- ✔ Voltage-regulator component feeding 33V line to the tuner.

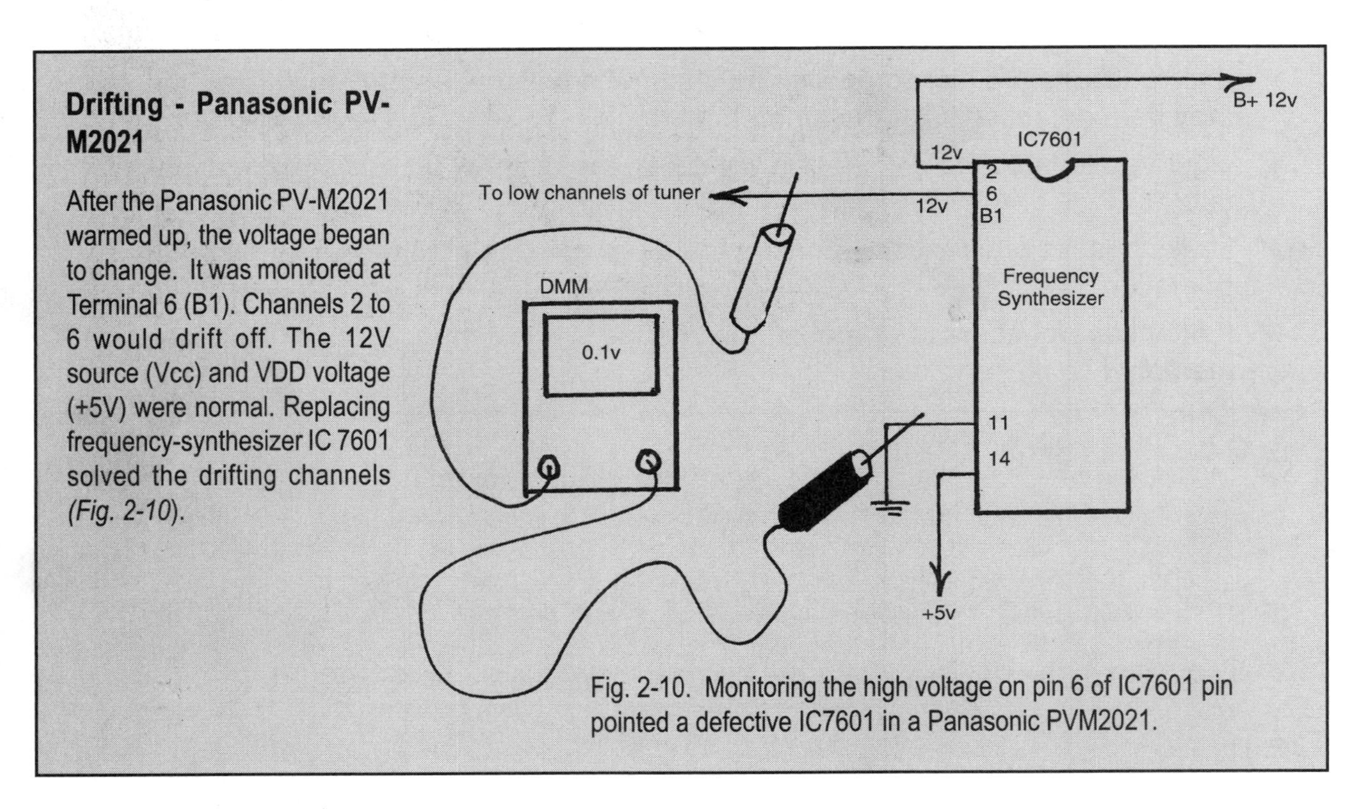

Drifting - Panasonic PV-M2021

After the Panasonic PV-M2021 warmed up, the voltage began to change. It was monitored at Terminal 6 (B1). Channels 2 to 6 would drift off. The 12V source (Vcc) and VDD voltage (+5V) were normal. Replacing frequency-synthesizer IC 7601 solved the drifting channels *(Fig. 2-10)*.

Fig. 2-10. Monitoring the high voltage on pin 6 of IC7601 pin pointed a defective IC7601 in a Panasonic PVM2021.

Snowy Picture

Possible Causes:

- ✔ Defective components within varactor tuner or IF stages
- ✔ Poor solder connections in tuner (also loss of audio and video)
- ✔ Bad antenna-isolation block
- ✔ Open RF transistor
- ✔ Defective IC in tuner (also no audio) *(Fig. 2-11)*
- ✔ Poorly soldered resistor joint on tuner-control board
- ✔ Poorly soldered ground connections
- ✔ Defective tuner (intermittent tuner voltage)

Fig. 2-11. A snowy picture with no audio can be caused by a defective tuner.

There are a few other problems you may encounter when looking for the culprit causing a snowy picture.

- ✔ When the picture is snowy, check the voltage going to the tuner control IC. If it is locked at a voltage below 1V and won't change, suspect a defective microprocessor IC.
- ✔ The screen can be extremely snowy with a change in channels and on-screen display operation. Monitor the tuning voltage applied to the varactor tuner and IC component; it may be a defective preamp transistor in the IF section.

Defective IC in Tuner

In many of the tuners constructed on a modular board or on the main PCB, an SMD or microprocessor inside the shielded-tuner area can cause many different problems. Any of the following symptoms may indicate a defective tuner IC.

- ✔ No tuner action
- ✔ An extremely snowy picture with low tuning voltage
- ✔ No video or audio on any channel with faint sound on Channel 2
- ✔ A dead tuner on most channels
- ✔ When tuning voltage stays at maximum (33V) and won't change
- ✔ Channel numbers change, but tuner gets only one snowy channel with no sound
- ✔ Snow on all channels
- ✔ Channel cycle does not stop
- ✔ Loss of low VHF channels when set is cold and, after warm-up, a loss of high VHF channels

Snow on High and Low Channels – RCA 13TVR60

On an RCA 13TVR60 TV/VCR, the tuner was snowy on all channels. The +5V source, 9V source, and 33V source were measured on the tuner N2001 *(Fig. 2-12)*. Since this type of tuner has both the tuner and IF stage enclosed, the sound was fairly normal with a snowy output. The tuner was defective and exchanged.

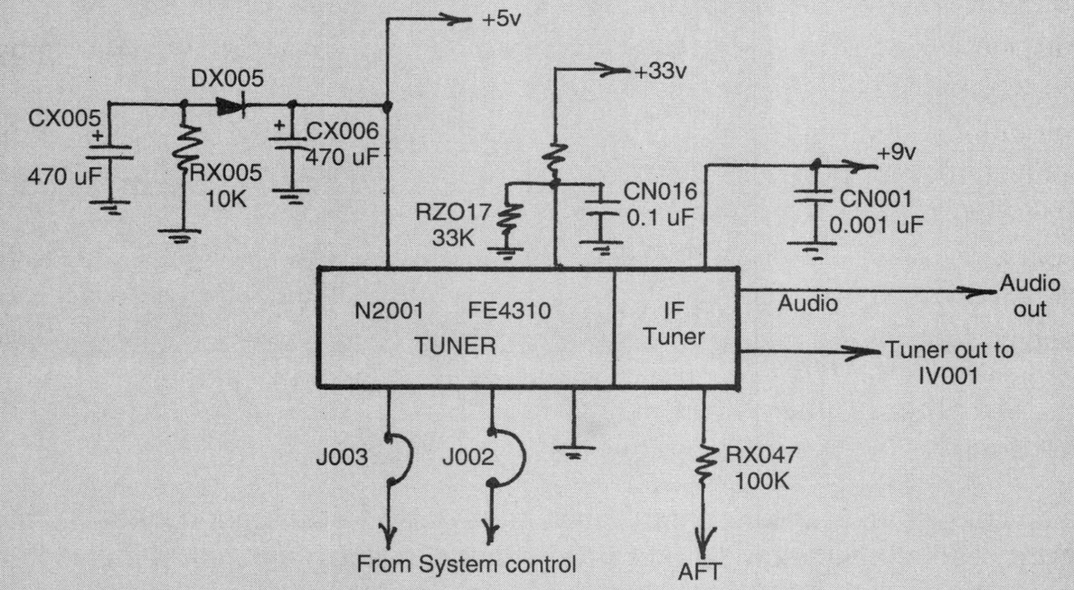

Fig. 2-12. A bad tuner caused a snowy picture on high and low channels in a RCA TV/VCR combination.

Defective EEPROM IC

EEPROM ICs are found in some TVs and can cause many different symptoms. If the channels will not change, the EEPROM IC may be bad. Replace the defective EEPROM IC and resolder all connections on the IC if there is no tuner operation. Resolder all grounds around the EEPROM terminals if intermittent tuner operation occurs. Other symptoms of a bad EEPROM IC include:

- ✔ Tuner does not receive all channels
- ✔ Channel numbers are off to the right of the screen and some are not readable
- ✔ No or intermittent reception of high channels (Channels 7 to 13) and cable channels

Storm Damage

When lightning strikes the TV through the antenna lead-in, cable, or power line, extensive damage may result in the varactor tuner, including weak video from a burned trace in the input line. When hit by lightning, the SMD IC in the tuner can cause a dead tuner symptom. An intermittent and snowy picture might be caused by burned or strip-foil ground connections in the tuner. The IC microprocessor that feeds tuning voltage to the tuner can be damaged in a lightning storm as well.

Unusual Tuner Problems

These are problems that are a bit out of the ordinary, but you may encounter them in the course of repairs.

- ✔ A squeaking noise in a tuner might be caused by defective capacitors in the microprocessor IC terminals.
- ✔ A hissing noise that sounds somewhat like a rushing waterfall accompanied by very poor sound may be caused by a bad tuner.
- ✔ Poor or intermittent reception and white hash lines throughout the picture on high channels may indicate bad grounds in the tuner. Resoldering the joints should cure the problem.
- ✔ Replace the tuner if intermittent lines or a band of snow a few inches high in the picture appears.
- ✔ Flashing horizontal lines across the entire picture can be caused by poor shield grounds in a tuner that has been constructed on the main chassis.

Outside the Tuner

Defective components outside the tuner, but connected to the tuning circuits can cause many different tuner problems.

ICs or Microprocessor

- ✔ A defective IC or microprocessor that supplies the tuning voltage can cause a dead-tuner symptom.
- ✔ If the TV has no tuner action, no channels up or down, and no remote action, it may seem that the problem is a bad tuner, but the problem may be a defective microprocessor where the clock and data pulse are okay.
- ✔ When the +33V can't be varied at the tuner, with a no-audio/no-video symptom, the cause may be a defective outside IC.
- ✔ A defective EEPROM IC can prevent the turner from receiving all channels.
- ✔ A defective IC or transistor regulator feeding the varactor tuner can cause the varactor tuner to drift off stations.
- ✔ A dead or drifting tuner also can be caused by a change in resistance or capacity of a resistor or capacitor tied to the line voltage supplied by an IC.

Bad Connections

- ✔ Resolder all contacts on the tuner plugs and ground terminals if the set comes on by itself at full audio volume.
- ✔ Burned or broken trace wiring in the input lines to the tuner can cause weak or snowy video.

Capacitors

- ✔ When the channel does not change and it looks like a bad tuner problem, check for leaky bypass capacitors or a change in resistance in the 33V line.

Electrolytics

- ✔ If there is no picture in the high channels and it looks like the problem is a bad tuner, the problem actually can be caused by defective electrolytics on the tuner module board.
- ✔ A dead tuner can be caused by a dried-up electrolytic on the B+ line. Use the ESR meter to check all capacitors in the tuner or outside.

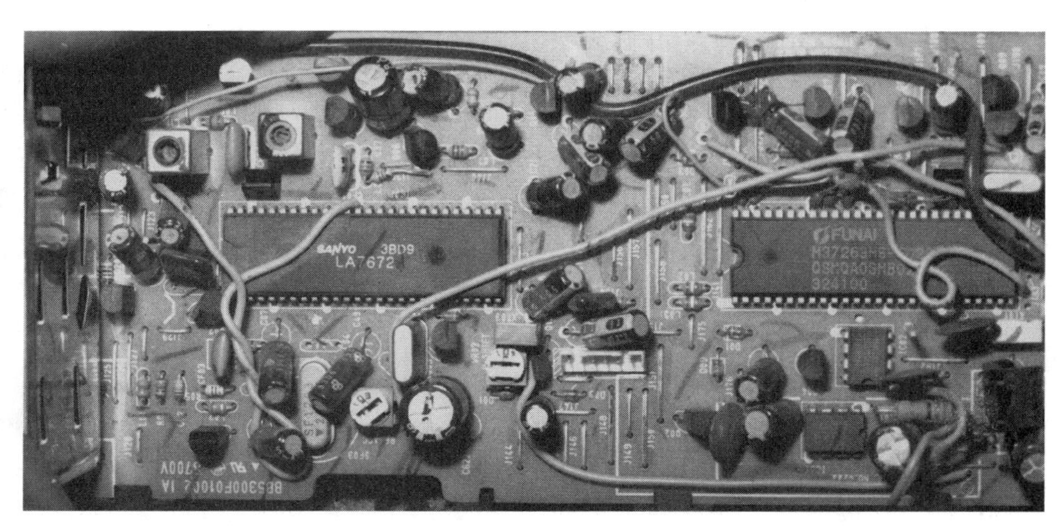

Fig. 2-13. The IF and varactor tuner are found on a module PCB in a Magnavox TV/VCR combination.

Tuner or IF

The dead or snowy TV symptom can be caused by the tuner or a defective IF stage. Determine if the tuner or IF circuits are defective by substituting another tuner. If the picture and sound return, then the tuner is defective. If injecting an IF signal at the IF cable has no effect, then the IF transistor or IF-video IC may be defective. Use the demodulator probe on the oscilloscope to check the IF signal at the output of the preamp transistor on the IF input terminal or IF video IC.

Scope the tuned signal after the preamp stage to the SAW filter or into the IF video IC. Most IF stages in the TV/VCR combo have a transistor as a buffer or preamplifier before the SAW filter. The SAW filter output is tied to Terminals 8 and 9 of video/chroma deflection/IF (IC301). The VIF amp and video detector are found inside IC301. Both the video and audio are separated within the IC.

IF Problems

Today, most IF stages are enclosed inside a large IC that might also contain video, chroma, sync, AGC, and deflection circuits. A defective IF IC with other stages might have a dead chassis, or only part of the IC might be defective. The sound might be present, for instance, but with no video or picture. When both the sound and picture are missing, check the video/IF IC. When all circuits (including deflection, IF, sound, and video) are missing, check for a defective IC and a low power-supply voltage or no power-supply voltage. The color-bar generator and scope are useful test instruments in servicing IF circuits.

All voltages in the IF/video circuits can be measured with the DMM. The sweep marker and color-dot bar generator can service the IF circuits with the oscilloscope. The tuner-subber can be used to insert a signal at the IF input and even at the base of the second IF amp or IC, with the picture tube as indicator. Since most IF circuits use a ceramic filter or SAW filter between IF stages, there is no need for IF alignment or the use of a sweeper-marker generator. Clamp the tuner AGC to determine if AGC circuits are normal. Intermittent, snowy, or no IF signal can occur in the IF stage.

Be very careful when removing metal shields that cover IF circuits. Always replace transistors and small components in the same spot as the originals. Do not replace IF components under the PCB; replace them in the original position. Carefully replace all shields after repairs are made.

Common IF problems can include the following:

Defective IF

- ✔ Snow on all channels can be caused by a defective IF amp transistor.
- ✔ A leaky or open IF transistor can cause a loss of signal and snowy picture.

Bad Connections

- ✔ A loss of signal, picture, or sound after the set has warmed up can be caused by poorly soldered joints in the IF circuits.
- ✔ A badly soldered joint in the IF module can cause intermittent loss of video.
- ✔ Intermittent loss of audio and video can be caused by poorly soldered joints in the IF or tuner module.
- ✔ Snow on all channels can result from badly soldered joints and connections in the IF module.
- ✔ When the TV picture shifts and becomes snowy after a few minutes of operation, you might suspect an AGC problem, but this is actually caused by poorly soldered joints under the IF shields.

Bad Grounds

- ✔ Intermittent tuner action may stem from poor grounds in the IF module.
- ✔ Intermittent tuner action that changes when the TV is tapped or moved might be caused by poor ground connections on the IF cable lead connected to the varactor tuner.
- ✔ Intermittent picture and sound can be caused by poor grounds surrounding the IF shields.

SAW Filter Circuits

The SAW filter circuit (surface acoustic wave network) is generally found between the preamp or buffer transistor and the IF/video IC. For example, in the Zenith TVSA1320 TV/VCR the SAW filter is located between the tuner IF plug and the IF/video IC, and in the Panasonic PV-M2021 TV the SAW filter network is found between Q4 in the tuner section and the VIF amp inside TV demodulator IC701 *(Fig. 2-14)*.

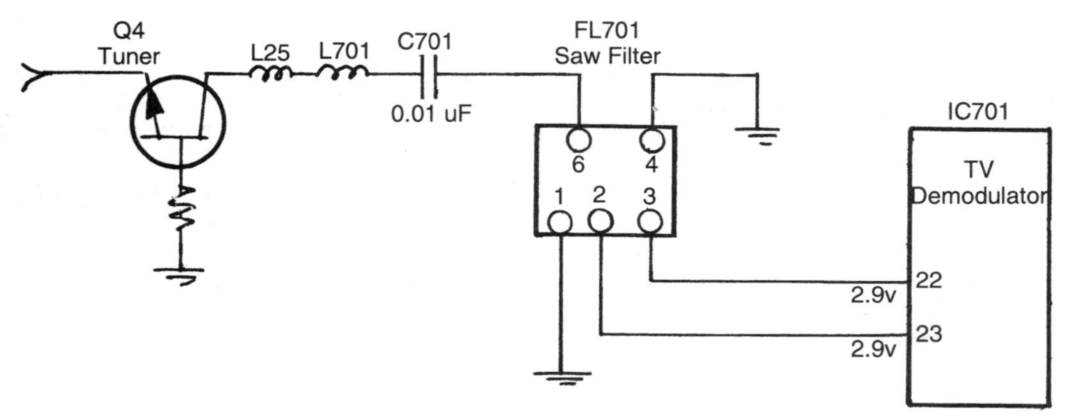

Fig. 2-14. The SAW filter network is located between buffer amp transistor (Q4) and IC701 in a Panasonic TV/VCR combination.

The surface-wave filter is made up of piezoelectric material with two pairs of transducer elements. When an AC input signal is applied to one set of elements, acoustic waves travel along the surface of the element to the other electrode generating an AC output voltage. A SAW filter network sets the proper IF frequency, which eliminates IF alignment.

Ceramic filters or resonators consist of piezoelectric material converting mechanical stress into electrical energy or vice versa. The ceramic filter network takes the place of an IF transformer found in the tube and early transistor IF stages. The ceramic filter contains a fixed frequency made up of piezoelectric ceramic material. The ceramic filter can be tested with the same methods when checking for a defective SAW filter network.

Chapter 3: Troubleshooting TV Horizontal Circuits

The horizontal circuits in the TV chassis might consist of a horizontal oscillator or deflection IC, horizontal driver, horizontal output, and flyback, or output transformer. The horizontal circuits provide horizontal sweep and high voltage to the CRT. Horizontal sweep is fed to the horizontal yoke, and high voltage is developed in the flyback. Most horizontal circuits are the same throughout any manufacturer's TV line, and these servicing problems are located in the horizontal circuits (*Fig. 3-1*).

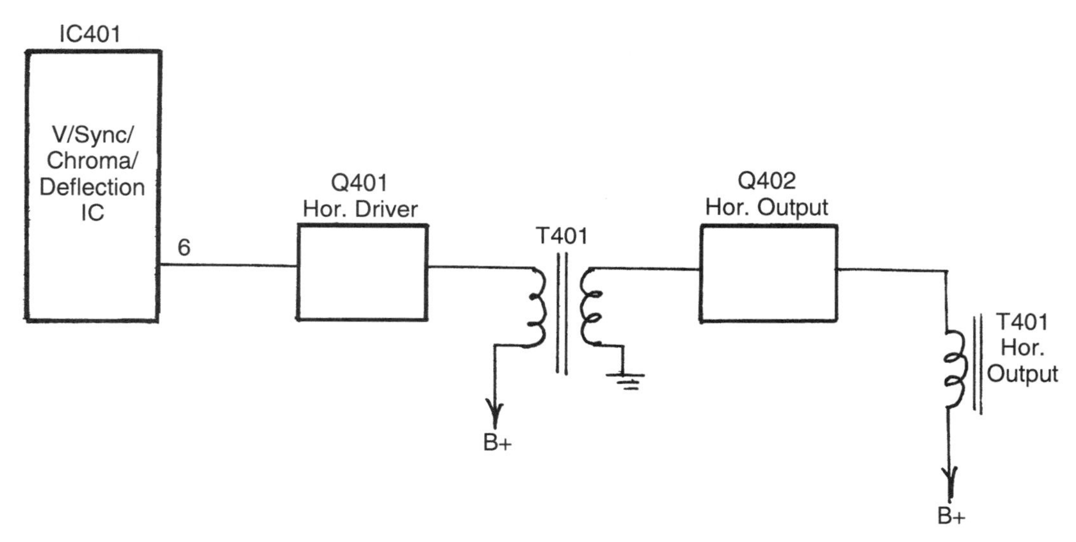

Fig. 3-1. Block diagram of the most critical parts in the horizontal circuits are a defective IC, horizontal driver transformer, output transistor, and flyback primary winding.

Troubleshooting Horizontal Circuits

Check the low-voltage circuits by testing across the positive and negative terminals of the main filter capacitor. This will indicate that the low-voltage sources are working even when the chassis is dead. Check the condition of the horizontal output transistor when the main fuse is open and there is no voltage across the main electrolytic. Shut the chassis down and take a quick resistance measurement from the collector terminal to the common ground. A leaky or shorted output transistor will have a low-ohm measurement from metal shield to ground.

Make sure the horizontal square-wave signal or horizontal pulse is coming from the deflection IC. Look up the correct pin terminal of the deflection IC and scope for both horizontal and vertical deflection waveforms. If both waveforms are missing, then check for a defective deflection IC, improper supply voltage, or defective components tied to the deflection IC. Look up the supply pin terminal (Vcc) or the highest voltage found on any IC terminal.

If the supply voltage is low or shows improper voltage, look for a leaky IC. Remove the supply pin (Vcc) from the PC wiring and take a resistance measurement from the unsoldered supply pin of the IC to the common ground. If a low resistance - below 500 ohms - is noted, check for a defective IC. Now, take a voltage test on the PC foil that was removed from the supply-pin terminal, and if the voltage is normal or a little higher, a leaky deflection IC is probably the cause. To be certain, take critical resistance measurements between all pin terminals and common ground. If the deflection IC is leaky, or if it could have an open circuit, replace the deflection IC.

Scope the base and collector terminals of the horizontal driver transistor. Suspect a defective driver transistor if no waveform is found at the collector terminal. Solder all terminals of the driver transformer to the board connections. Some technicians, when seeing intermittent startup and shutdown symptoms, resolder these terminals right away. Whenever a shorted driver transistor is located, you should check the resistance of the primary winding. Measure the resistance of an isolation-dropping or voltage-dropping resistor in series with voltage source and primary winding. An open primary winding will not provide adequate drive or a correct waveform to the horizontal output transistor.

Trace the drive waveform to the base terminal of the horizontal output transistor. Now, check the output waveform with a scope test on the flyback (*Fig. 3-2*). Sometimes, a defective yoke or a bypass capacitor in the yoke circuit will break down and overload the output circuit. Remove the red yoke horizontal lead if the supply voltage is low at the

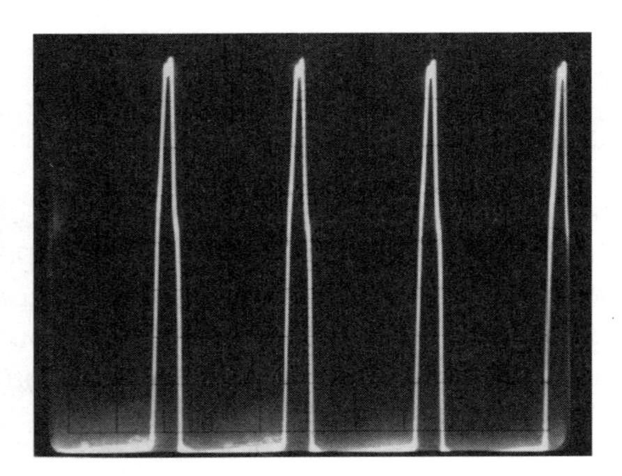

Fig. 3-2. Check the waveform on the horizontal output transistor to determine if the transistor, yoke, and flyback are normal.

flyback. Often, the voltage will rise and the output waveform will be quite low when the yoke is removed from the circuit indicating a defective yoke assembly. Check for a defective flyback when the horizontal output transistor appears warm as the line voltage is raised when using a variable isolation transformer and no waveform is found at the output transistor.

Deflection IC

Most of today's TV chassis' have the horizontal and vertical deflection-drive circuits inside an IC component, which includes video/sync/chroma and IF circuits. Often, the deflection circuits are crystal controlled and fed to a VCO and a vertical and horizontal countdown circuit. The horizontal drive signal might include a pre-driver amplifier inside the deflection IC. The horizontal output drive waveform is found at Pin 20 of IC301 in a Sears TV/VCR (*Fig. 3-3*).

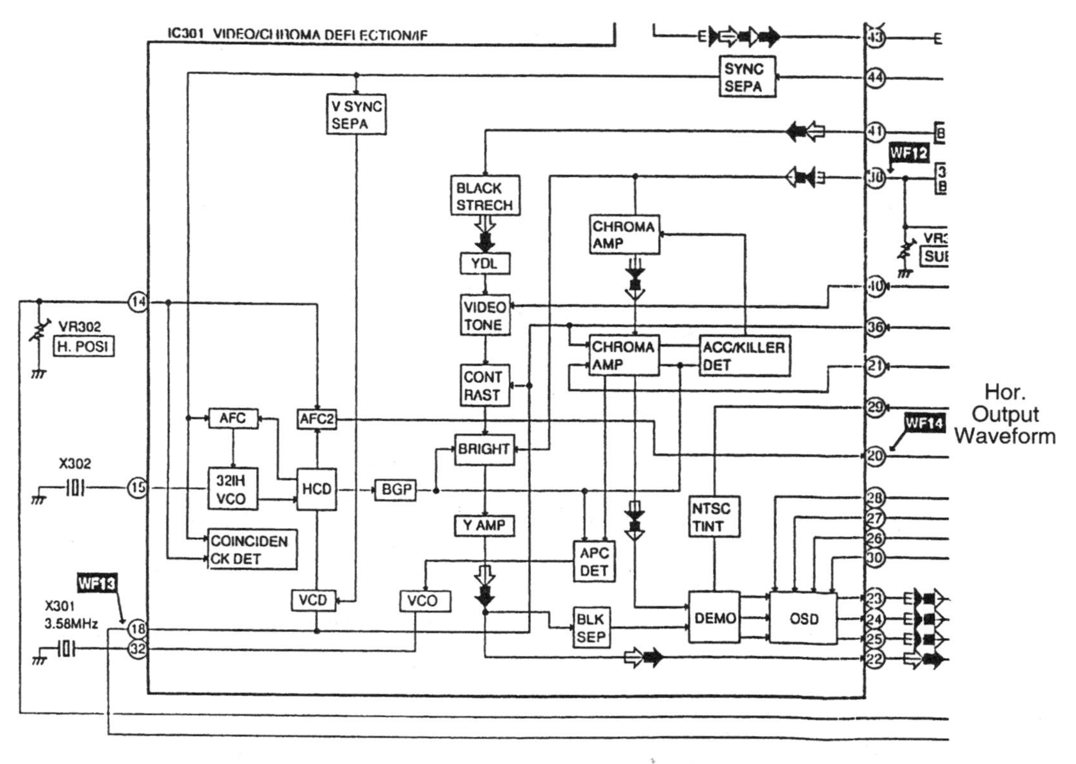

Fig. 3-3. The horizontal deflection driver signal is found at pin 20 of deflection IC301.

In a Panasonic 13-inch TV, the horizontal circuits are developed in a luminance/ chrominance/signal processing IC. The horizontal oscillator circuits are developed inside IC301 with a first countdown stage. The horizontal oscillator is crystal controlled by X501, and the horizontal signal from the countdown stage is fed to a horizontal pre-driver stage and out Pin 17 of IC 301. The horizontal square-wave drive signal is directly coupled through a driver transformer (T502).

A quick waveform test of the horizontal drive terminal pin of the deflection IC can determine if the oscillator deflection circuits are functioning (*Fig. 3-4*). If there is a problem, check for a defective oscillator crystal, deflection IC, and improper supply voltage to the deflection IC. Also check the supply voltage of the deflection IC with the DMM when no drive waveform is at the deflection IC output pin terminal. This supply voltage is powered from the switching power supply. The deflection IC horizontal waveform is coupled to the horizontal driver stage through a coil, resistor, or coupling transformer.

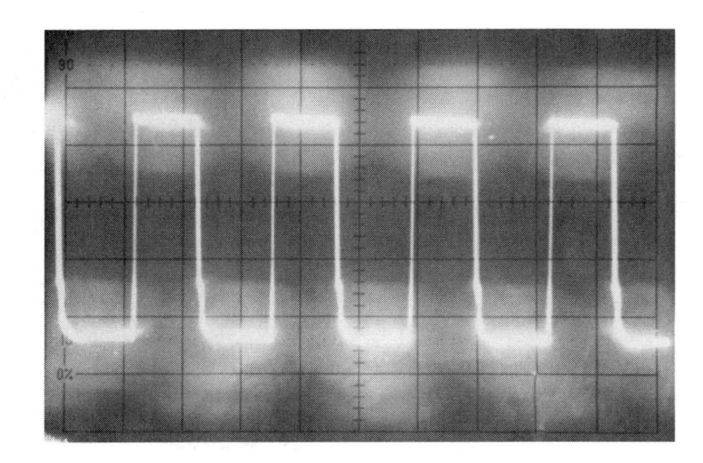

Fig. 3-4. A square waveform applied to the horizontal driver transistor with horizontal amp inside deflection IC.

The horizontal driver transistor greatly amplifies the drive waveform and is transformer-coupled to the horizontal output (*Fig. 3-5*). A defective horizontal driver transistor, drive transformer, and supply voltage can cause several different service problems in the drive circuits. The horizontal driver and output circuit cause most of the severe problems found in the TV chassis. The driver and output transistor are fed from the same power source.

A driver transformer couples the horizontal drive signal to the base of the horizontal output transistor (*Fig. 3-6*). If the drive waveform is missing on the base of the output transistor, then the transistor becomes hot and is damaged. Often the horizontal output transistor becomes shorted between collector and emitter terminals. When the main fuse is blown, check for a leaky or shorted output transistor. More horizontal output transistors are replaced than any other transistor in the TV chassis.

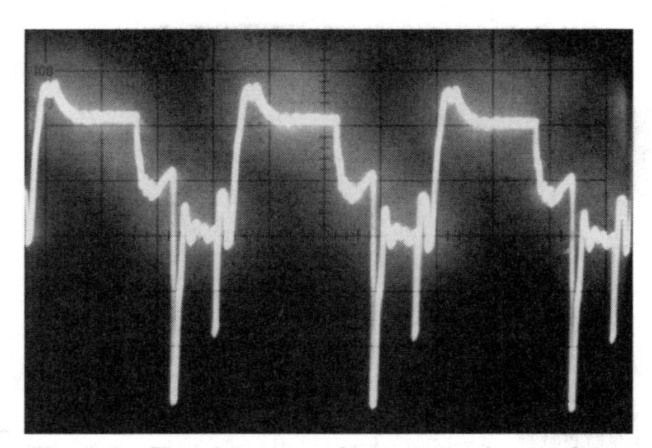

Fig. 3-6. The drive waveform on the base of the horizontal output transistor.

The horizontal output transistor provides a sweep drive for the primary winding of the flyback or horizontal output transformer and horizontal deflection yoke. Do a quick output waveform test by placing the scope

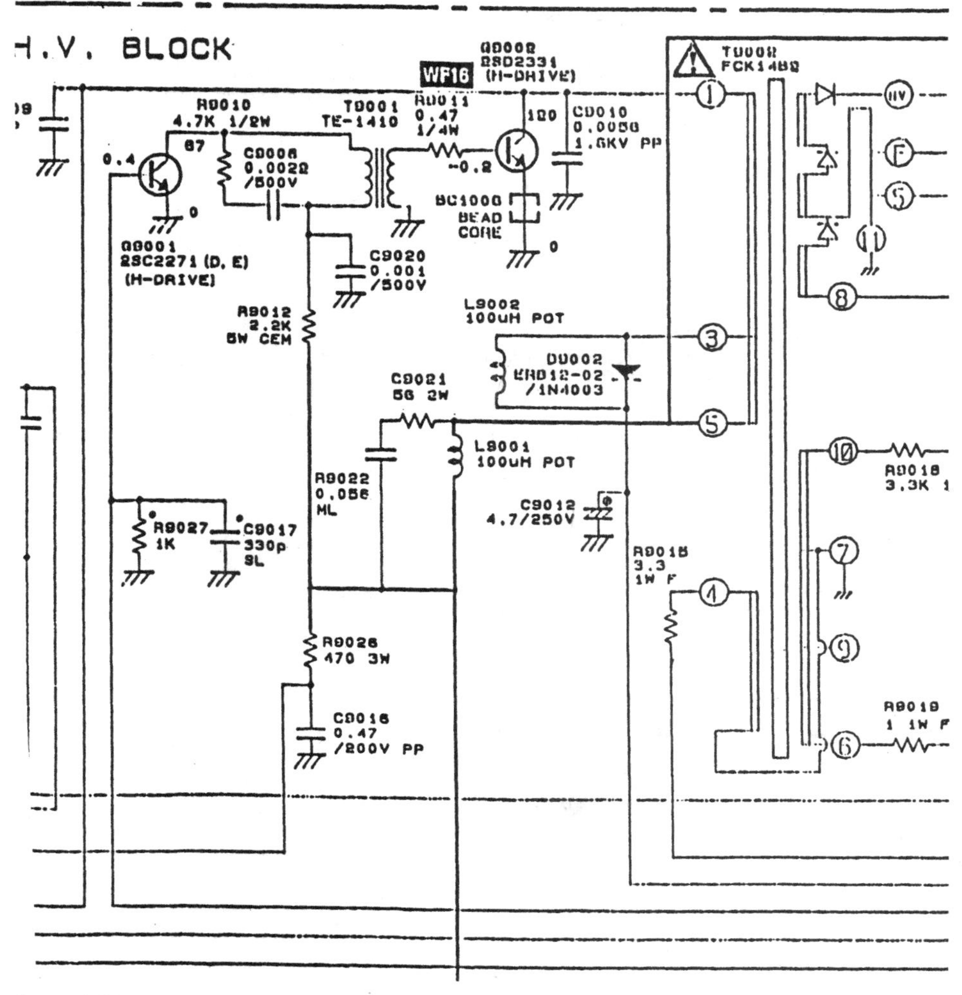

Fig. 3-5. The horizontal driver circuits of a 13-inch Sears TV/VCR.

probe alongside the flyback to determine if the horizontal stages are working (*Fig. 3-7*). If no waveform is found here, then perform a quick waveform test with the scope at the drive pin of the deflection IC and on the collector terminal of the horizontal driver, and also do a base waveform test at the horizontal output transistor.

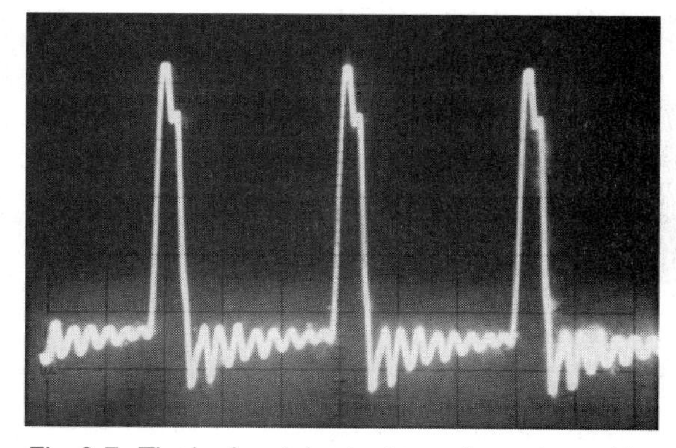

Fig. 3-7. The horizontal output waveform along side the flyback or horizontal output transformer.

When no voltages are found on the deflection IC, horizontal driver, and horizontal output transistor, check the low-voltage power supply. Usually, the horizontal driver and output-transistor power source have the highest voltage within the switching transformer power sources.

An isolation resistor or secondary fuse might be found between the power supply and the flyback primary winding to protect horizontal circuits. The horizontal output transistor receives its voltage source through the primary winding of the flyback. Although some DMMs can measure the collector voltage on the horizontal output, it's best to take a waveform measurement instead because the meter can be damaged.

To keep the high voltage from increasing to a dangerous high arcover voltage, a hold-down or safety capacitor is found on the collector terminal of the output transistor. This safety capacitor causes many service problems in the high-voltage circuits (*Fig. 3-8*). An open or defective safety capacitor may be to blame when the high voltage arcs over at the anode connection of the picture tube and the CRT spark gaps. In addition to high voltage, the open safety capacitor can damage the output transistor. A leaky safety capacitor can blow the secondary or line fuse. (For more information on the flyback and high-voltage circuits, see Chapter 5.)

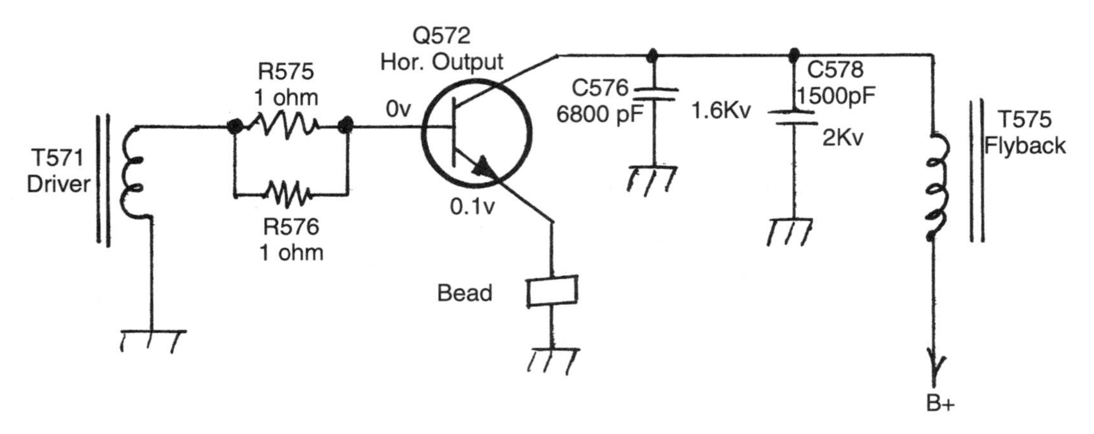

Fig. 3-8. The safety capacitor holds down the HV generated by the output transistor.

Horizontal Circuit Problems

Problems in the horizontal circuits can cause a burning smell, the picture to pull in on the sides, and pulsing on and off. The most common service symptoms found in the horizontal circuits are:

Open B+ fuse

A blown or open fuse might occur with a defective flyback, horizontal output transistor, or horizontal yoke winding.

Dead chassis

A dead chassis can occur with about any component breakdown in the horizontal circuits.

Intermittent operation

Intermittent horizontal circuits can result from a defective output transistor, flyback, or safety capacitor. Poorly soldered connections on the driver transformer can cause intermittent startup and shutdown problems. Poorly soldered connections on the flyback can cause intermittent shutdown or intermittent chassis operation. A broken safety capacitor lead or internal break can cause the fuse to open and intermittent horizontal output voltage.

Horizontal lines in the picture

Horizontal lines in the picture might be caused by a defective safety capacitor, while numerous black horizontal lines can result from a bad zener diode in the horizontal supply voltage to the deflection IC. Horizontal bars in the picture can be caused by a dried-up or defective electrolytic in the power sources (*Fig. 3-9*). Horizontal tearing might be caused by a defective safety capacitor and zener diodes. Check the horizontal deflection for possible horizontal tearing and a defective electrolytic tied to the deflection circuits.

Fig. 3-9. Heavy, dark horizontal lines are an indicator of the horizontal out of sync and off frequency.

Ticking sound

A ticking or squealing noise might be caused by a defective horizontal output transistor and poor grounds in the driver-transistor circuits. The defective output transistor might cause a chunk-chunk sound instead of a ticking noise. The ticking noise can be caused by a defective safety capacitor. With a dead-ticking symptom, check for open resistance or a change in resistance of the resistors tied to the flyback. Also, poor horizontal driver transformer connections can cause a ticking noise.

Deflection IC Problems

A defective deflection IC can cause a dead chassis and chassis shutdown problems. Scope the deflection-drive signal to determine if the horizontal oscillator and countdown circuits are functioning. Check the supply voltage of the deflection IC to determine if voltage from the power supply is at the deflection IC. Low or no voltage can cause intermittent or a dead horizontal section. Low voltage from the supply source might indicate a defective voltage-regulator transistor or IC. A raster pulled in 1/4" on the sides can result from poor regulation and a low-voltage regulator. A defective crystal in the horizontal countdown circuits can cause symptoms such as no startup, no horizontal sweep, and shutdown.

If the TV is slow to come on and the channels sometimes change automatically, it may be caused by poor solder joints and connections on the deflection IC terminals. Horizontal tearing can result from a defective deflection IC. For horizontal tearing or horizontal lines in the picture, check the electrolytic capacitors in the horizontal deflection circuits. Deflection IC damage can be caused by storm damage or power-line surges.

Driver Transistor Problems

When no driver voltage is at the base terminal, the driver transistor can run warm with low collector voltage. An intermittent driver transistor can cause intermittent power-up and shutdown problems. The leaky driver transistor can damage the output transistor with no drive voltage at the base terminal. An open coupling resistor to the base of the driver transistor can result in a dead chassis.

For a dead chassis, check for a burned resistor or an increase in resistance of the isolation resistor to the primary winding of the driver transformer. A leaky driver transistor can cause a dead-relay click symptom and cause the isolation resistor to run hot and burn (*Fig. 3-10*). An intermittent driver transistor can cause a dead-relay click symptom as well as causing the TV to intermittently come up with HV and then shutdown.

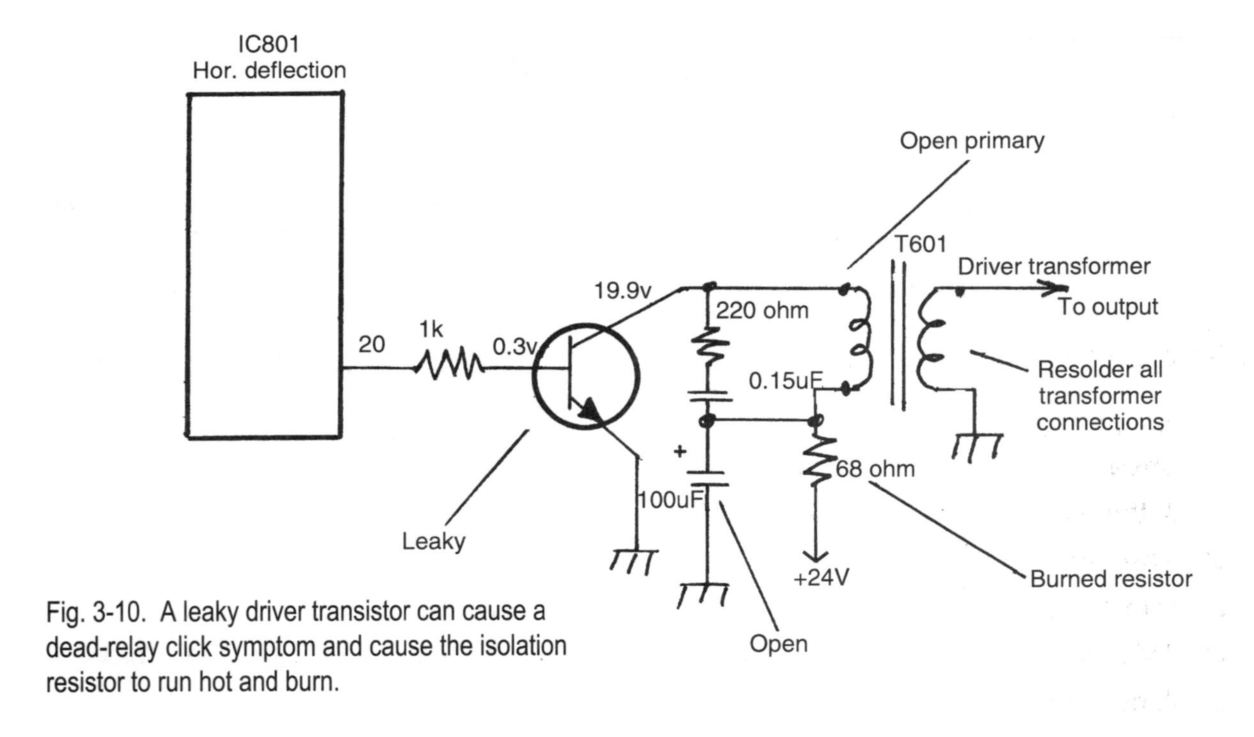

Fig. 3-10. A leaky driver transistor can cause a dead-relay click symptom and cause the isolation resistor to run hot and burn.

A dead chassis can be the result of no voltage on the collector terminal of the horizontal driver transistor with a burned isolation resistor and open primary winding of the driver transformer. A dead relay click symptom might result from an open resistor feeding the base terminal of the driver transistor. Poorly soldered connections on the driver transistor can cause a dead relay with low or no voltage applied to the driver transistor.

Horizontal Driver–Transformer Problems

The horizontal driver-transformer circuits can cause many different problems in the horizontal circuits.

Poorly Soldered Joints on Transformer Connections

Symptoms:

- ✓ TV shuts off (not shut down) and immediately comes on again and after playing for several minutes, shuts off again
- ✓ Dead with ticking from the flyback
- ✓ Intermittent shrinkage on both sides of the raster
- ✓ Intermittent collapse of the horizontal sweep circuits (*Fig. 3-11*) (resolder connections)

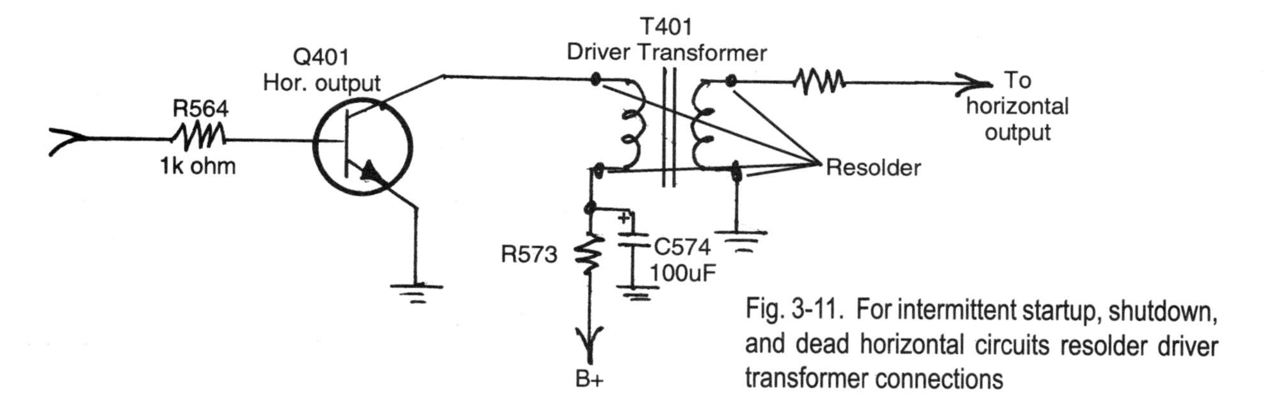

Fig. 3-11. For intermittent startup, shutdown, and dead horizontal circuits resolder driver transformer connections

Other problems resulting from poor driver-transformer connections include:

- ✓ An intermittently dead chassis, which also can be caused by an open primary on the driver transformer.
- ✓ A thin vertical line in the middle of the raster with no horizontal sweep.
- ✓ The output transistor can run hot.
- ✓ Intermittent startup with shutdown after 30 minutes.
- ✓ TV immediately shuts off and sometimes blows the horizontal output transistor.
- ✓ A pulsating noise in the flyback.

Bad transformer joints can also produce problems such as an intermittent picture, the picture pulling in on the right-hand side, and the TV shutting off. A dead chassis/fuse okay situation can be the result of poor board connections of the driver transformer.

Intermittently Dead - Emerson VT1920

In an Emerson VT1920 TV/VCR, the intermittently dead symptom was caused by poorly soldered joints on the primary winding of T401. Always resolder all driver transformer connections when there is intermittent operation, the set fails to start up, or the chassis is dead.

If after the set is turned on, it intermittently goes off and the relay clicks off and on, the Pin 3 board connection of the driver transformer may be bad. If the TV experiences a repeating cycle in which it plays only two or three minutes, goes dead, and can't be turned on again for five minutes, poor driver transformer connections may be to blame.

Horizontal Output Transistor Problems

Keep in mind when looking at horizontal output transistor problems, if the horizontal output transistor is shorted, there may be an open 2.5-amp or 4-amp fuse as a result. A quick low-ohmmeter test from the collector terminal of the horizontal output transistor to common ground will identify a leaky output transistor. If there is excess thermal grease on pins of the horizontal output transistor, it can shut off the chassis immediately. If internal arcing of the flyback occurs, it can damage a horizontal output transistor.

Dead – Panasonic PV-M2021

In a Panasonic PV-M2021 TV/VCR, the TV chassis was dead with no high voltage. A quick check of the horizontal output transistor indicated a short between the emitter and collector terminals so the horizontal output transistor (Q551) was replaced. L552 in the emitter circuit was checked for continuity and was found to be good (*Fig. 3-12*). Before firing up the chassis, safety capacitor C554 indicated leakage and was also replaced.

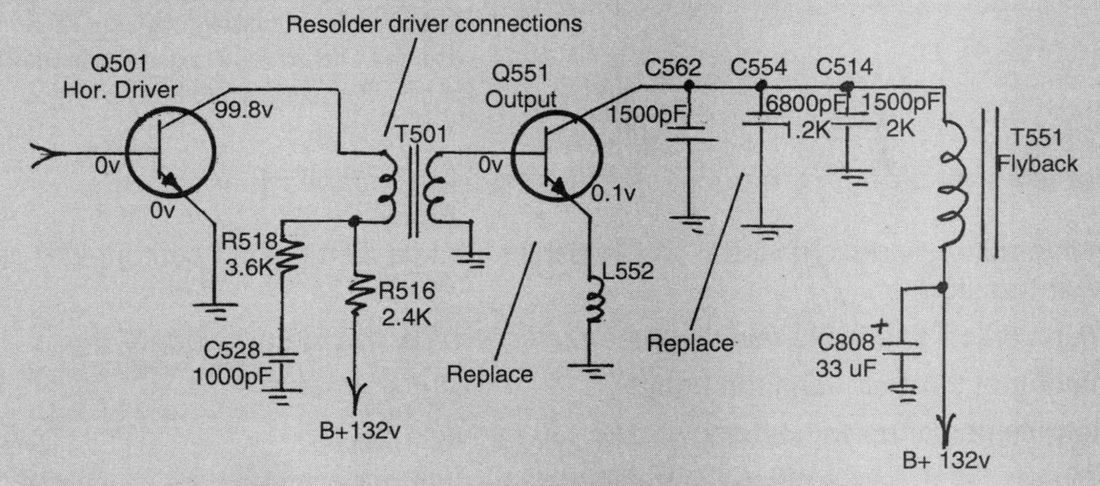

Fig. 3-12. Replace output Q551 and C554 in a dead Panasonic PVM2021 TV/VCR.

Defective, Leaky, or Shorted Horizontal Output Transistor

Symptoms:

- ✓ TV power pulsates and picture pulls in from right side
- ✓ Relay clicks, power indicator comes on, and set goes off immediately
- ✓ Set appears dead, the relay clicks, and the flyback makes a high-pitched squeal
- ✓ Intermittent collapse of horizontal sweep
- ✓ Dead chassis with ticking from the flyback (*Fig. 3-13*)
- ✓ Dead chassis accompanied by a severe squealing noise
- ✓ Dead chassis-fuse okay symptom with channel LEDs lit and a 130V source that goes down to 62V.
- ✓ Open 2.5-amp fuse (caused by horizontal output transistor and defective electrolytic capacitor in voltage source)
- ✓ Dead chassis-fuse blown symptom
- ✓ Dead chassis-fuse okay symptom (short from base to emitter or a defective horizontal output transistor and an open isolation resistor)
- ✓ Set tries to turn on, flyback squeals
- ✓ Dead chassis-relay click with a blown fuse
- ✓ Set intermittently shuts off after two or three minutes
- ✓ Dead chassis with an open 4-amp fuse (also check voltage-regulator IC)

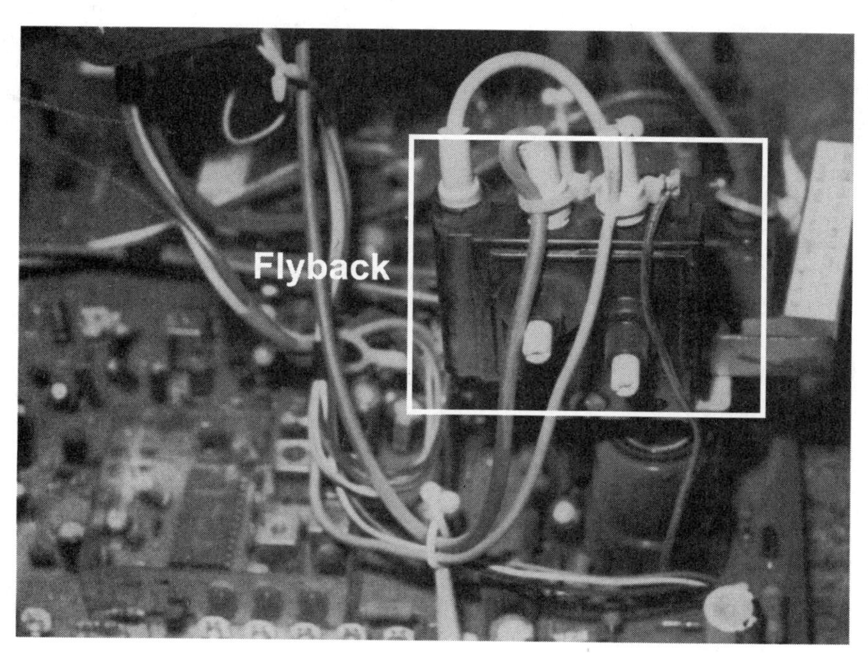

Fig. 3-13. When there was no startup, it was a leaky horizontal output transistor that caused a ticking noise in the flyback.

Horizontal Output Transistor Socket

Symptoms:

- ✓ Set goes completely dead after sides of picture pull in
- ✓ Set operates for several minutes and goes off, resulting in a ticking noise

Horizontal Foldover

Horizontal foldover is a vertical line or picture that is rolled into a small space. The picture rolls or folds into a thin area and has no horizontal width action. Most horizontal foldover problems occur in the horizontal and high-voltage circuits. Horizontal foldover can be caused by the following defective parts or conditions:

- ✓ Damper diodes
- ✓ Safety capacitors
- ✓ Horizontal output transistors
- ✓ SCRs
- ✓ Flybacks
- ✓ Poorly soldered connections
- ✓ Yokes

A poor coil connection at the emitter terminal of the horizontal output transistor can cause the raster to go black or dim and the side to pull into a horizontal line after warm-up. A thin vertical line in the center of the screen also results from a leaky bypass capacitor in the yoke circuits and a defective safety capacitor (*Fig. 3-14*). Horizontal foldover in the center of the screen, from top to bottom, can occur if there is a defective 1-ohm resistor in the emitter circuit of the horizontal output transistor. Poor horizontal linearity and foldover problems can result from badly soldered terminals in the reactors and regulation transformers.

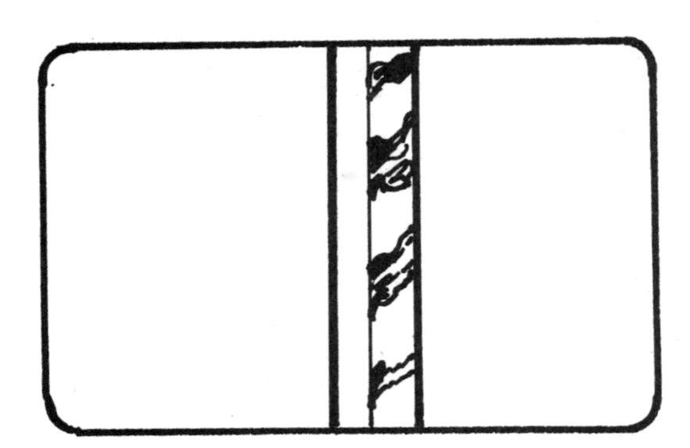

Fig. 3-14. Horizontal foldover might be caused by a defective bypass capacitor in the yoke circuit or a safety capacitor.

Horizontal Output Transistor Keeps Going Bad

- ✓ A shorted yoke or bypass capacitor in the yoke circuits can damage the horizontal output transistor. Sometimes, after replacing the open fuse and horizontal output transistor, the output transistor will be ruined again.
- ✓ Use the ESR meter to check all electrolytics in the voltage source of the horizontal driver transformer for horizontal output transistor damage then resolder all connections on the driver transformer. If repeated failure of the horizontal output transistor occurs, check for open bypass capacitors in the horizontal output circuits. If after replacing the output transistor an unusual dead condition occurs and the transistor blows after five seconds making a loud noise, the cause may be a defective CRT.
- ✓ If the horizontal output transistor is continually knocked out it may be a 2.7-ohm, 7-watt isolation resistor that is burned or open. Replace the flyback and then the horizontal output transistor.
- ✓ If the horizontal output transistor fails after only one or two minutes and keeps destroying the horizontal output, replace the safety capacitors (*Fig. 3.15*).
- ✓ If the chassis is dead and the output transistor is hot, use the ESR meter to check the small electrolytic capacitors in the driver-transformer circuit. The problem can be caused by small electrolytic capacitors in the driver-transformer voltage source.
- ✓ To prevent damaging another horizontal output transistor, plug the TV cord into a variable-isolation transformer. Slowly raise the voltage on the line transformer and notice if the output transistor starts to get warm. Most horizontal output transistors run warm after operating for several hours, but not red hot. Scope the collector-output pin for a waveform. When low DC voltage and no waveform are found at the collector terminal of the output transistor, check for a leaky flyback. Notice if the transistor starts to heat up before any waveform appears. Repair the defective circuit before installing a new output transistor.

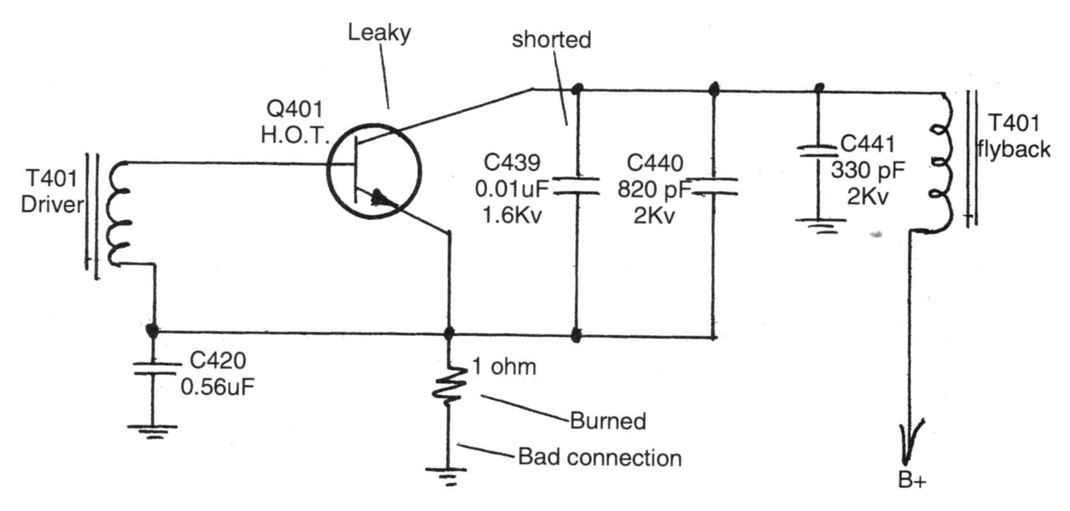

Fig. 3-15. A shorted safety capacitor can damage the horizontal output transistor.

Dead Chassis, Fuse Okay – Emerson VT1920

In an Emerson VT1920 TV/VCR, the main fuse was intact but the chassis was dead. Horizontal output transistor (Q402) was found shorted and was replaced. The output transistor has a damper diode inside. Emitter resistor R430 appeared to be burned and was replaced. By using the ESR meter, the C441 (1 μF, 160V) electrolytic in the leg of the driver transformer showed signs of ESR problems and was also replaced (*Fig. 3-16*). All driver-transformer connections were resoldered.

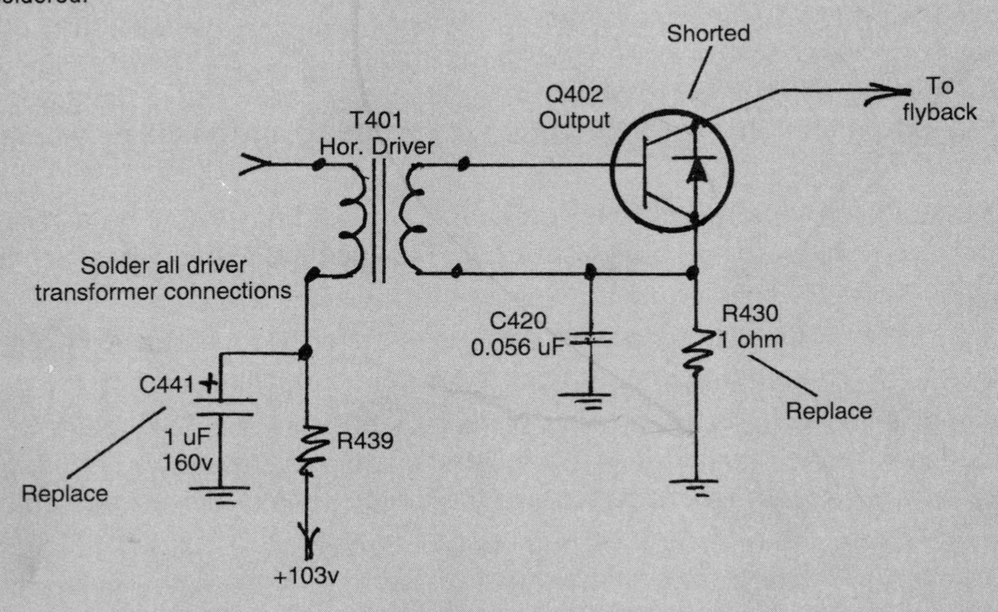

Fig. 3-16. A dead chassis with an okay fuse in an Emerson VT1920 was caused by Q402, R430, and C441.

Defective Safety Capacitors

Always replace the safety capacitors with original part numbers or exact capacity and working voltage. If a safety capacitor has a 6800 pF, 1.2 kV rating and is defective, replace it with a 6800 pF, 1.6 kV or 2 kV voltage rating. The safety capacitor is a critical component found in the HV hold-down circuit. A shorted safety capacitor can blow the fuse and damage the horizontal output transistor. When the safety capacitor becomes open or has a broken internal lead, the voltage goes very high and may cause the picture tube to arcover and fire at the CRT spark gaps (*Fig. 3-17*). A dead or smoking symptom can be caused by a shorted safety capacitor. If the set has a dead chassis and a ticking noise, replace the leaky picofarad (pF) capacitor alongside the safety capacitor. Other symptoms of a defective safety capacitor include intermittent horizontal tearing of the picture and horizontal lines in the picture. Always use the ESR meter to check the safety capacitor whenever a leaky or shorted output transistor is found.

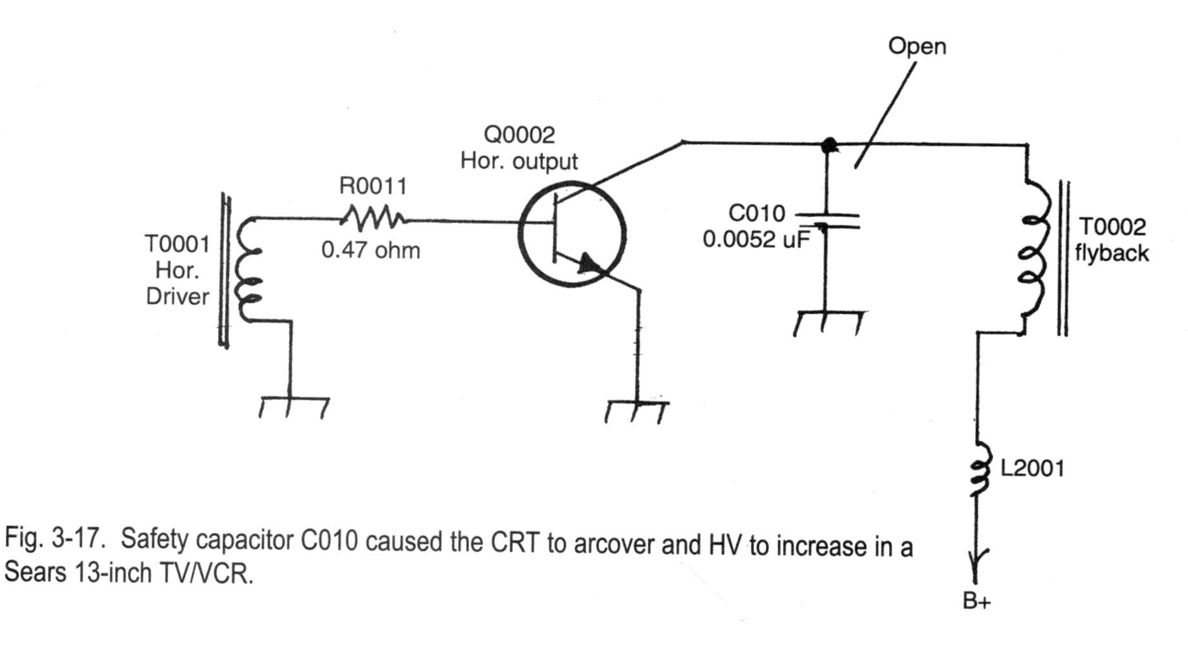

Fig. 3-17. Safety capacitor C010 caused the CRT to arcover and HV to increase in a Sears 13-inch TV/VCR.

Replace the safety capacitor if:

- ✓ The set has a dark raster.
- ✓ Only the menu is visible.
- ✓ The audio is okay.
- ✓ The set has a change in stations.
- ✓ The chassis shuts down after 10 or 15 minutes of operation.
- ✓ The chassis shuts down after five minutes (the dead chassis, open fuse symptom can be caused by a shorted safety capacitor)
- ✓ The set is dead and the horizontal output transistor fails after one or two minutes of operation.
- ✓ The set has the problem where the HV comes up momentarily and the relay clicks off and then shuts down (the dead chassis symptom can result from a shorted horizontal output transistor and safety capacitor located in the yoke or pincushion circuits).
- ✓ The set is dead and has a blown fuse.

Dead Chassis, Fuse Okay

Possible Causes:

- ✓ Leaky horizontal output transistor and isolation resistor
- ✓ Defective horizontal driver transistor.

Dead Chassis, Fuse Okay, Relay Clicks

When replacing and/or resoldering to correct this problem, also check for a defective diode and resistor at Pin 1 of the horizontal driver transformer. Although the diode might look good, if this set of symptoms occurs, replace it anyway.

Possible Causes:

- ✓ Badly soldered joint on a coil in emitter terminal of horizontal output transistor
- ✓ Defective horizontal output transistor
- ✓ Defective voltage regulator
- ✓ Burned isolation resistor
- ✓ Badly soldered joint on the driver transistor (*Fig. 3.18*)
- ✓ Defective deflection IC
- ✓ Defective safety capacitor
- ✓ Defective flyback

Fig. 3-18. A badly soldered joint on the driver transistor can cause intermittent startup.

Dead, Open, or Blown Fuse

Possible Causes:

- ✓ Leaky or shorted horizontal output transistor
- ✓ Shorted flyback
- ✓ Defective yoke assembly
- ✓ Line voltage regulator
- ✓ Defective switching or chopper transistor
- ✓ Defective zener diodes
- ✓ Shorted main-filter capacitor
- ✓ Shorted bypass capacitors in horizontal deflection yoke circuits

Poor Width

This symptom can vary in its form and causes.

- ✓ Shrinking on both sides of the raster and repeated cycling can be caused by a dried-up filter capacitor. Use the ESR meter to check all electrolytic capacitors in the voltage sources.
- ✓ Shrinking on both sides of the screen can be the result of a defective yoke or flyback.
- ✓ Suspect a defective safety capacitor when the raster is shrunk on both sides of the screen.
- ✓ A defective voltage-regulator IC may be the cause when the raster is pulled 1/4" on both sides.
- ✓ Check the low-voltage fusible resistors when the raster is pulled in on both sides.
- ✓ A defective voltage regulator and fusible resistors may be the cause when the picture is pulled in on both sides and the top and bottom (*Fig. 3-19*).

Fig. 3-19. Suspect a defective line voltage regulator IC if the screen pulls in on all sides.

- ✓ A damaged emitter resistor in the emitter circuit of the horizontal output transistor may cause a white driveline and partial collapse on both sides of the screen.
- ✓ A defective horizontal output transistor and emitter resistor can cause the sides to pull in with a pulsating raster. Replace the emitter resistor even if it looks good and tests good.

Intermittent or No Startup/Shutdown

Causes Include:

- ✓ Poorly soldered joints on the driver transformer and low resistance feeding the voltage source to the transformer.
- ✓ A defective horizontal output transistor (with high voltage and audio).
- ✓ Poorly soldered horizontal driver transformer connections, horizontal output transformer terminals, and flybacks.
- ✓ A poorly soldered joint on a coil or resistor in the base of the horizontal output transistor.
- ✓ Bad solder connections on Pins 1 and 2 on the flyback
- ✓ Intermittently functioning zener diodes in the secondary voltage sources.
- ✓ Poorly soldered connections on the deflection IC terminals.

Dead, No Startup – Magnavox CCRO95ATO4

The TV section was dead and wouldn't start up in a Magnavox CCRO95ATO4, but the fuse was okay. Badly soldered joints were found on connector CN6001, Q02, and on the AC input connections.

TV Shuts Down

Horizontal tearing begins across the screen; the chassis shutdown is caused by a deflective zener diode in the failsafe circuit. If the TV has problems such as the horizontal becoming intermittent, going off frequency, and shutting down, it may be the result of a badly soldered joint on the flyback. When the TV chassis shuts down after 30 minutes of operation, the likely cause is badly soldered joints on the horizontal driver transformer. Badly soldered joints on the horizontal driver transformer sometimes can also cause the fuse to blow or the TV to shut down intermittently and then come back on when the chassis is moved.

A leaky output transistor, burned emitter resistor, or voltage regulator all can cause a problem in which the relay clicks, the power indicator comes on, and the set goes off at once. A bad yoke will cause the set to smoke, come on momentarily with the horizontal sweep collapsed, and then shut down. If, after the set operates for a few minutes, the relay clicks on and off intermittently and the set shuts down, look for a badly soldered joint on Pin 3 of the horizontal driver transformer. A poor collector-terminal connection

No Startup - 19-Inch Emerson

In a 19-inch Emerson that wouldn't start up automatically, the horizontal driver-transformer (T401) terminals were soldered. A quick-drive waveform was taken on the base of Q402 (horizontal output transistor) with no results. Locating the horizontal deflection (IC 401), a horizontal waveform was taken at Pin 6 of the horizontal output pin, but still no waveform. The voltage-supply pin (24) of IC 401 measured only 3.1V. The supply voltage normally should measure around 11.1V.

The 11V source was developed through a silicon diode from the 12V source in the switching transformer secondary. The 12V source was normal. After taking resistance tests on each pin of IC 401, the deflection IC was removed and replaced. Thus, replacing IC 401 restored the no-start up chassis (*Fig. 3-20*).

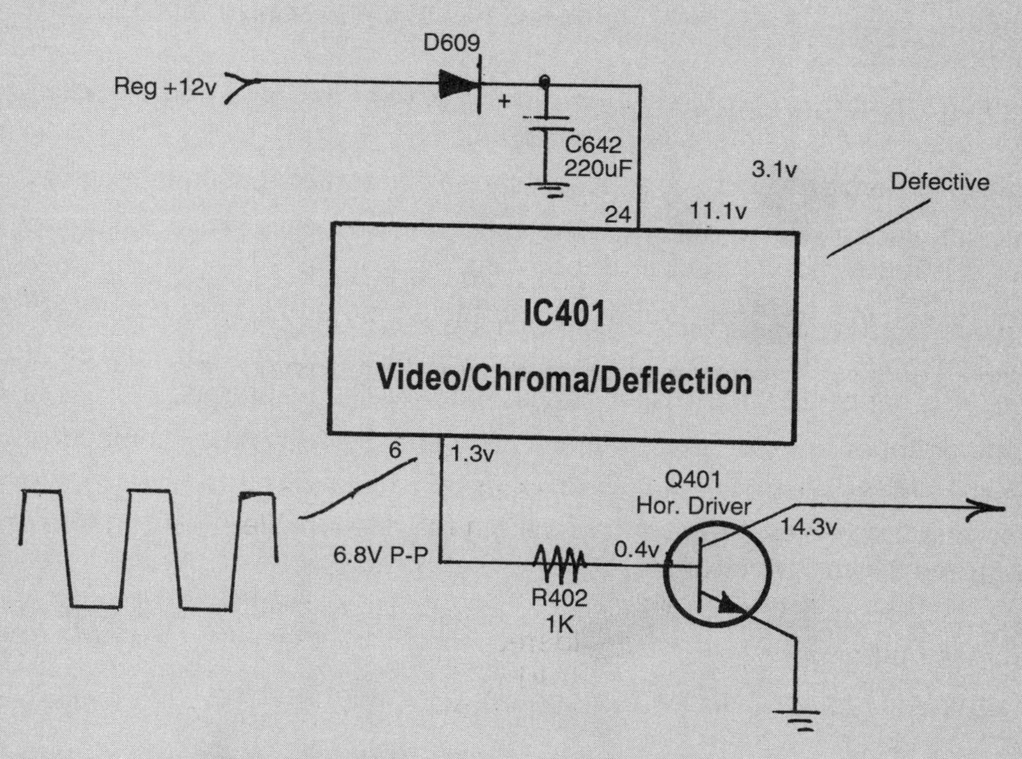

Fig. 3-20. An Emerson 19-inch TV would not start up because of a defective deflection IC.

on the driver transistor also causes intermittent shutdown. When the chassis shuts down intermittently after two or three minutes, replace a leaky driver and horizontal output transistors.

When the TV comes on and shuts down, replace the 220 µF electrolytic in the voltage source of the deflection IC. When the TV intermittently turns off and then back on, double-check the zener and fixed diodes in the voltage source.

Horizontal Lines

Horizontal lines in the picture often occur in conjunction with any of these other symptoms or alone. Use the ESR meter to check all capacitors.

- ✓ Horizontal lines in the picture might occur when the horizontal circuit drifts off frequency. To fix this problem, check capacitors in the horizontal deflection countdown circuits.
- ✓ The intermittent-dead problem, with clicking of the relay, and horizontal lines on the raster, results from a defective deflection IC.
- ✓ A defective horizontal hold control can cause horizontal lines in the picture.
- ✓ Defective electrolytic capacitors in the deflection and driver voltage sources can cause the horizontal off-frequency problem.
- ✓ Check the main filter capacitor for horizontal tearing and loss of horizontal sync.
- ✓ If the horizontal circuits are off frequency, accompanied by horizontal bars, check the electrolytic capacitors in the driver and horizontal output stages.
- ✓ When horizontal lines appear on the screen and a smoky smell comes from the chassis, check the safety capacitors.
- ✓ Numerous black lines that overload the picture and look like AGC can be caused by a defective 33-volt zener diode.
- ✓ Check for a badly soldered joint on the flyback when the horizontal is intermittent and goes off frequency.
- ✓ If the tuner is mounted on the main board, horizontal flashing lines can be caused by poor grounds on the tuner shields.
- ✓ A defective EEPROM can cause horizontal flashing lines in the picture.

Ticking Noise

A ticking noise may occur alone or with other symptoms. Here are some of the situations where it might be encountered.

- ✓ If the TV is dead but has a ticking noise, check for a badly soldered joint on the horizontal driver transformer.

- ✓ Replace the horizontal output-emitter resistor for a ticking noise in the flyback (a defective horizontal output transistor can also cause a chunk-chunk noise instead of a ticking noise in the flyback).
- ✓ If a ticking noise is heard, a poor ground connection of the emitter resistor or a leaky horizontal output transistor may be the cause.
- ✓ A bad horizontal output transistor can cause the set to try to start up making a ticking sound.
- ✓ If the TV plays for several minutes, goes off, and then tries to start up making a ticking sound, it may be the result of a bad horizontal output transistor socket.
- ✓ Badly soldered joints on a resistor feeding the driver primary winding may cause a ticking noise with a dead TV.
- ✓ A poor connection between the driver transformer and horizontal output transistor can result in ticking.
- ✓ Replace leaky picofarad (pF) safety capacitors in the collector circuit of the horizontal output transistor in order to get rid of the ticking noise, and also look for burned resistors or increased resistance off of the deflection IC terminals.

Tearing in the Picture

Intermittent, horizontal tearing of the picture can be caused by a defective safety capacitor. Loss of horizontal sync and tearing in the picture can also be caused by a defective filter capacitor. Low-voltage decoupling electrolytics in the horizontal voltage sources can cause horizontal bars and tearing in the picture. Horizontal tearing followed by shutdown can result from a defective 6-volt zener diode in the failsafe circuit. When horizontal tearing occurs, check all components in the horizontal deflection and output circuits.

Squealing, Buzzing, and Whining Noises

If there is a high-pitched sound when the TV tries to turn on, it may be due to a defective horizontal output transistor. A badly soldered joint on the coupling coil to the base terminal of the output transistor can create an intermittent high-pitched squeal. A TV that is dead but has a whining sound can be the result of a leaky horizontal output and poorly soldered joints on Pins 1, 2, and 10 of the flyback. A loud squeal can result from a badly soldered joint on a coil in the horizontal deflection-yoke circuit.

Check for a defective horizontal output transistor and yoke if the TV squeals but is dead with a relay that clicks. If there is a faint squealing noise when the set is turned on, replace the horizontal output transistor. A buzzing sound inside the chassis may be a bad yoke.

Noisy – Zenith TVSA1320

In a Zenith TVSA1320 TV/VCR a terrible noise was heard from inside the chassis when it was fired up. At first, the flyback was suspected. After prodding with an insulated tool, it was determined not to be the problem. It turned out to be a very noisy horizontal driver transformer (T571), which was replaced (*Fig. 3-21*).

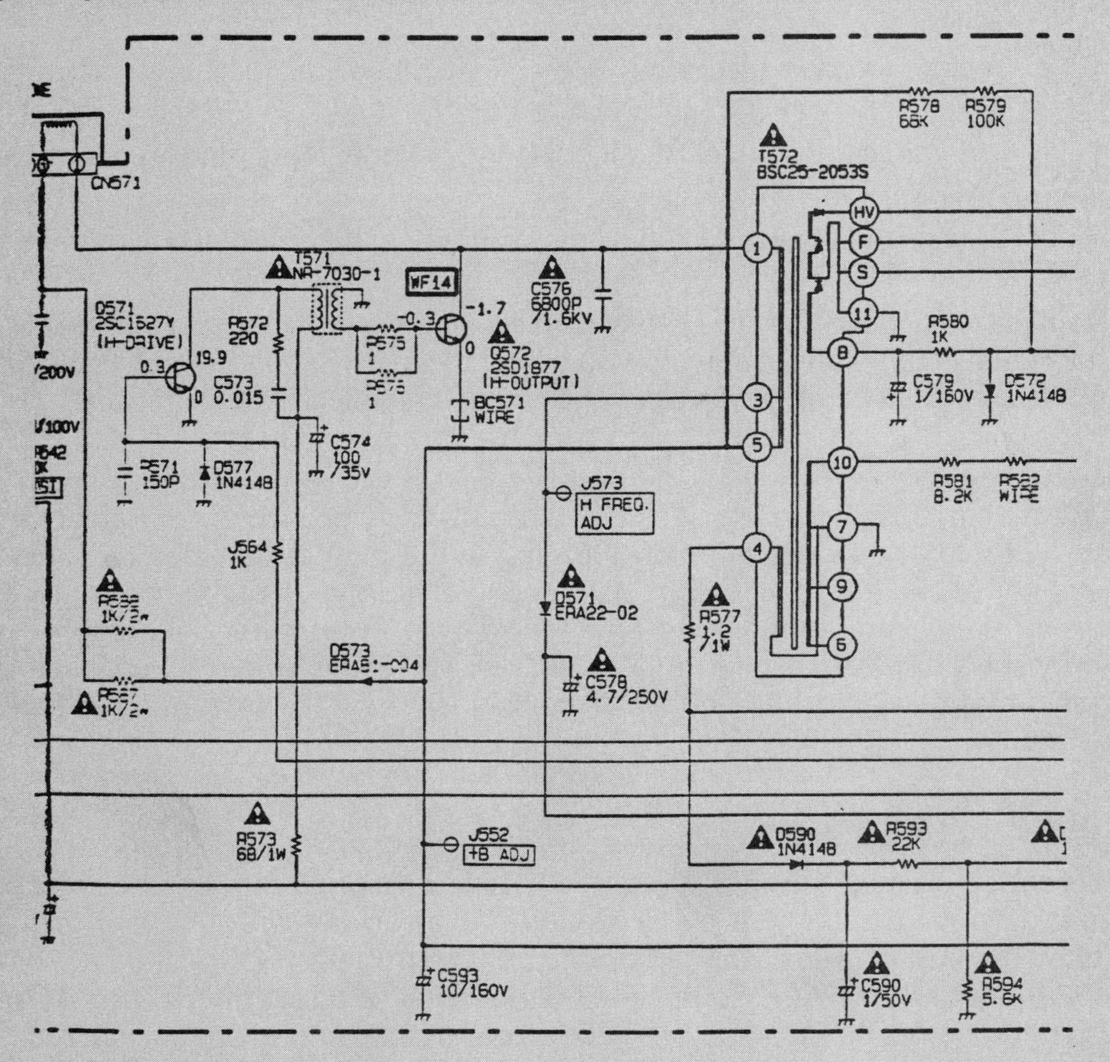

Fig. 3-21. A noisy T571 horizontal driver transformer was replaced in a Zenith TVSA1320 TV/VCR.

Defective Yoke Circuits

If the set goes dead, a buzzing sound comes from inside the chassis, intermediate shutdown occurs, or the B+ fuse blows after operating 30 minutes, the yoke circuit may be defective.

If there is a defective yoke circuit, be aware of the problems it can cause.

- ✓ A bad yoke assembly can cause the set to come on momentarily and begin to smoke.
- ✓ An open horizontal winding or poor yoke connection can cause a vertical bright line with no horizontal sweep.
- ✓ A defective yoke can cause a dead chassis with an open fuse.
- ✓ Leaky or shorted capacitors in the yoke circuits can cause a dead chassis.
- ✓ Poor contacts or corroded contacts on the yoke socket can cause shrinkage from both sides of the picture.
- ✓ A bad solder joint in an inductance coil within the horizontal yoke circuit can produce a loud squealing noise.
- ✓ Bad connections of the yoke socket can result in arcing lines in the picture.
- ✓ A defective yoke may have burned spots, a bad smell, and create noise in the speaker at startup.

Flyback Problems

When soldering the horizontal driver transformer terminals, go directly to the flyback and solder all connections to the PCB. Replace the flyback if the horizontal output transistor and isolation resistor are constantly being knocked out.

A chassis that is dead with a good fuse – with a 52V source feeding the horizontal driver transistor and horizontal output transistor – can be caused by a shorted horizontal output transistor and flyback. Additional information on flyback problems is addressed in Chapter 5.

- ✓ A defective flyback or horizontal output transformer can keep damaging the horizontal output transistor.
- ✓ A leaky flyback winding to ground can open the fuse and show a short between the collector and ground of the horizontal output transistor.
- ✓ A flyback can be damaged by a defective or open safety capacitor, causing the flyback to arc over with excessively high voltage. And, while arcing internally, the flyback can destroy the output transistor.
- ✓ Poorly soldered connections at Pins 1 and 2 of the flyback can cause symptoms including intermittent horizontal sweep, a vertical white line, and cycling off and on.

- ✓ A leaky output transistor and bad solder joints on Pins 1, 2, and 10 of the flyback can cause whining and a dead chassis.
- ✓ Poorly soldered joints on the flyback may cause the horizontal sweep to go off frequency intermittently.

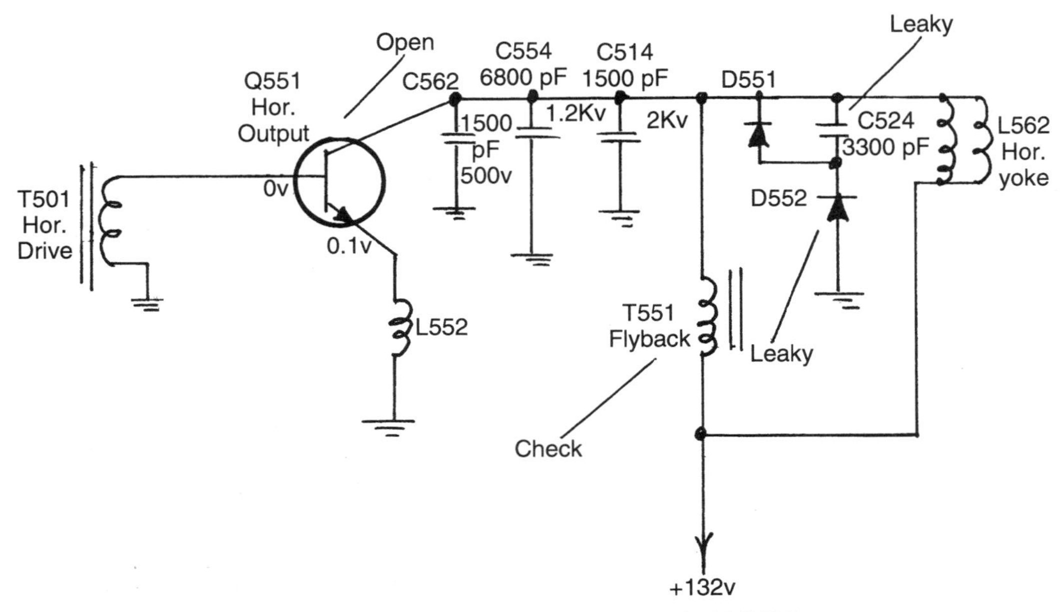

Fig. 3-22. Several components can be defective in the horizontal circuits of a TV/VCR.

Storm Damage and Power Surges

Although most of the problems resulting from lightning or power-surge damage are found in the tuner and power supply, other components in the horizontal circuits also can be damaged. After storm damage has occurred, these are some of the problems to look for:

- ✓ A damaged deflection IC.
- ✓ Blown fusible resistors.
- ✓ A dead chassis with an okay fuse.
- ✓ A damaged isolation resistor.
- ✓ A damaged horizontal output transistor.

Replace the horizontal output transistor and isolation resistor and also check for:

- ✓ Stripped PC wiring to the tuner assembly.
- ✓ A damaged horizontal driver transistor.
- ✓ Stripped foil around the horizontal driver transistor.
- ✓ Badly soldered connections on the driver transformer.
- ✓ A burned 1-ohm emitter resistor.

Chapter 4: Troubleshooting the TV Vertical Section

The vertical deflection IC in the TV/VCR chassis develops a vertical drive waveform that is fed to a vertical deflection-control IC, IC541. This vertical deflection-control IC might include a vertical trigger input, one-shot multiplier, ram generator, vertical size-control switch, vertical drive, protection, and vertical output. The vertical output is fed directly to the vertical yoke winding (*Fig. 4-1*).

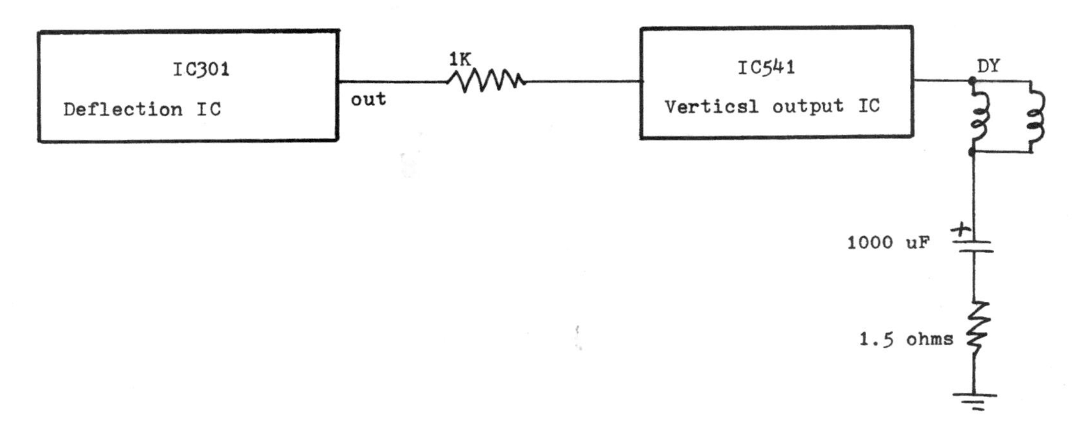

Fig. 4-1. Block diagram of a deflection IC301 and vertical output IC541 that generates the vertical circuits.

In the early TV chassis', vertical output circuits were made up of two transistors operating in a push-pull operation with a deflection IC. Today, in the vertical section in the TV/VCR chassis, a vertical countdown circuit is included in the horizontal defection IC. In a Sears 9-inch TV/VCR, the vertical drive waveform is taken from a VCO terminal Pin 18. This vertical drive signal is connected to a connector and plug (CN575 to CN576) that is joined to the vertical output IC (*Fig. 4-2*).

Vertical Problems

Scope the vertical output pin from the vertical deflection IC to determine if the deflection IC is functioning. No vertical sweep might display a bright horizontal line on the screen. Insufficient vertical sweep is when the vertical raster is only two or three inches high or cannot fill the entire screen. If the picture is pulled up from the bottom, the cause might be a defective vertical circuit. Intermittent vertical sweep is when the vertical raster collapses and then fills out the picture tube again. Improper height is a lack of complete vertical sweep.

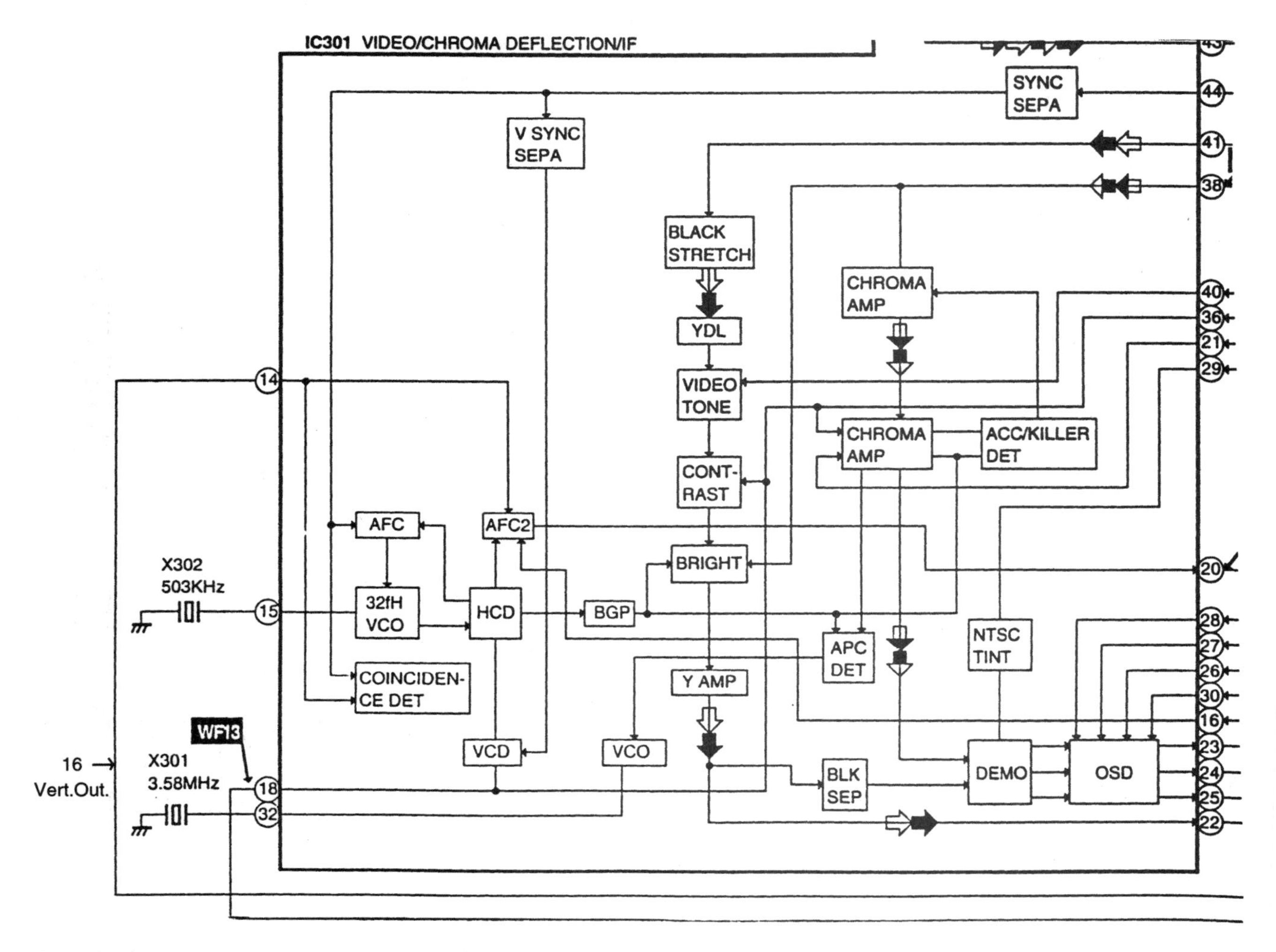

Fig. 4-2. Scope the vertical output pin 18 of vertical deflection IC301 to determine if deflection circuits are functioning.

Improper vertical sweep is indicated when a portion of the raster is black. Likewise, a partial loss of vertical sweep also constitutes improper vertical sweep. Poor vertical linearity results when the scanning lines are not linear across the entire screen – that is, the lines are wider or pressed too tightly together. Poor vertical linearity causes scanning lines to be compressed at the top of the screen and stretched over the remainder of the raster. If there is no vertical sweep at the top of the screen, a defective top output transistor or IC may be the cause.

Severe vertical jumping and bouncing occurs when the vertical raster is unsteady and moves up and down. Vertical crawling can be caused by a defective electrolytic in the vertical voltage source. Vertical foldover happens when the picture is rolled up in the center or at the top or bottom of the raster. When the picture begins to move or roll up or down, it is out of vertical sync or lock. Bad solder joints and foil connections can cause intermittent and partial loss of the vertical sweep. Use the ESR meter to help locate bad connections and broken trace or foil in PC wiring.

Vertical Output IC

The vertical IC output circuit might consist of a vertical trigger input at Terminal 2 of deflection IC551. A one-shot multiplier, ramp-amp generator, vertical size control SW, vertical drive, protection, pump-up, and vertical output are included in the output IC of a Sears 13-inch TV/VCR. The amplified output signal at Pin 12 is fed directly to the vertical deflection yoke. Vertical size and vertical positioning controls are included in the vertical output-deflection control circuit (*Fig. 4-3*).

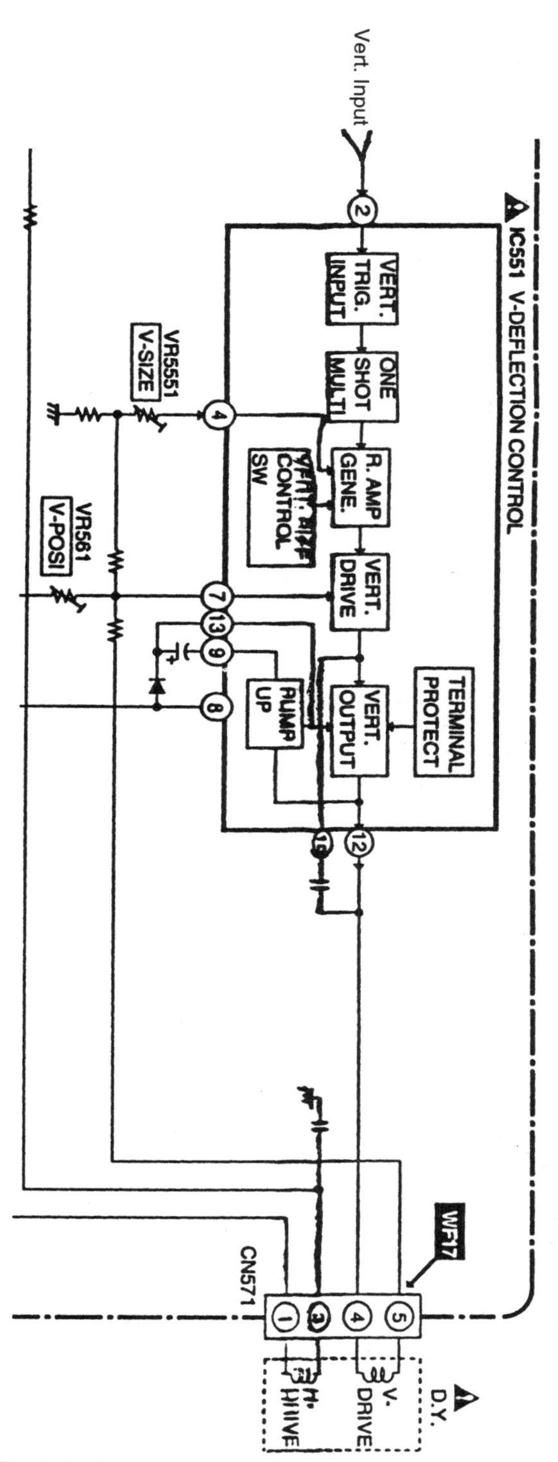

Fig. 4-3. Block diagram of the vertical deflection control IC551 in a Sears 13-inch TV/VCR combination.

Signal Tracing Vertical Circuits

Start at the vertical output pin of the vertical deflection IC and scope the oscillator-drive waveform. If there is no waveform or there is an improper waveform at the output pin, check for a defective vertical deflection IC or improper voltage source (*Fig. 4-4*). Trace the deflection waveform to the input terminal of the vertical output IC. Often, the vertical drive waveform goes through a plug and socket before reaching the output IC. Go directly to the output pin of the vertical output IC for a more amplified vertical signal.

The vertical output signal can be traced directly to the vertical deflection-yoke winding. If the signal is going into the vertical output IC but not out, a defective output IC or improper supply voltage may be the problem.

Many different vertical problems can occur in the vertical output circuits. Common problems include no vertical sweep and a white horizontal line. Improper or partial loss of vertical sweep can be caused by a defective electrolytic, a defective diode, or improper voltage to the output transistor.

Fig. 4-4. Locate the vertical deflection and driver circuits found with the video/chroma/deflection/IF (IC3301).

Coupling and decoupling electrolytics within the output-sweep circuit can cause improper vertical sweep or foldover problems, while dried-up electrolytics or those with ESR measurements can cause incomplete vertical sweep at the bottom of the picture, as well as intermittent sweep, foldover, and lines in the top of the picture. Use the ESR meter to check each electrolytic for correct capacitance and ESR problems.

Vertical Waveforms

The vertical deflection-IC output waveform may differ from other TVs when the waveform is taken from the vertical frequency oscillator (VFO) or with an internal vertical driver stage. The deflection output may appear as a vertical pip or a sawtooth waveform (*Fig. 4-5*). The vertical deflection output is fed through a 1 kilohm or 1.5 kilohm resistor to the input terminal of the vertical deflection-control IC.

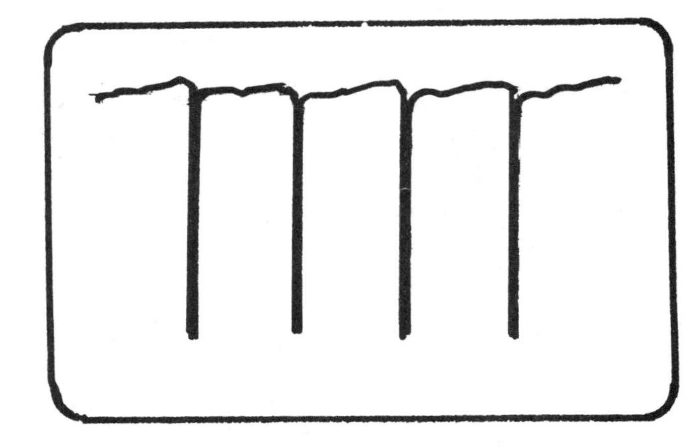

Fig. 4-5. The vertical input waveform might look like a pip or inverted line waveform to the output IC.

Take a critical vertical waveform at the input and output terminals of the deflection-output IC. Most vertical waveforms are very unstable compared to other TV waveforms. The vertical output waveform is coupled directly to or through a large electrolytic between the output and yoke terminal or after the yoke winding (*Fig. 4-6*). A normal output waveform at the vertical yoke winding indicates that the vertical circuits are functioning. Notice that a high supply voltage is fed to the vertical output IC (25V to 33V).

Fig. 4-6. The vertical waveform of vertical output IC.

Vertical IC Output Circuits

If the vertical output IC (*Fig. 4-7*) is defective, any of the following conditions might occur.

- ✔ No vertical sweep
- ✔ Insufficient vertical sweep
- ✔ Intermittent vertical sweep
- ✔ Improper supply voltage
- ✔ No vertical sweep on the bottom half of the picture
- ✔ No sweep on the top half of the raster
- ✔ Poor vertical linearity
- ✔ A bouncing raster
- ✔ Vertical foldover

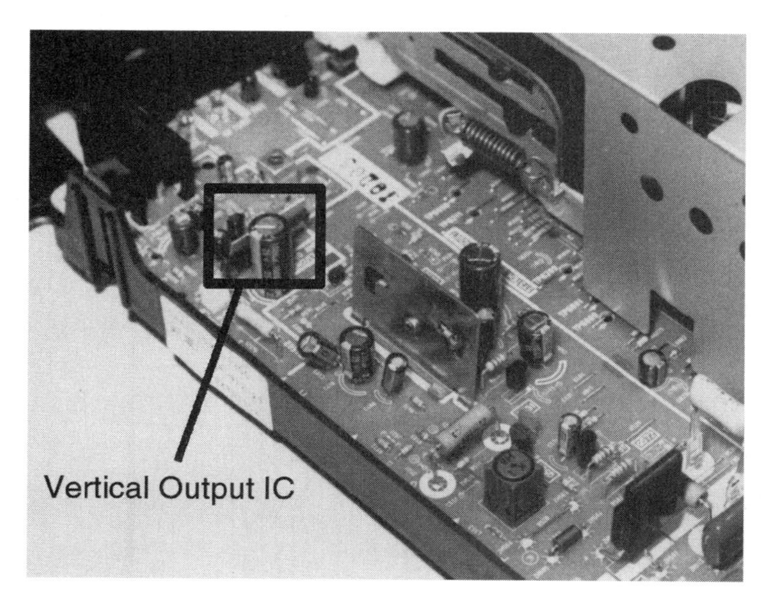

Fig. 4-7. The vertical output IC on a TV/VCR chassis.

The leaky or shorted output IC can lower the supply voltage at Pin 13. When low voltage is measured at the supply pin, suspect a leaky or shorted output IC or improper voltage applied to the supply pin.

Remove the supply-voltage pin by applying solder-wick around the terminal and removing the excess solder. Flip the IC pin with the blade of a knife or a small screwdriver to make sure the pin is loose. Take a resistance measurement between the removed IC pin and the common ground. A low resistance (under 500 ohms) indicates a leaky IC. If the resistance is under 10 ohms, the IC is internally shorted.

To determine if the IC is leaky, take a voltage measurement on the foil that was soldered to Pin 13 and notice if the voltage returns to normal. Replace the leaky or shorted IC if you get a low-resistance-to-ground measurement and a normal voltage measurement on the voltage supply trace or PC wiring (*Fig. 4-8*).

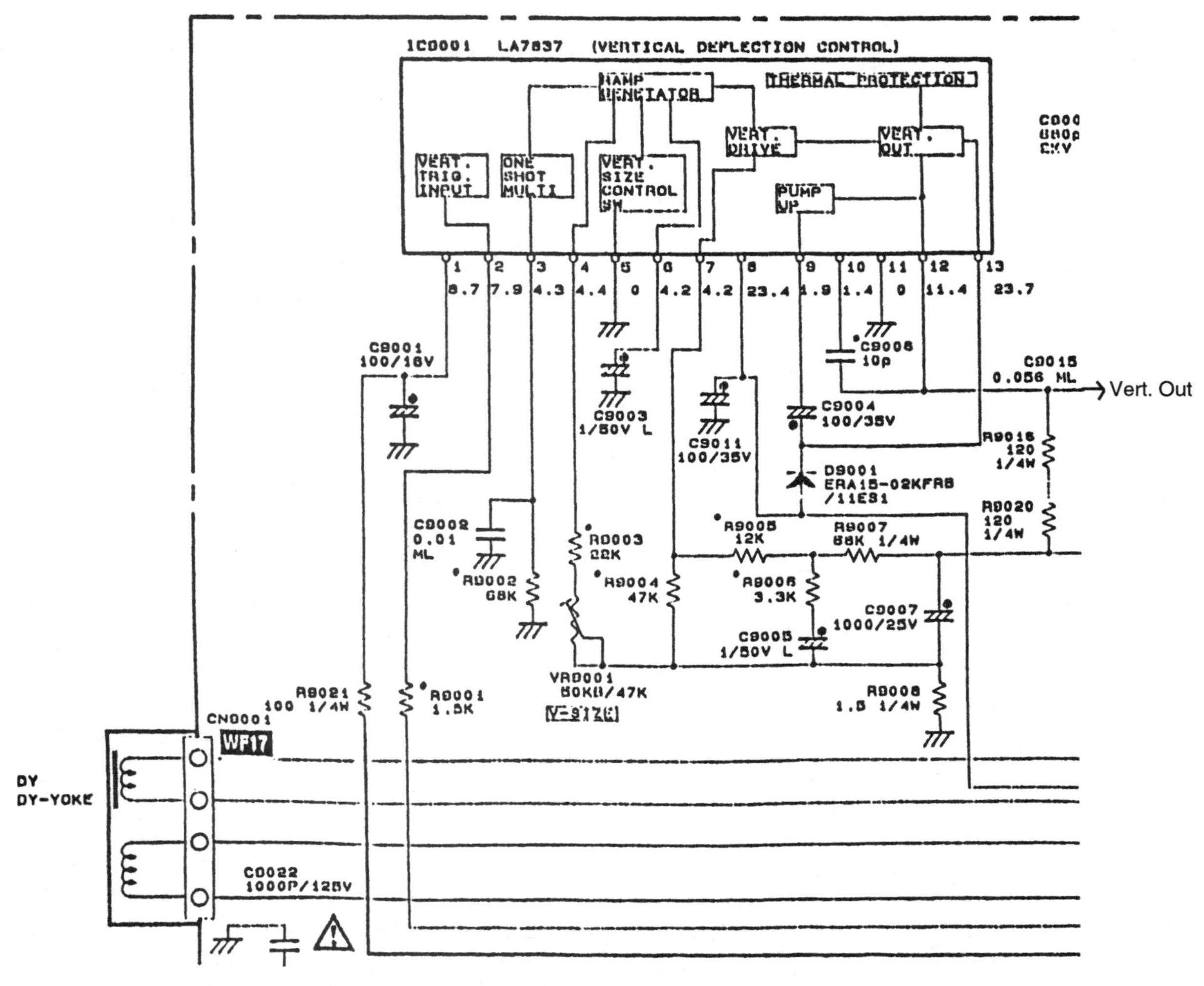

Fig. 4-8. A vertical deflection IC IC0001 as found in a Sears 9-inch TV.

Sometimes, the vertical output IC might be defective with a normal or near normal supply voltage. Check for an input and output waveform to determine if the signal is coming in and if there is no output waveform. Take a voltage measurement on each pin terminal of the output IC to the common ground and compare with the schematic. Measure each pin resistance to the common ground.

Check for a defective output IC if the pin voltages do not measure up to the schematic. A low resistance from one or more pins might indicate a leaky output IC. Use the ESR meter to check each electrolytic connected to an IC output pin. If all components tied to each pin have normal resistance and voltage measurements and you have no output waveform, then replace the defective IC. The defective output IC tests could be normal, but the IC still may need to be replaced in order to restore the vertical output signal.

No Vertical Sweep

Sometimes, by turning up the screen control, the raster will show up as a no-raster or vertical sweep. Often, a white horizontal line across the middle of the screen indicates a problem of no vertical sweep. No vertical sweep can be caused by:

- ✔ A defective deflection IC
- ✔ A bad solder joints on the deflection IC terminals
- ✔ Open traces between resistors tied to the vertical size control
- ✔ A 1-ohm, ½-watt or 0.47-ohm, ½-watt fusible resistors in the voltage supply source
- ✔ A defective vertical output IC

Check the correct vertical voltage source that comes from a tap on the secondary winding of the flyback. Also, check all electrolytics in the vertical output IC, testing them with the ESR meter. Check electrolytic capacitors C541 (47 μF), C543 (1 μF), and C544 (100 μF) on pin Terminals 1, 6, 9, and 13 of IC541 (*Fig. 4-10*).

Leaky or open diodes in the vertical deflection-control IC circuits also can cause a no vertical sweep problem. Resolder all vertical output-IC pins for a no vertical sweep symptom. Also, check for a badly soldered joint on Pin 1 of IC541 with this symptom.

Sometimes, defective electrolytic capacitors with ESR problems can cause the vertical output IC to fail, and this problem cannot be located with a regular capacity tester; instead, use an ESR meter. These electrolytics might be low in capacitance and high in ESR.

A shorted vertical output IC can destroy the secondary voltage source and cause other transistor and diode problems. A defective zener diode or voltage regulator in the supply-voltage source of the output IC can cause a no-sweep symptom. Check each diode out of the circuit; remove one end of the diode for a normal test. When a defective diode is located, always check small, low-ohm or fusible resistors tied to the diode.

No Vertical Sweep - Sears 934.44727790

A Sears 9-inch TV/VCR (934.44727790) came in with only a horizontal white line. A quick scope test on Pin 18 of IC301 indicated no vertical sweep. Since both horizontal and vertical sweep are developed in the same deflection IC, a defective component in the vertical drive circuits could be at fault. The horizontal drive waveform was present with no vertical deflection. All voltages on the IC were quite close, except that Pin 18 was low at 1.5V. All electrolytics were checked in the vertical circuits and appeared normal. The problem of no vertical sweep was solved by replacing IC301 (*Fig. 4-9*).

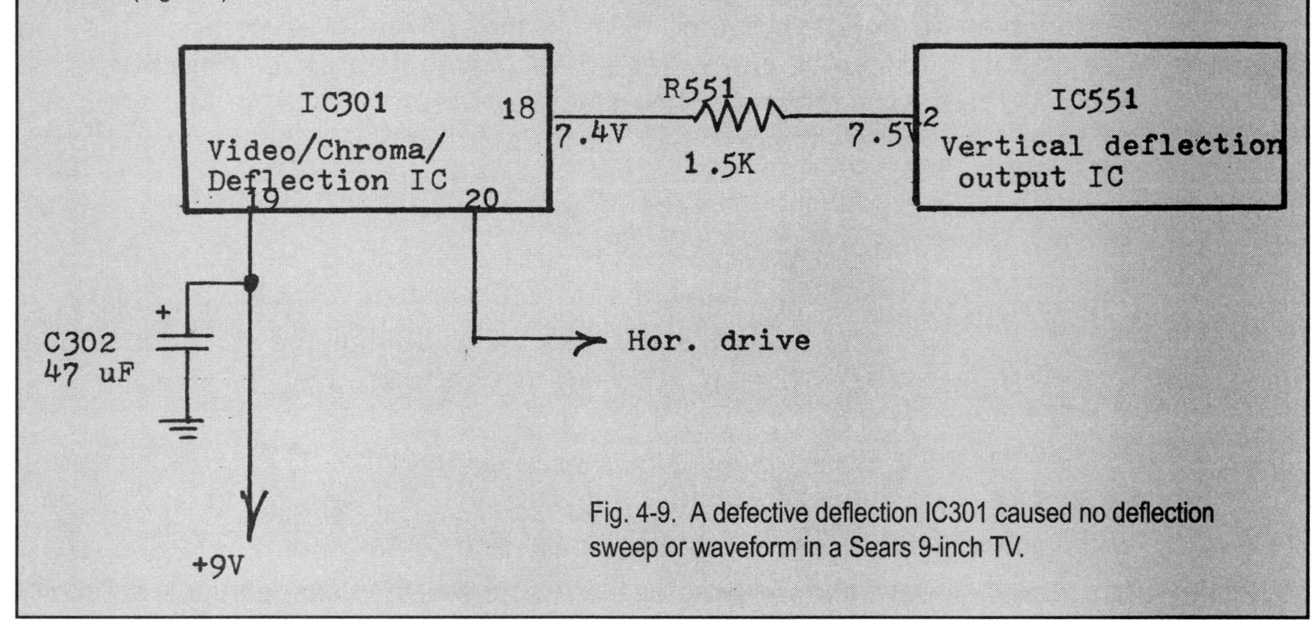

Fig. 4-9. A defective deflection IC301 caused no deflection sweep or waveform in a Sears 9-inch TV.

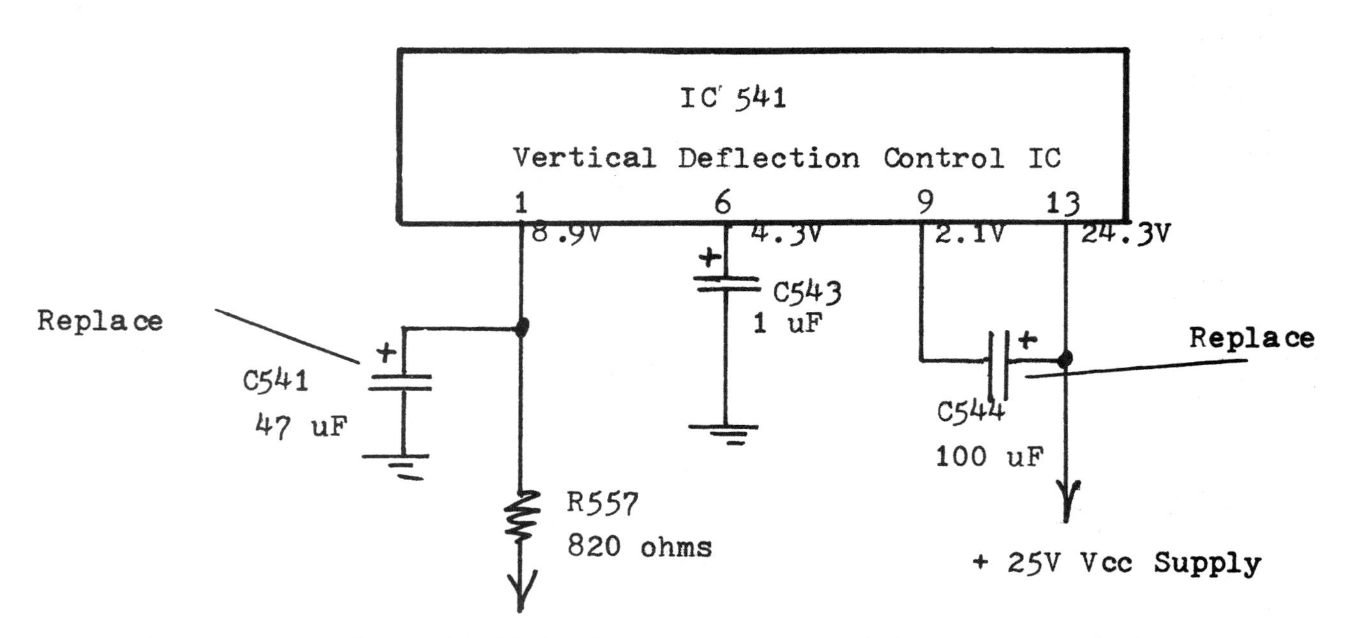

Fig. 4-10. Check electrolytics C541 and C544 when there is no vertical sweep in the vertical output circuits.

Some other causes of a no vertical sweep condition include:

- ✔ An open coupling inductance or low-ohm resistor between the deflection and output ICs.
- ✔ An open coupling capacitor (1000 µF) between the vertical output IC and yoke.
- ✔ A badly soldered joint on the coupling capacitor (1000 µF).
- ✔ An open vertical yoke winding or open return resistor in the leg of the deflection yoke.
- ✔ A defective EEPROM found in the chassis of some TVs.

Loss of Vertical Sweep at the Bottom

- ✔ If there is no vertical sweep on the bottom half of the picture and the top half has only lines, replace the vertical output IC.
- ✔ If there is incomplete vertical sweep at the bottom with poor linearity, check the 1 µF electrolytics.
- ✔ If there is a loss of one to two inches of vertical sweep at the bottom of the screen, and if adjustment of the height control only stretches the linearity of the raster, then check for leaky zener diodes and a low supply voltage to the vertical circuits.
- ✔ Partial loss of vertical sweep at the bottom of the screen with poor linearity at the top can be caused by defective 1 µF and 2.2 µF electrolytics. These capacitors might check okay on the capacitor tester, but not with the ESR meter (*Fig. 4-11*).

Fig. 4-11. Loss of vertical sweep at the bottom of the picture or raster can be caused by a defective vertical output IC.

- ✔ Check all 100 µF electrolytics in the vertical circuits that can cause a vertical sweep loss of 1½ inches at the bottom and 3½ inches at the top of the screen. These capacitors seem to dry up and have a very low capacity.
- ✔ A defective 47 µF electrolytic in the vertical circuits can cause a vertical sweep loss of 3 inches at the bottom of the screen, a 2-inch band of white, and dark lines at the bottom of the remaining raster.
- ✔ When there is no vertical sweep on the bottom half of the screen and only six inches at the top with foldover, use the ESR meter to check all 1 µF electrolytics in the vertical output circuits. A bad solder joint on the 1 µF capacitors can cause a loss of vertical sweep at the bottom of the screen.
- ✔ No vertical sweep at the bottom also can result from a defective 1000 µF coupling capacitor in the vertical output circuits.

Loss of Vertical Sweep – Panasonic PVM2021

In a Panasonic PVM2021 TV/VCR, a loss of vertical sweep at the bottom of the screen was found. All voltages were checked on the vertical output deflection IC451 (*Fig. 4-12*). The supply voltage on Pin 7 was quite close (+26.5V). Electrolytic capacitors C402 (470 µF), C408 (1 µF), and C409 (100 µF) were checked with the ESR meter. ESR measurements solved the problem and C402 and C408 were replaced.

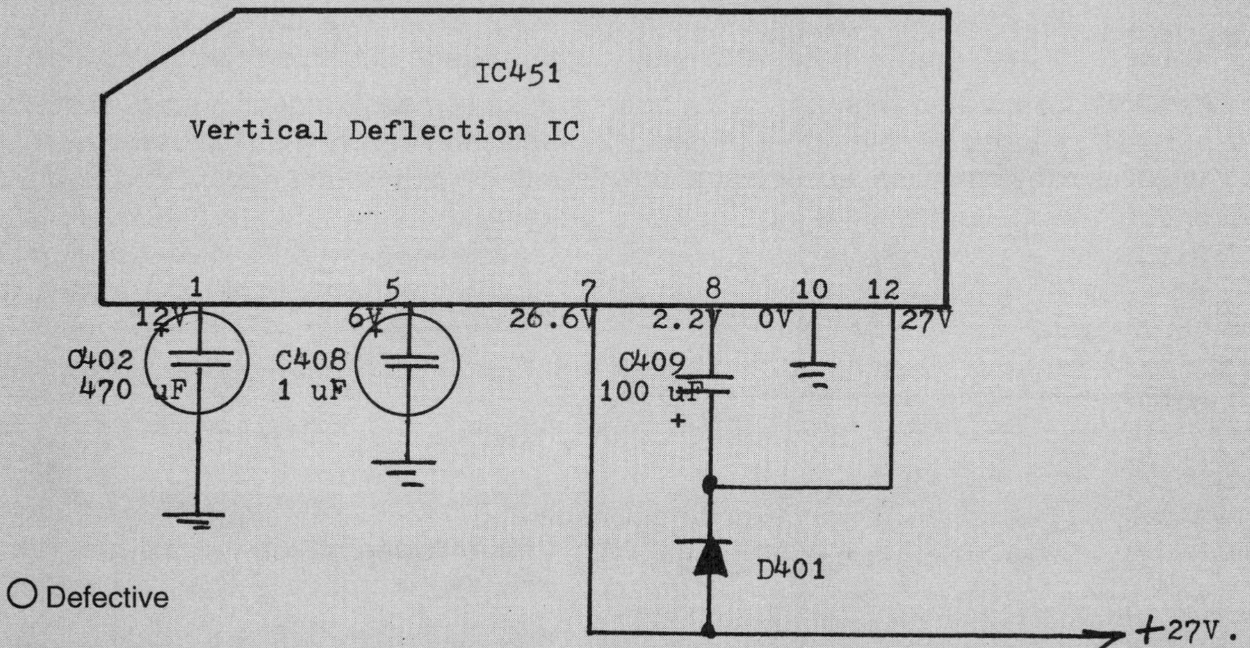

Fig. 4-12. C402 and C408 were replaced when there was a loss of vertical sweep or picture in a Panasonic PVM2021A.

Improper Vertical Sweep

Improper vertical sweep occurs when the raster cannot be fully extended without stretching the linearity at the top or bottom of the picture. A raster that lacks one or two inches of vertical sweep can be called improper vertical sweep. The picture that is only two or three inches in the center of the screen might be called poor vertical sweep. Improper vertical sweep can occur at either the top or bottom of the picture.

Most sweep problems are developed in the vertical output circuits. A defective EEPROM found in some TVs can also cause improper vertical sweep.

If there are only three inches of vertical sweep, check for poorly soldered joints on the vertical output-IC pins. If there is insufficient vertical sweep, check for defective diodes in the vertical output circuits. A loss of one inch of vertical sweep at the top might be caused by 33 μF and 47 μF electrolytics in the vertical deflection and output circuits. A defective vertical output IC can cause five or six inches of vertical sweep.

Improper vertical sweep can be seen when advancing the screen control. Incomplete vertical sweep can be caused by defective electrolytics in the voltage-supply line. For two or three inches of vertical sweep, check the large 1000 μF coupling capacitor in the vertical output and yoke circuit. For improper vertical sweep, substitute all 1 μF electrolytics.

In older TVs experiencing insufficient vertical sweep, check the vertical output transistor. Resolder all terminal leads of the output transistor for improper vertical sweep. If there is partial loss of vertical sweep, check all electrolytics in the vertical output and feedback circuits. If there is insufficient vertical sweep, readjustment of the height control may be necessary in order to return to full sweep. Use the ESR meter to check the 1 μF to 2.2 μF capacitors connected to the height control. Sometimes, in order to solve the insufficient vertical sweep problem, you might have to replace two or three electrolytics in the output and feedback circuits. If there is improper vertical sweep, you should replace the large filter capacitors on the voltage supply source.

Insufficient vertical sweep at the top and bottom of the screen can result from a defective diode tied to the vertical output IC and fusible resistor in line from a pin terminal on the flyback. To resolve an intermittent and partial loss of sweep, replace both the diode and the electrolytic tied together in the vertical output IC. Replace the vertical size control in order to resolve only two inches of vertical sweep. Only two to three inches of vertical sweep can also result from a defective deflection yoke.

If the TV is dead but the fuse checks out okay, there could be several causes, including:

- ✔ A leaky vertical output IC.
- ✔ A fusible resistor.
- ✔ A 100 µf electrolytic tied to the vertical output pins.
- ✔ Shorted bypass capacitors off of the vertical output IC.

Check for correct resistances of the small resistors in series with the yoke winding to the common ground, and for defective resistors and capacitors in the pincushion circuits outside the vertical stages to cause improper vertical sweep. In order to repair insufficient vertical sweep, always check circuits outside the original vertical circuits that might be tied to the vertical output.

Loss of Vertical Sweep at the Top

For partial loss of the top half of the screen and a stretched portion of the remaining raster, check the vertical output circuits for defective electrolytics (100 µF) tied to the vertical output IC. If there is a loss of three inches at the top, resolder all terminals on the vertical output IC (*Fig. 4-13*). For loss of a few inches of vertical sweep at the top of the raster, substitute small 33 µF to 4 µF electrolytics and check for badly soldered joints of resistors in the vertical output.

Fig. 4-13. Loss of vertical sweep at the top of the screen can be caused by a defective output IC and electrolytic capacitors.

For insufficient vertical sweep at the top, replace the vertical output IC. Remember, it is possible that a new replacement IC might be defective. Incomplete vertical sweep at the top with a dark 2-inch band on the left side and tearing of the picture on certain channels can be caused by 220 to 470 µF electrolytics in the voltage supply source.

For no vertical sweep on the top half of the screen, check for defective silicon diodes in the vertical output IC circuits. Loss of vertical sweep at the top four inches of the raster can be caused by a defective 33 µF, 160V electrolytic. Loss of the entire top half of the vertical sweep can result from a defective 47 µF, 63V electrolytic with ESR problems.

Improper Vertical Sweep - Panasonic PVM20201

In this Panasonic PVM2021 TV/VCR, the vertical waveform was normal at Pins 2 and 6 of vertical output IC451. Improper vertical output waveform was found at Pin 11. All voltages on IC451 were checked and compared to the schematic. Pin 11 had zero voltage and Pin 7 was low (15.3V). Replacing the vertical output IC451 and electrolytic C409 (100 μF) solved the improper vertical sweep symptom (*Fig. 4-14*). IC451 (LA7835) output IC can be replaced with a universal NTE 1855 IC

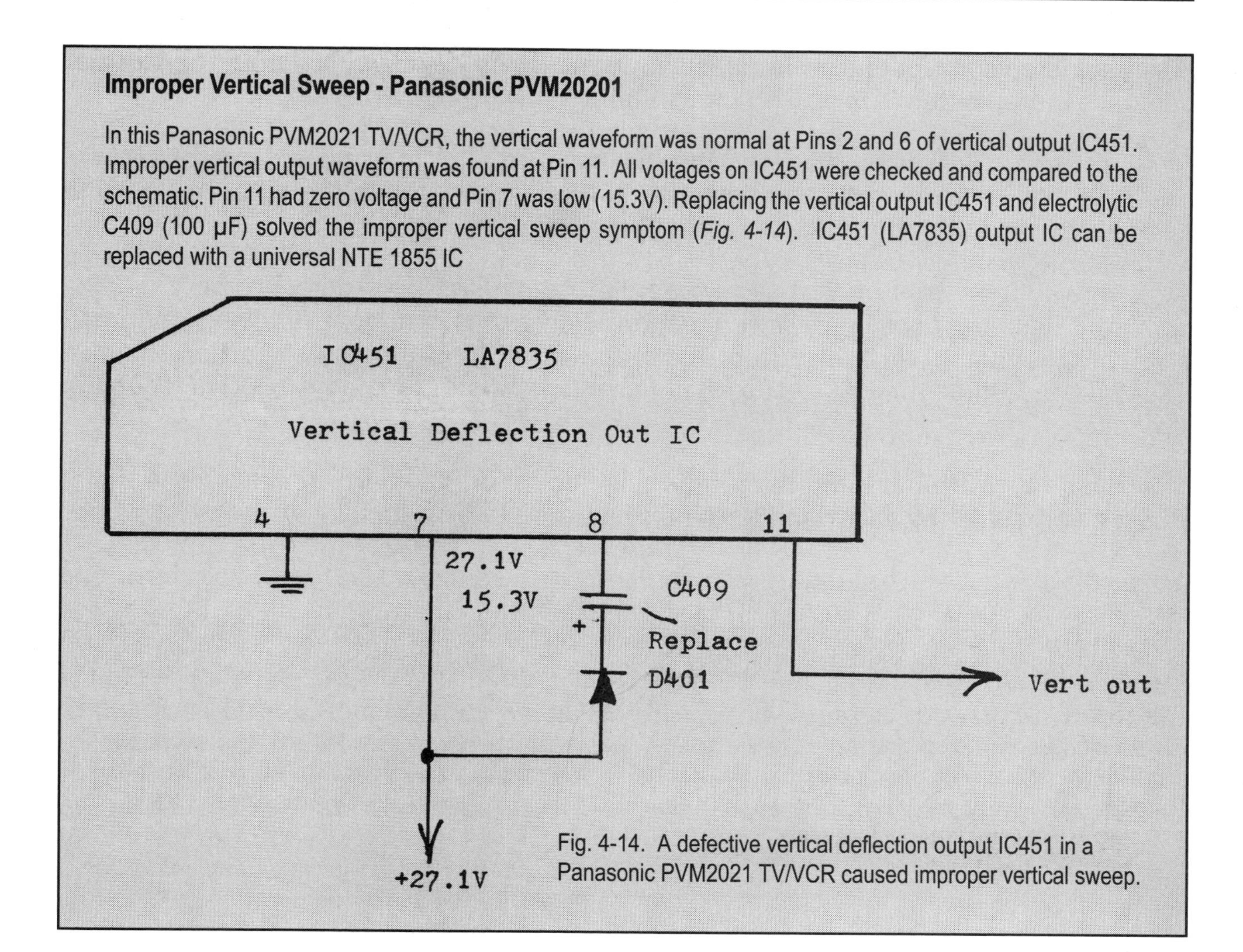

Fig. 4-14. A defective vertical deflection output IC451 in a Panasonic PVM2021 TV/VCR caused improper vertical sweep.

Intermittent Vertical Sweep

Intermittent vertical sweep can be the result of:

- ✔ A defective vertical deflection IC or vertical output IC.
- ✔ Poor yoke socket connections or a defective yoke winding.
- ✔ Bad foil or traces where electrolytics and resistors tie into pins of the output IC.
- ✔ Defective vertical output transistors.
- ✔ Poorly soldered transistor-board terminals.
- ✔ Poorly soldered joints on SMD resistors and capacitors.
- ✔ Poorly soldered connections or defective 100 μF electrolytics in the vertical output.

Intermittent loss of vertical sweep at the top can result from an intermittent top vertical output transistor, while a poorly soldered connection of pin terminals on the deflection or output IC can cause intermittent collapse of the vertical raster.

Badly soldered connections on Terminals 1, 2, and 3 of the horizontal output transformer can cause intermittent sweep and intermittent snow in the raster.

Check for defective silicon diodes in the vertical output for intermittent loss of vertical sweep. A leaky silicon diode and defective 100 µF electrolytic in the output-IC circuits can cause intermittent and partial loss of vertical sweep.

Also, intermittent loss of vertical sweep can be caused by a defective vertical size control. Check for poor resistor connections tied to the size control. Check all small electrolytics in the vertical output-IC circuit for ESR problems. In order to solve intermittent loss of vertical sweep after warm-up, substitute a large 1000 µF coupling capacitor in the vertical output IC to the yoke circuit.

In some TVs with tuner circuits mounted on the main PC board, removing the metal shield and resoldering all ground terminals can solve intermittent loss of vertical sweep.

Intermittent Vertical Sweep – Emerson VT1920

In an Emerson VT1920 the vertical sweep would bounce in and out. The deflection IC out of Terminal 15 was monitored with the scope and was found to be normal. The vertical input waveform at Pin 3 of vertical deflection IC402 was good with intermittent sweep out of output Pin 5 (*Fig. 4-15*). The supply voltage on Pin 7 (25V) was fairly normal and did not change when the sweep collapsed. C423, C422, C456, and C426 were checked with the ESR meter. At that point, there was no doubt IC402 was the intermittent component; it was replaced with an original replacement (VPC1488H).

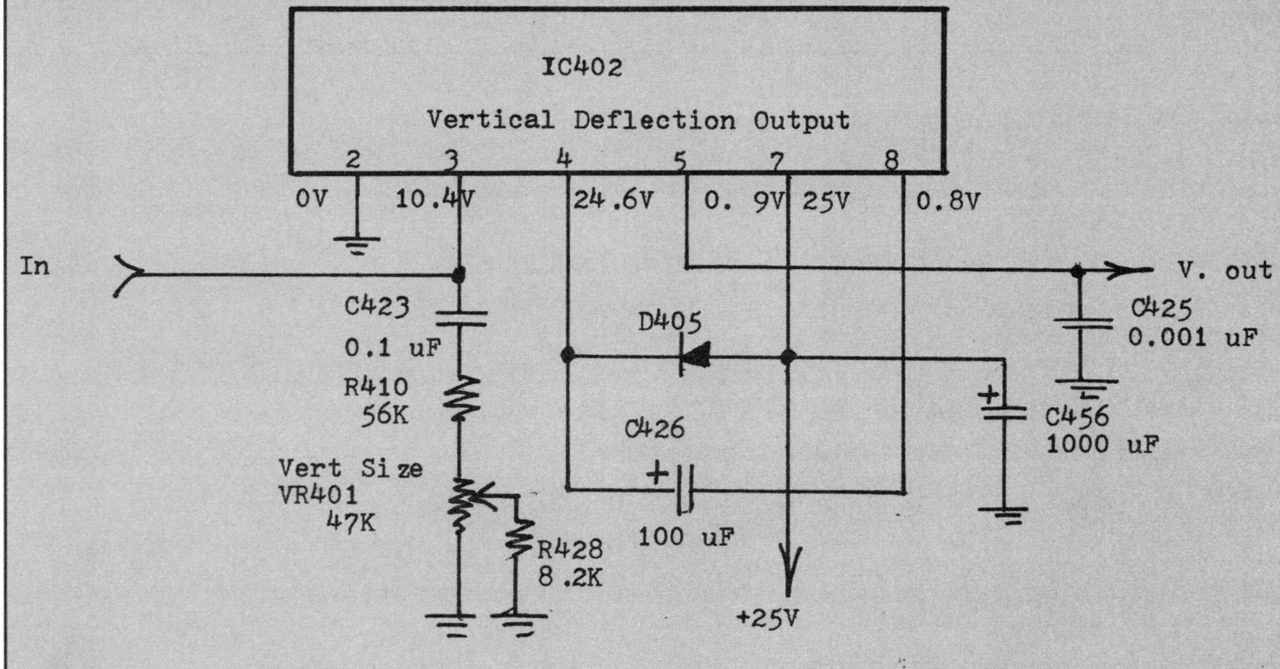

Fig. 4-15. IC402 was replaced in an Emerson VT1920 because it was causing intermittent vertical sweep.

Vertical Foldover

Vertical foldover problems are caused by components in the vertical output or pincushion circuits. Vertical foldover at the top of the screen can be caused by a 100 µF electrolytic tied to the vertical IC output pin terminals. Substitute a 1000 µF coupling yoke capacitor. Use the ESR meter to check all feedback electrolytics for foldover symptoms. One inch of vertical foldover can result from a defective 470-µF, 63-volt electrolytic in the voltage supply source.

A defective vertical output IC can produce a vertical foldover picture at the top with flipping vertical roll symptoms. A slight vertical foldover with poor regulation can be caused by a defective IC regulator. Check for a defective diode in the vertical output circuits. No vertical sweep on the bottom half of the screen and only six inches of vertical sweep at the top with foldover in the remaining raster, is caused by a 1 µF electrolytic tied to the deflection output control IC.

Poor Linearity

Poorly soldered connections on the vertical output IC pins can result in poor linearity of the raster (*Fig. 4-16*). Use the ESR meter to check all electrolytic capacitors in the vertical output-voltage sources for poor linearity of vertical sweep at the top of the picture. Defective zener diodes in the vertical output and voltage sources can cause insufficient vertical sweep, and adjustment of height or size control only stretches linearity of vertical sweep. A small electrolytic (1 µF to 2.2 µF) in the vertical output

Fig. 4-16. Stretched lines or poor linearity at the top of the picture or raster.

stages can cause partial loss of vertical sweep at the bottom of the screen, as well as cause the picture to stretch at the top of screen. These capacitors might check out okay with a capacitor tester, but will reveal problems when tested with the ESR meter.

For severely stretched vertical sweep, check for an increase or decrease in the large megohm resistors of the vertical circuits. Remove one end from the PCB and check it again. For a weaving picture and a stretching of vertical height due to poor voltage regulation, don't overlook a defective IC regulator.

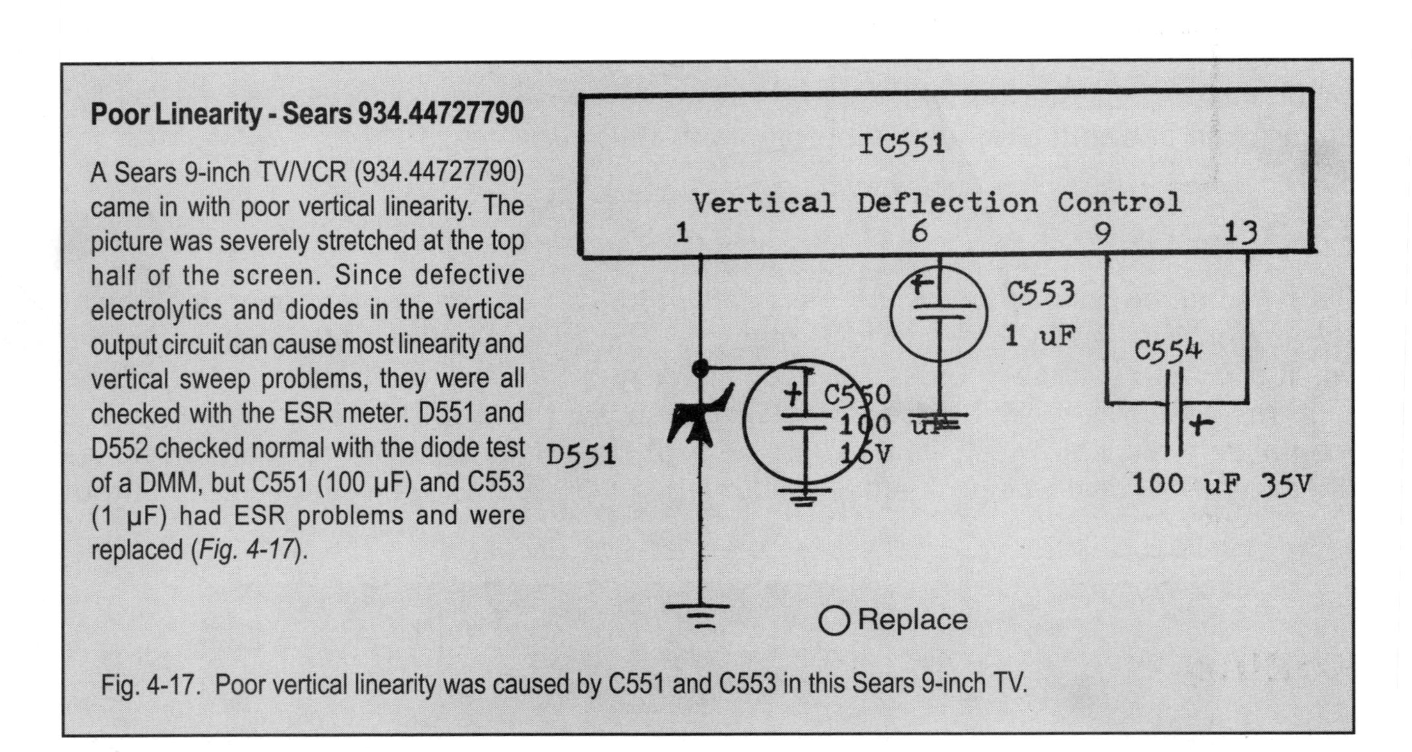

Poor Linearity - Sears 934.44727790

A Sears 9-inch TV/VCR (934.44727790) came in with poor vertical linearity. The picture was severely stretched at the top half of the screen. Since defective electrolytics and diodes in the vertical output circuit can cause most linearity and vertical sweep problems, they were all checked with the ESR meter. D551 and D552 checked normal with the diode test of a DMM, but C551 (100 µF) and C553 (1 µF) had ESR problems and were replaced (*Fig. 4-17*).

Fig. 4-17. Poor vertical linearity was caused by C551 and C553 in this Sears 9-inch TV.

Vertical Yoke Problems

An open yoke winding or poor yoke socket contacts can cause a no vertical sweep problem. Shorted turns in the vertical yoke winding can cause linearity and improper vertical sweep. Arcing within the yoke winding can produce intermittent vertical sweep and shutdown of the TV chassis. Internal firing of the yoke winding can cause lines in the picture and improper vertical sweep. A thermal yoke problem can cause intermittent loss of vertical sweep.

The large, coupling electrolytic (1000 μF) from output to yoke winding could cause improper sweep, bunching of lines in the raster, and loss of vertical sweep (*Fig. 4-18*). A bad yoke can cause:

- ✔ Only two or three inches of vertical sweep.
- ✔ Insufficient vertical sweep
- ✔ Retrace lines at the top.
- ✔ A singing yoke noise.

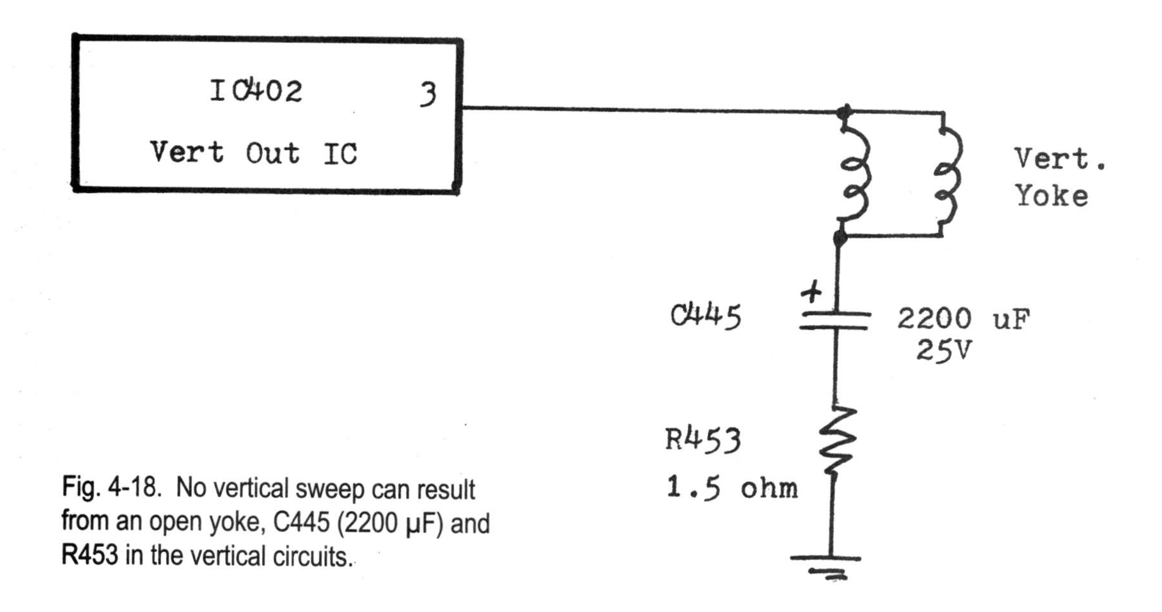

Fig. 4-18. No vertical sweep can result from an open yoke, C445 (2200 μF) and R453 in the vertical circuits.

Rolling Pictures

Poor vertical sync problems within the vertical deflection IC can cause a rolling picture or no vertical lock. Scope the vertical input circuits for correct vertical sync waveform. Check for low or improper voltage sources feeding to the vertical circuits for rolling and crawling pictures, and also check for a change in resistance of large resistors within the vertical deflection and driver circuits. A defective vertical size control can cause a rolling picture.

Vertical jumping and collapse of vertical sweep can be caused by a defective deflection vertical IC. Vertical roll can result from a defective 1 μF electrolytic in the deflection or output vertical circuits. Replace the vertical output IC for vertical roll and foldover at the top on some channels. Also, a defective diode might cause the picture to bounce or roll.

Vertical Crawling and Jitters

Scanning lines moving up the picture with a dark section can be called a vertical crawling symptom. Dark bars and lines at either the top or bottom of the picture might appear with a crawling symptom. Substitute another large filter capacitor across the terminals of the suspected main filter capacitor, observing correct polarity. Always remove the power plug, discharge the main filter capacitor, and clip another electrolytic with the same or higher capacity and working voltage. Replace the original capacitor if the crawling picture or raster disappears.

For crawling problems, check the 470 μF, 220 μF, and 100 μF electrolytics within the vertical voltage source of the output IC. A slight weaving in the picture and stretched vertical height can be due to a poor voltage regulator. When lines and bars appear in the picture, check for a defective output IC.

Vertical jitter and loss of vertical sweep can be caused by a defective output IC and a 100 μF capacitor tied to a pin terminal of the output IC. If vertical jitter occurs when the channels are changed and the jittery picture becomes worse on strong signals, you should replace the voltage regulator IC. Severe vertical jitter and intermittent vertical jumping can be caused by a defective deflection IC. Triple images, vertical jitter, and incomplete vertical sweep at the bottom are caused by a large electrolytic in the supply voltage of the vertical circuits. Severe vertical jumping and a loss of a quarter of the vertical sweep can be caused by a large (1000 μF) output coupling capacitor.

ESR Meter Capacitor Tests

Use the ESR meter to test all electrolytic capacitors in the vertical deflection and vertical output circuits. A lot of vertical loss and insufficient vertical sweep is caused by electrolytic capacitors.

Defective capacitors also can be responsible for:

- ✔ Reduced vertical sweep at the top and bottom.
- ✔ Poor vertical linearity.
- ✔ Stretched vertical sweep.
- ✔ Tearing in the picture.
- ✔ Vertical foldover.
- ✔ Vertical jitter.
- ✔ Rolling pictures (*Fig. 4-19*).

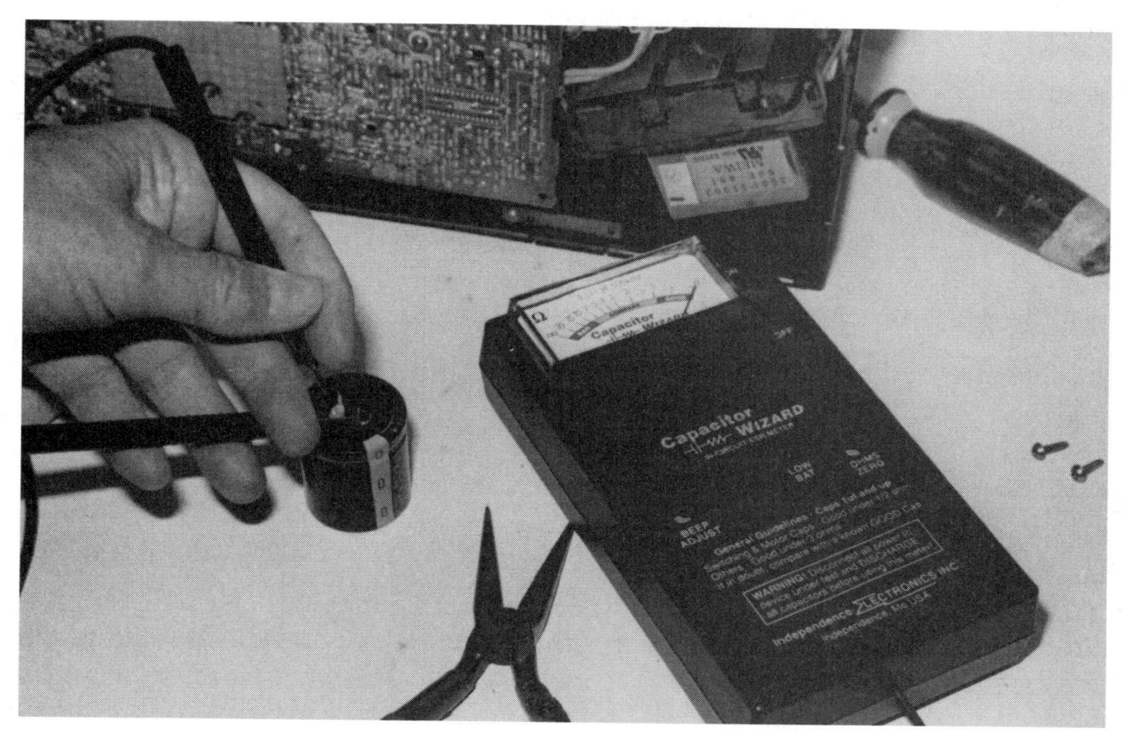

Fig. 4-19. Check for defective electrolytics in vertical circuits for vertical jitter and a rolling picture.

Loss of vertical sweep by a defective output IC can be caused when a defective capacitor damages the IC component.

Double-check all electrolytic capacitors with the ESR meter when taking in circuit tests. Voltage and capacitor tests may show a bad capacitor to be normal. The regular capacitor tester might test for correct capacitance, but the electrolytic might have ESR problems. The tested electrolytic might test normal, but be very high in ESR problems. Remove and replace all electrolytics that have a bad ESR measurement. You can quickly test all electrolytics in the vertical circuits.

Retrace Lines

Sometimes, several horizontal lines are found at the top of the raster or picture. These retrace lines may be spread widely or bunched closely together – that is, more closely or more wider than normal. Numerous black lines throughout the picture, loss of 1-inch vertical sweep, and vertical foldover at the top of the picture all can be the result of a defective capacitor in the output IC circuits. With vertical retrace lines at the top of the raster, replace both vertical output transistors. A defective vertical-control output IC also can cause retrace lines at the top. A defective yoke assembly can cause insufficient vertical sweep with retrace lines at the top. When bunching lines are found in the picture, replace the vertical output IC.

Retrace Lines at the Top – Zenith TVSA1320

In a Zenith TVSA1320 TV/VCR insufficient vertical sweep and retrace lines at the top were found. Sometimes, a singing sound could be heard inside the cabinet. All voltages on the vertical output IC541 were quite close. All electrolytic capacitors were checked with the ESR meter and were good. D541 and D542 checked normal with in-circuit diode tests by a DMM. When prodding the yoke, the sound would stop and the vertical sweep acted up. Replacing the DY yoke removed the lines at the top (*Fig. 4-20*).

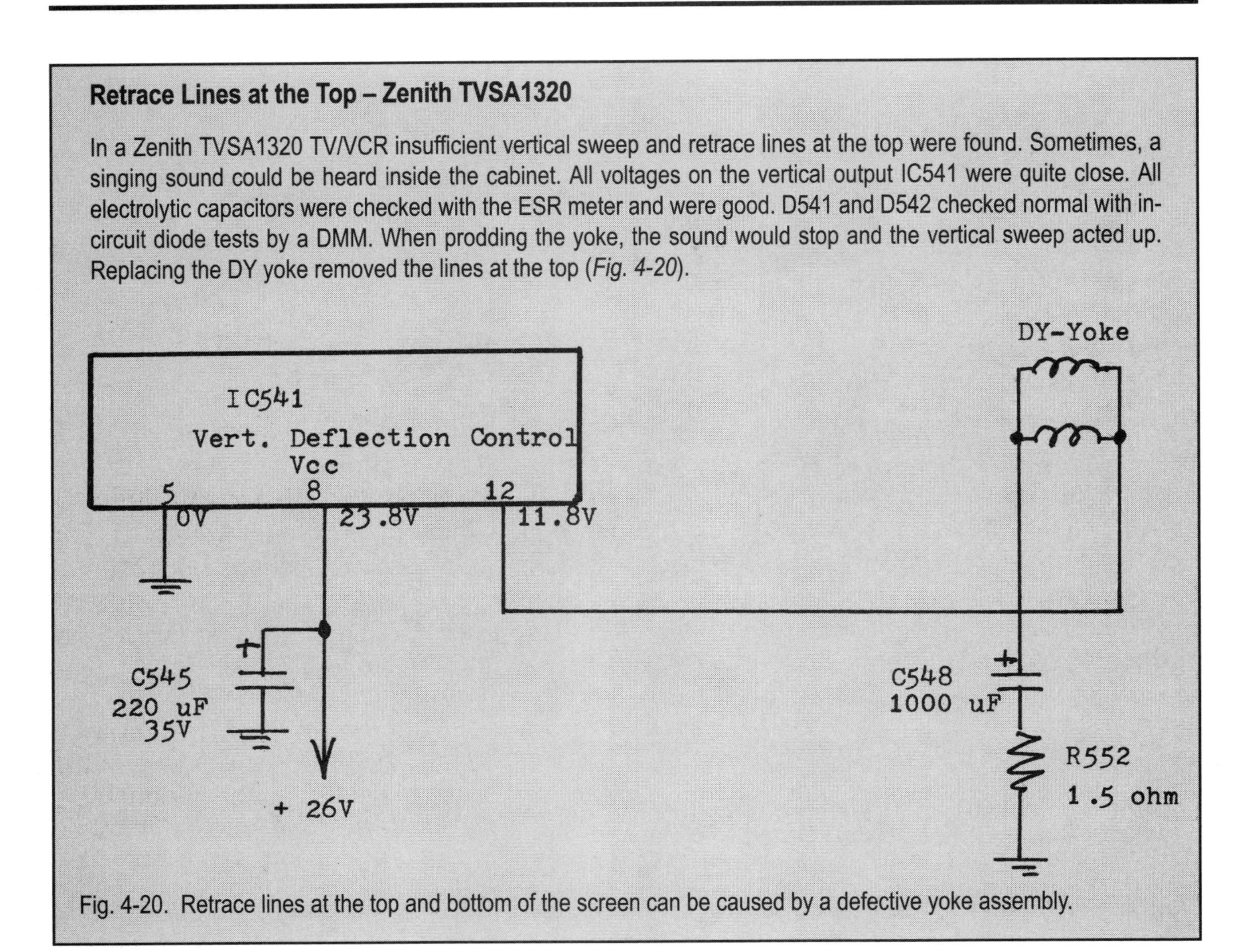

Fig. 4-20. Retrace lines at the top and bottom of the screen can be caused by a defective yoke assembly.

Vertical Pincushion Problems

Because larger screen sizes are now being manufactured with VCRs in the TV chassis, vertical pincushion troubles can appear. The pincushion circuit prevents the picture from sagging in the middle with straight scanning lines at the end of the picture tube and in the corners. Pincushion circuits are found in 27-inch TVs and larger.

When the sides are bowed inward, check for a defective component in the pincushion circuits (*Fig. 4-21*). A defective pin driver or output transistor can cause extreme bowing at the sides. Remove the output transistor and test out of the circuit. The intermittent output transistor may test okay when removed, but if in doubt, replace the output transistor. Test the pin buffer or pin transistor with in-circuit tests. For intermittent bowing of the raster or picture, check for poorly soldered connections in the pincushion circuits.

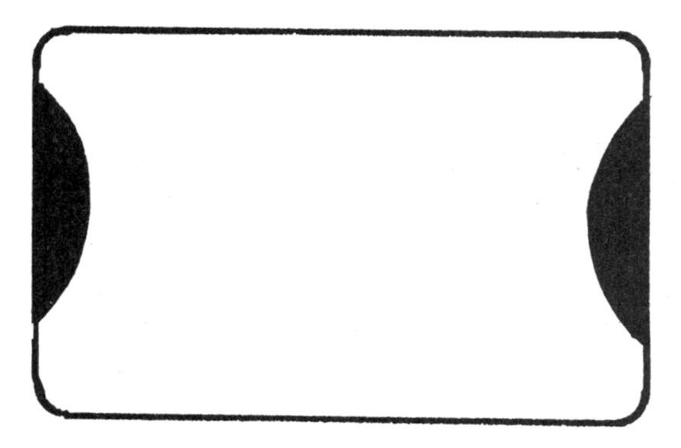

Fig. 4-21. Check components in the vertical pincushion circuits when the picture is bowed at the sides.

The pincushion output transistor can cause the picture to be too wide and the sides to bow in, while intermittent width can result from Q1 in the pincushion circuits. If the right side of the screen is bowed in and there is a flickering picture, replace both the Q1 and Q2 pincushion transistors. If after warm-up the sides begin to bow inward, replace Q1. A bad solder joint on Q2 of the pincushion board can cause insufficient width on both sides of the raster.

A high-pitched squeal inside the set can result from a pincushion transformer; replace it. The same can be said when flexing the pincushion board and the raster bows in on both sides; suspect a bad pincushion transformer. Open resistors in the pincushion circuits can cause only 3 inches of vertical sweep in the center of the screen. A badly soldered joint on the pincushion transformer can cause the vertical raster on the right side to be compressed into a pattern similar to a trapezoid.

For bowed pictures and pincushion problems, check both transistors on the pincushion PCB. A PNP-pincushion transistor can cause straight lines bowed outward. When the width is curved and narrow, suspect a bad solder terminal on the driver-output transistor. When both sides are really curved in with poor pincushion action, check for a loose core in the coil on the pincushion board.

Solder all pincushion transformers and coils in the pincushion circuits. Check for foil or trace breaks within the pincushion circuits. A shorted component within the pincushion circuits can damage the horizontal output transistor. When locating a leaky or shorted horizontal output transistor, always check the pincushion circuits.

Unusual Vertical Problems

Check components outside of the vertical circuits that might be tied to the vertical circuits and result in a vertical problem. A defective pincushion-output transistor could cause the sides to bow inward (*Fig. 4-22*). In some chassis', poorly shielded grounds in the PCB tuner can cause a lack of vertical sweep. A shorted vertical output IC, leaky

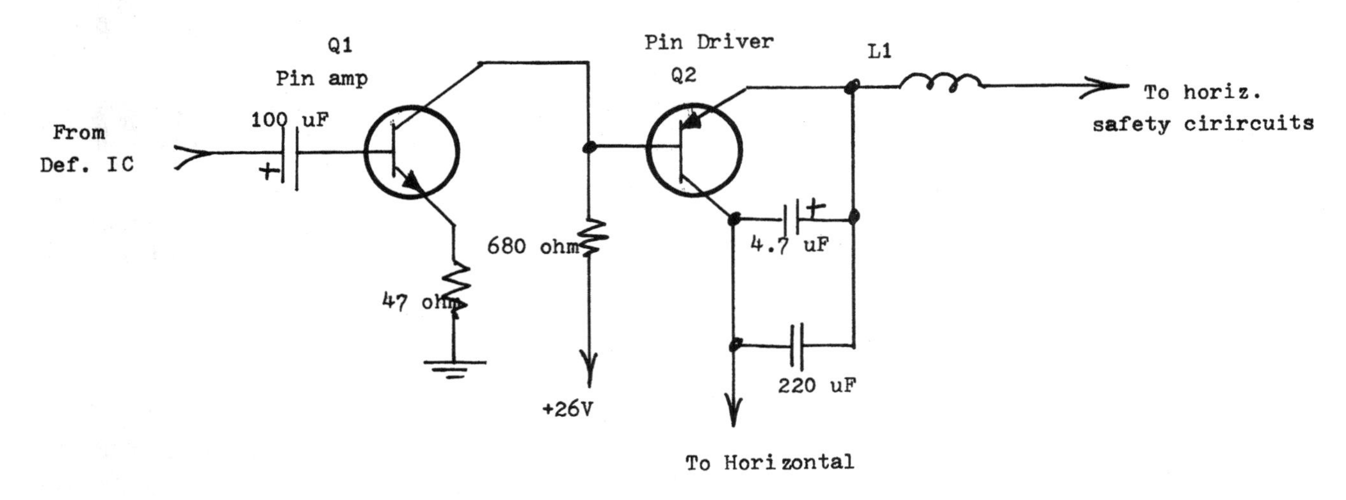

Fig. 4-22. Check the pin output transistor when the picture or raster is bowed in on the sides.

diode, and 100 µF electrolytic with ESR problems produce horizontal lines over the entire screen, a severe thump-thump noise in the speaker, and failure by the set to start.

A defective component within the pincushion circuits also may cause vertical foldover to occur. A defective EEPROM found in a TV chassis can cause symptoms such as the set failing to come on, a blue screen with 1 inch of vertical sweep up from the bottom, the customer menu unable to be straightened, and no audio.

Chapter 5: Servicing High-Voltage and CRT Problems

The horizontal output circuit provides horizontal sweep to the horizontal deflection yoke, and high voltage is developed in the flyback circuits. High-voltage diodes molded inside the flyback rectify the high voltage created by the horizontal output transformer and fed to the anode terminal of the picture tube. The high voltage pulls the electrons from the cathode terminal in the CRT to the front of the screen at a high rate of speed creating a raster or picture.

Besides applying high voltage to the anode terminal of the picture tube, focus, screen, and heater voltage are also created in the flyback circuit (*Fig. 5-1*). The screen and focus controls are often included inside the flyback component.

In a TV chassis that has a line-voltage regulator operating from the power line, additional scan-derived voltages developed in the secondary of the flyback are fed to other circuits in the TV. The TV/VCR chassis with a switching power supply has several different low-voltage sources developed in the secondary of the switching transformer.

The horizontal collector-terminal voltage is supplied through the primary winding of the flyback, which provides horizontal sweep and high voltage to the picture tube circuits. Hold-down or safety capacitors in the collector and primary flyback circuits keep the high voltage from rising excessively at the anode terminal of the CRT. When a safety capacitor goes open, the high voltage severely increases and can cause arcover problems in the picture tube circuits. Most components in the horizontal and high voltage circuits are critical parts and must be substituted with original or equivalent replacements.

The Flyback

The primary winding of the flyback might have a voltage between +117V and +160V applied to the horizontal output transistor. A secondary fuse or isolation resistor might be found between the primary winding and power supply to protect the flyback and high voltage circuits. When the horizontal output transistor becomes shorted, the fuse or resistor will open, removing the B+ voltage from the flyback circuits. A no raster problem with an open fuse might be caused by a leaky flyback, a horizontal output transistor, or both (*Fig. 5-2*).

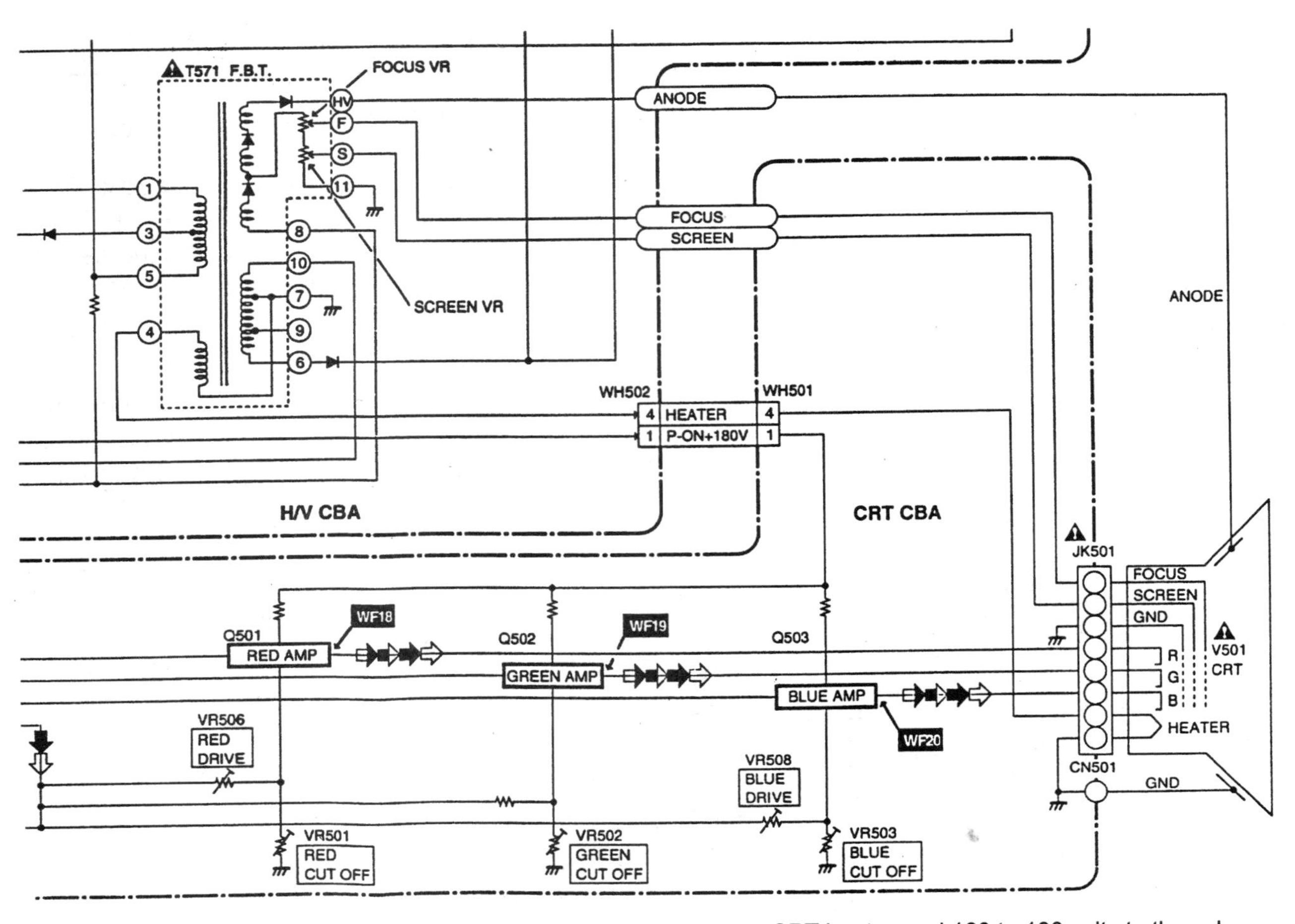

Fig. 5-1. The flyback supplies high voltage to the anode, focus, screen, CRT heater, and 160 to 180 volts to the color output transistors on the CRT board.

Fig. 5-2. The flyback and high-voltage components are located at the rear or the chassis or along the side in a TV/VCR.

High voltage in the secondary winding of the flyback is developed and rectified by molded diodes inside the flyback. A heavy-duty, high-voltage wire and plug are attached to the flyback in order to connect the high voltage to the anode terminal of the CRT. The integrated flyback secondary winding might arcover or the silicon diodes inside the flyback might become leaky and shorted, causing arcover or a smoking chassis. A leaky flyback can destroy the horizontal output transistor and cause the fuse to open. High-voltage arcover inside the flyback or high-voltage arcover to the common ground shield can easily be seen and heard from the rear of the TV chassis.

Dead Chassis, Open Fuse – Hitachi 13VR8B

In a Hitachi 13VR8B TV/VCR the chassis was dead and the main fuse was open. Replacing D605, D607 and the horizontal output transistor (Q572) solved the dead chassis with open-fuse problem.

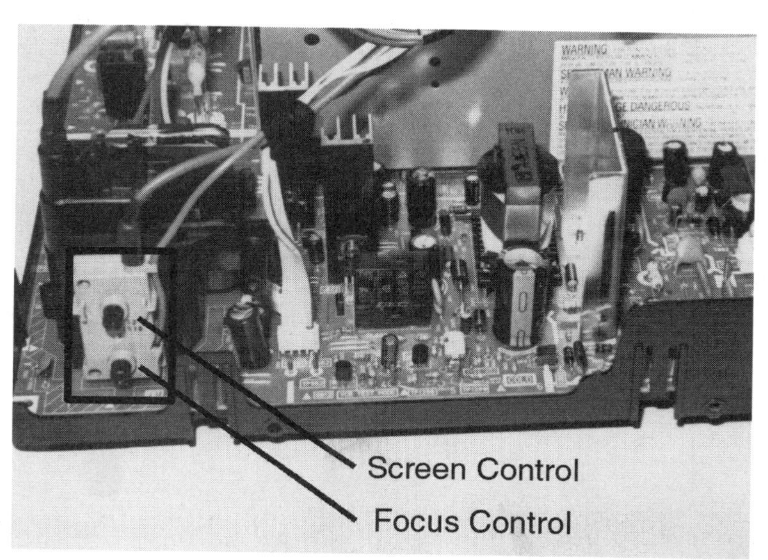

Fig. 5-3. The focus and screen controls are connected to the flyback component.

In addition to providing high voltage to the anode terminal of the CRT or picture tube, the flyback provides a focus and screen voltage from the high-voltage winding (Fig. 5-3). A correct focus voltage keeps the scanning lines sharp and clean. The DC screen voltage is fed to the screen-grid control of the picture tube. A separate winding around the flyback core allows the heater voltage to light up the gun assembly in the CRT. Another separate winding on the primary, or a tap in the primary winding, provides a high voltage (+180V) to the color-output transistors mounted on the CRT board.

Any breakdown of high voltage components – including resistors, diodes, and electrolytic capacitors within these high voltage circuits – can prevent the CRT from operating. Open lines or foil traces can prevent voltage from reaching the picture tube circuits. Poorly soldered connections in the high-voltage circuits can cause an intermittent picture or raster.

High–Voltage Problems

High-voltage problems can be caused by many different sources, including the breakdown of horizontal and flyback components. Use a high-voltage probe to measure the high voltage at the anode socket of the picture tube (*Fig. 5-4*). Make sure the ground clip of the meter is attached to the common ground, ground spring, or ground strap of the picture tube. If it isn't, you can receive a terrible shock just by holding the plastic meter. This may cause you to drop it resulting in meter damage.

Improper adjustment of the high voltage can produce X-ray radiation. When the high voltage is excessive, radiation can penetrate the shell of the CRT. High voltage should always be kept at the rated value and no higher. Excessive high voltage can cause failure of the picture tube and high voltage circuits.

When the high voltage is adjusted properly, there is no possibility of a radiation problem. The brightness control must be set while monitoring the high voltage with an high-voltage meter in order to be certain that the high voltage does not exceed the ratings set by the manufacturer. Usually, when the brightness is raised, the high voltage will lower, and when the brightness is reduced, the high voltage increases.

Red Hot Horizontal Output Transistor – Emerson VT1920

After replacing a horizontal output transistor (Q402) in an Emerson VT1920 chassis, the transistor became warm when the variable line voltage approached 85V AC. A quick waveform on the horizontal driver transistor (Q401) and the base of Q402 looked fairly normal. A quick test of the flyback indicated a leaky horizontal output transformer (T401). Replacing the flyback solved the overheated output-transistor (Q402) problem.

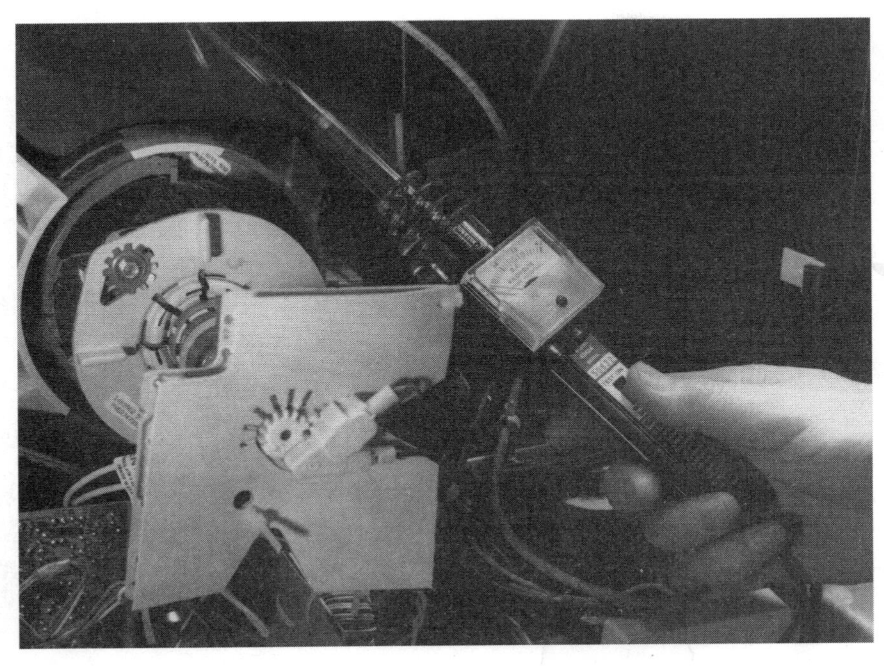

Fig. 5-4. Check the high voltage at the anode button on the bell of the picture tube.

When servicing and doing tests because of excessive high voltage, be careful not to get too close to the TV chassis. If high-voltage arcing occurs, do not operate the chassis any longer than necessary to repair the chassis or to locate the cause of excessive high voltage. Go directly to the safety capacitors and check for an open condition or broken lead connections.

A really low-high voltage symptom can be caused by a defective line-voltage regulator, while low or weak high voltage can cause a dim or weak raster.

No Raster, No High Voltage

Possible Causes:

- ✔ Defective horizontal output transistor
- ✔ Defective fuse
- ✔ Defective flyback

Raster Failure With Good High Voltage

Possible Causes:

- ✔ Defective picture tube circuit or picture tube
- ✔ Defects in the different voltage sources feeding the picture tube
- ✔ Badly soldered joints on CRT pins or the CRT board
- ✔ Defective high-voltage bypass capacitors on CRT board
- ✔ No screen voltage or poor socket contacts of the picture tube
- ✔ Defective video circuits
- ✔ Open filament
- ✔ Poor CRT socket contacts

Some other symptoms you may encounter when servicing high-voltage problems include:

- ✔ A dead chassis, but the relay clicks and the fuse is okay. The cause may be a hairline crack from Pin 2 of the flyback, causing an open circuit to the horizontal output transistor.
- ✔ When high voltage comes up and then instantly shuts off for a moment. Look for a defective diode in the standby power supply.
- ✔ Poor focus resulting from improper focus voltage or a defective CRT.
- ✔ When high voltage comes up momentarily when power is applied and then goes off or shuts down. Check for a bad safety capacitor. A defective safety capacitor can cause the high voltage to come up and shut down at once.
- ✔ Intermediate shutdown caused by a badly soldered joint on the yoke socket.
- ✔ High-voltage shutdown due to a defective high voltage-shutdown circuit.

High Voltage, Immediate Shutdown – Symphonic TVCR13E1

In a Symphonic TVCR13E1 the high voltage came up and then immediately shut down. A 5V-regulator IC was suspected to be the problem. After IC603 was replaced, the 13-inch TV stayed on and operated normally.

- ✔ Excessive arcing inside the flyback or spark gaps caused by extremely high-voltage arcover, flyback windings, or arcing between flyback and chassis.
- ✔ An arcing noise with intermittent loss of focus resulting from a defective focus control.
- ✔ Excessive arcing inside the gun assembly of the CRT caused by an open heater or cracked glass of the CRT neck assembly.
- ✔ Smoke rising up from the chassis caused by arcing inside the flyback or yoke assembly.

Different Screen Sizes

The larger the size of the picture tube, the greater the high voltage. Today, VCRs can be included in anything from 9-inch to 27-inch TV/VCR combo units. Improper adjustment of the B+ or 12V DC adjustments can inversely change the high voltage applied to the picture tube. The picture tube anode voltage might be quite close to the size of the CRT. A 19-inch TV might have an anode high voltage from 19 kV to 24 kV, whereas a 25-inch to 26-inch TV might have one from 25 kV to 27 kV and a 27-inch TV might have one from 27 kV to 30 kV. Check the manufacturer's service literature and Sams Technical Publishing's PHOTOFACT® schematics for the correct anode voltage. These anode voltages seem to vary with each TV manufacturer.

Flyback Problems

A defective flyback can cause many problems including, a dead chassis with an open fuse or a dead chassis with an okay fuse and a dead TV with relay click and an okay fuse. While a leaky flyback can cause its own set of problems including, relay clicks with a blown fuse and a dead chassis with a shorted horizontal output transistor. Low-high voltage and no raster problems also indicate a defective flyback.

Many flyback problems can be fixed by resoldering joints.

- ✔ If the chassis is dead but the fuse is okay, resolder all flyback pin terminals.
- ✔ Resolder Pins 1, 2, and 10 of the flyback in order to correct problems such as intermittent picture, vertical sweep, and a snowy picture. Poorly soldered connections on Pins 1 and 2 of the flyback can cause a loss of vertical sweep, a vertical line down the center of the screen, and may cause the set to cycle off and on.

- An intermittently dead chassis might come on when first turned on, shut down, and then operate for two or three hours. This can be caused by a poorly soldered connection on the flyback.
- A poorly soldered joint on the flyback can cause a noise in the audio and the sides to pull in intermittently.
- When the chassis works for 20 minutes and goes dead, resolder bad joints on the flyback.
- Poorly soldered connections on the flyback can lead to intermittent loss of the CRT filaments, intermittent loss of vertical sync, and an intermittent dead symptom.
- Intermittent loss of vertical sweep might be caused by a defective flyback or poorly soldered connections.
- Badly soldered connections of the driver transformer can cause intermittent and shutdown problems. Resolder the driver-transformer connections at the same time the flyback connections are resoldered.

Other problems involving the flyback include:

- A dead chassis with a singing noise can be caused by a shorted horizontal output transistor and plastic oozing out of the molded flyback (*Fig. 5-5*).
- A defective flyback can cause a dead chassis with a hum.
- A completely dead chassis can be caused by an open flyback. Check the CRT filaments before replacing the suspected flyback.
- A flyback shorted between Pins 2 and 3 and ground can cause a dead chassis, fuse okay symptom.
- If the set has intermittent horizontal lines and goes off horizontal frequency, this can be caused by a defective flyback.
- If the set experiences intermittent, partial loss of horizontal sweep, plus the loss of horizontal sync that becomes worse when the board or flyback is moved or flexed, this can be the result of a bad flyback.

Arcing Flyback

An arcing flyback can produce a pungent odor and a loud cracking noise in a chassis shutdown situation. For intermittent shutdown, resolder all joints on the flyback. For a no startup or shutdown problem, check for cracked or broken traces next to the flyback terminal connections.

Check for different high voltage and flyback symptoms that might be tied to the flyback circuits. Inspect the flyback circuits for an open isolation resistor that might cause a thin, blue horizontal line and no vertical sweep. If the set has no raster and no high voltage with the CRT filament lit, a badly soldered joint at the collector of the horizontal output transistor on the line pad that connects to Pin 8 of the flyback, might be the cause.

Fig. 5-5. An arcing flyback might have a cracked body or plastic oozing out of the molded flyback.

Some other problems you might encounter include:

- ✔ When the set comes on momentarily with high voltage and shuts down at once, check for a defective safety capacitor.
- ✔ A bad yoke socket connection can cause high voltage shutdown.
- ✔ If the set sometimes comes on and operates for a while before shutting down, resolder all flyback pin connections.
- ✔ Severe arcing in the flyback might cause the chassis to remain in shutdown.
- ✔ If the set fails to start and there is a ticking noise in the flyback, look for cracks or split-open sides in the molded flyback.
- ✔ Check for a defective flyback if the chassis is dead, the fuse checks out okay, but the set shuts down, or if there is no raster and no audio, and there is no high voltage crackling when you feel the face of the picture tube. (High voltage can be detected when your forearm is placed near the TV screen; the hair on your arm stands up.)
- ✔ The defective flyback can cause immediate shutdown.
- ✔ Leaky or shorted picofarad (pF) capacitors off of Pin 10 of the flyback might cause extremely bright raster with retrace lines.
- ✔ A dead chassis might be caused by a leaky silicon rectifier and burned low-ohm resistor on Pin 9 of the flyback.
- ✔ If the set has no raster and shows no vertical sweep when turning up the screen control, check for a burned low-ohm resistor on Pin 9 of the flyback.

Removing and Replacing the Flyback

After determining that the flyback is defective, remove each terminal with solder wick and a soldering iron. First, discharge the high voltage at the anode button to a ground wire around the bell of the picture tube. Next, remove the anode plug from the picture tube. Be very careful not to lift or break foil and traces tied to each flyback connection. Next, remove all solder from each connection then carefully lift the flyback from the board. Some TV/VCR flyback and high-voltage circuits are found in a module mounted alongside of the cabinet.

Always replace the flyback with an exact factory replacement. Solder all eyelets or flyback connections. Clean up the solder flux and excess solder from the board. Finally, double-check each connection with an ohmmeter or ESR meter.

ESR meters are ideal for checking for broken traces and poor connections. Test each soldered connection to a corresponding foil or trace that connects to the first soldered connection of a component. Check the resistance between the two points with the ESR meter. The ESR meter hand should hit the peg and sound off at each good connection. A poor connection will show a small resistance with an erratic sound from the meter. An open foil will show no needle movement on the ESR meter.

Bad Flyback Terminals – Emerson VT-1322

In an Emerson VT-1322 TV/VCR an intermittently dead problem was caused by poorly soldered contacts on the flyback. All flyback soldered board connections were resoldered. Also, the horizontal driver-transistor transformer was located and all the transformer terminals were soldered. Simply soldering the driver transformer and flyback connections can cure a lot of service problems in TV/VCRs.

Arcing Noises

High-voltage arcover can be caused by an internal shorted winding or leaky high-voltage diode. Arcing inside the flyback can show burn marks and a smoking flyback. Severe arcing in the flyback can lead to symptoms such as a dead TV with an okay fuse, shutdown, and a popping noise inside the set. A badly soldered joint on the flyback, focus, or screen terminal can burn the board around the bad connection. A red picture with a cracking and popping noise from inside the cabinet can be the result of an arcing flyback (*Fig. 5-6*).

Internal arcing of the flyback can cause poor focus with an intermittent popping noise. Also check the flyback for other problems such as immediate shutdown with a severe arcing noise just after the set is turned on. If the set is dead but the fuse checks okay followed by a thumping noise, check for a crack in the molded flyback. The horizontal output transformer that keeps arcing internally can keep destroying the horizontal output

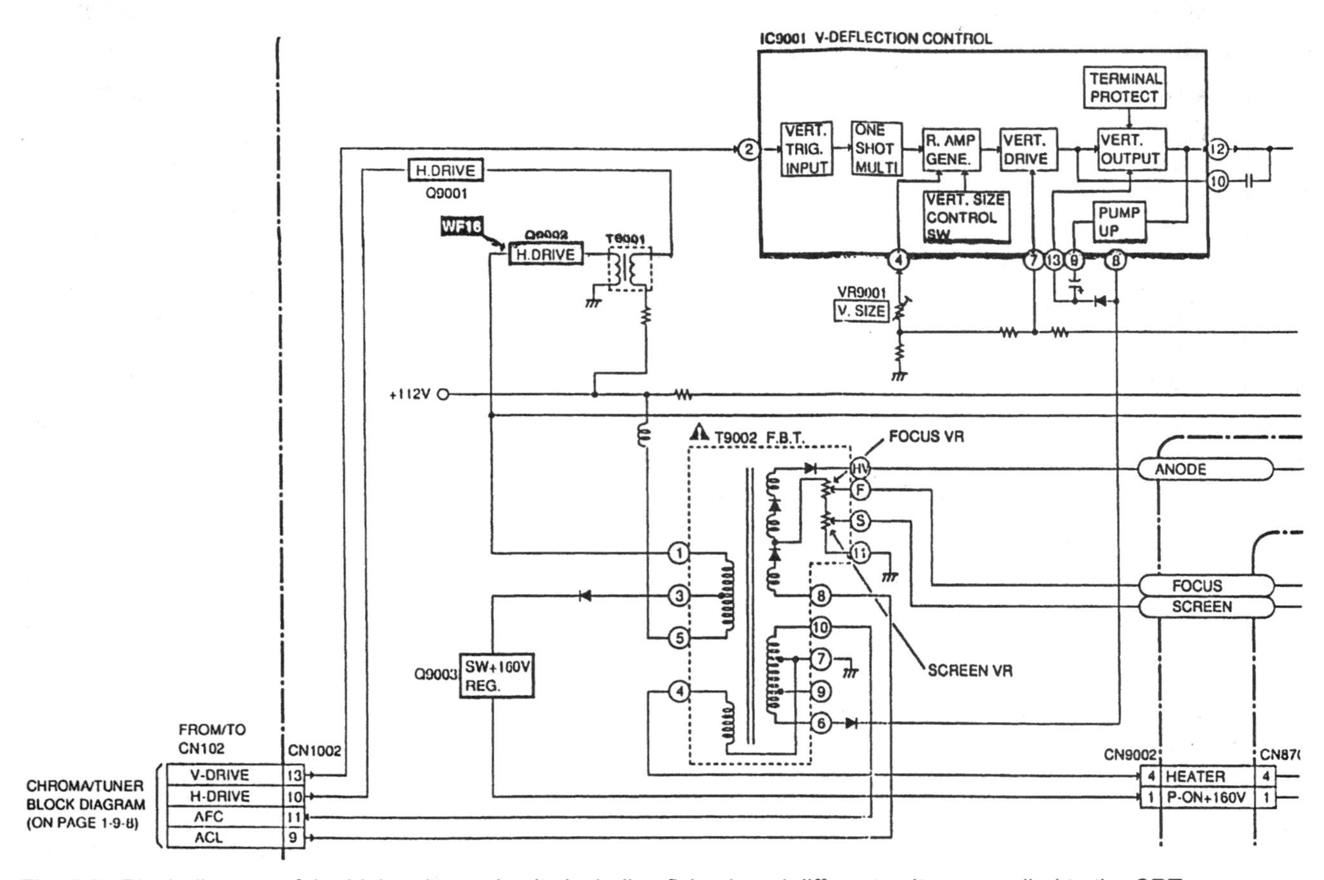

Fig. 5-6. Block diagram of the high-voltage circuits including flyback and different voltages applied to the CRT.

transistor. If, after replacing an arcing flyback, the set is still dead with an okay fuse, this is caused by a defective deflection IC.

The arcing flyback can destroy the green CRT amp, the red and green driver transistors, and the collector load resistors with a red-picture and cracking-sound symptom.

If there is a hissing noise in the flyback and the set attempts to turn on but won't, this can be the result of a defective electrolytic in the supply-voltage circuits.

A dead set with a ticking noise can be caused by a short between Pin Terminals 1 and 2 of the flyback. This also can destroy the horizontal output transistor.

Other problems that may be encountered with an arcing flyback include:

- ✔ Intermittent snippets in the picture with an arcing noise caused by the arcing flyback and deflection IC.
- ✔ A popping noise with a blue light in the back of the chassis can result from an arcing flyback.

- ✔ No startup accompanied by severe thumping and groaning noises in the speaker can be caused by a bad flyback.
- ✔ A badly soldered joint on the flyback can cause intermittent shrinkage from both sides of the screen, severe noise in the screen, and noise in the speakers.

Scan-Derived Secondary Voltages

Besides the high voltage, screen, focus, heater, and color output-transistor voltages, several other derived voltages might be developed with additional windings in the flyback, silicon-diode rectifiers, and electrolytic-filter capacitors. The high voltage, screen, focus, and heater voltages are fed directly to the picture tube circuits. Another +180V source might feed the color and driver output transistors. The +25V source might supply voltage to the vertical output circuits (*Fig. 5-7*).

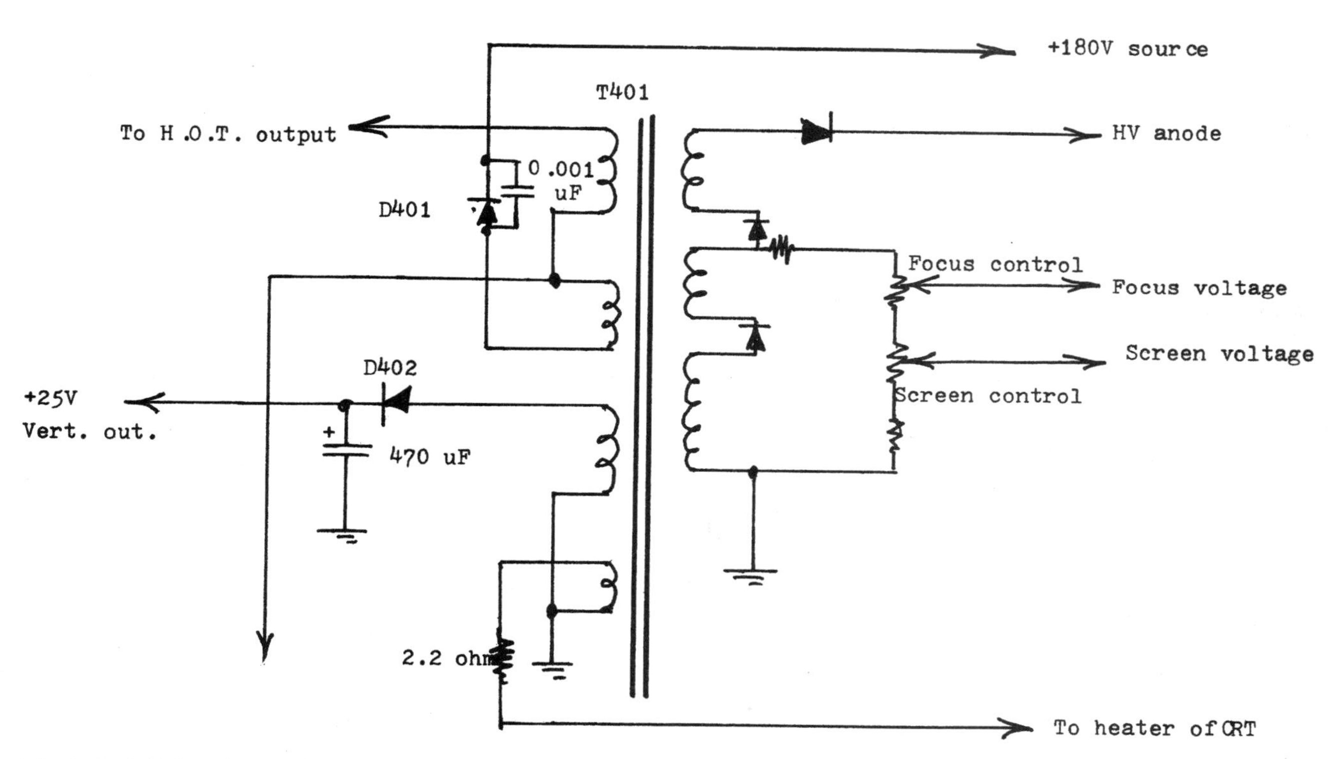

Fig. 5-7. Additional scan-derived windings on the flyback can provide different voltage sources for other TV circuits.

Within the same chassis, the low-voltage switching power supply might provide a +13V to 26V source to the sound-output circuits. A separate, switching-derived secondary winding might provide a +12V source for several other circuits. Another higher-voltage winding can provide a +103V to 120V source to the horizontal driver and output circuits.

In TV/VCR power supply circuits that have a line-voltage IC and work directly from the power line, the regulated +130V is fed directly to the horizontal driver and output transistor. The flyback-derived secondary voltages feed the high voltage, screen, focus,

Squealing Noise – Funai F19TRBIC

In a Funai F19TRBIC TV/VCR a faint squealing noise was heard. The horizontal output transistor was shorted so it was replaced. Shutdown occurred very shortly after the chassis was fired up. Transistor Q02 was once again damaged. A leaky flyback caused the repeated damage of the horizontal output transistor. Replacing the leaky flyback solved the repeated damage of Q02.

and heater voltage of the picture tube. A +160V to 180V source winding provides voltage to the driver and color amps on the CRT socket board.

Another winding provides a +27V to 33V source for the vertical output IC. The +12V-source winding provides a voltage source to the deflection IC, switching transistors, buffer transistors, and other circuits. In some TV/VCR units, the VCR circuits might be powered by another switching power supply.

CRT Heater Circuits

The heater voltage for the picture tube is developed in a separate winding wound around the core of the horizontal output transformer. In older TVs, the winding was loosely threaded through the open core of the flyback. Now the heater winding is found molded inside the body of the flyback. You might find a low-ohm voltage-dropping resistor in series with one heater lead to the CRT tube socket (*Fig. 5-8*).

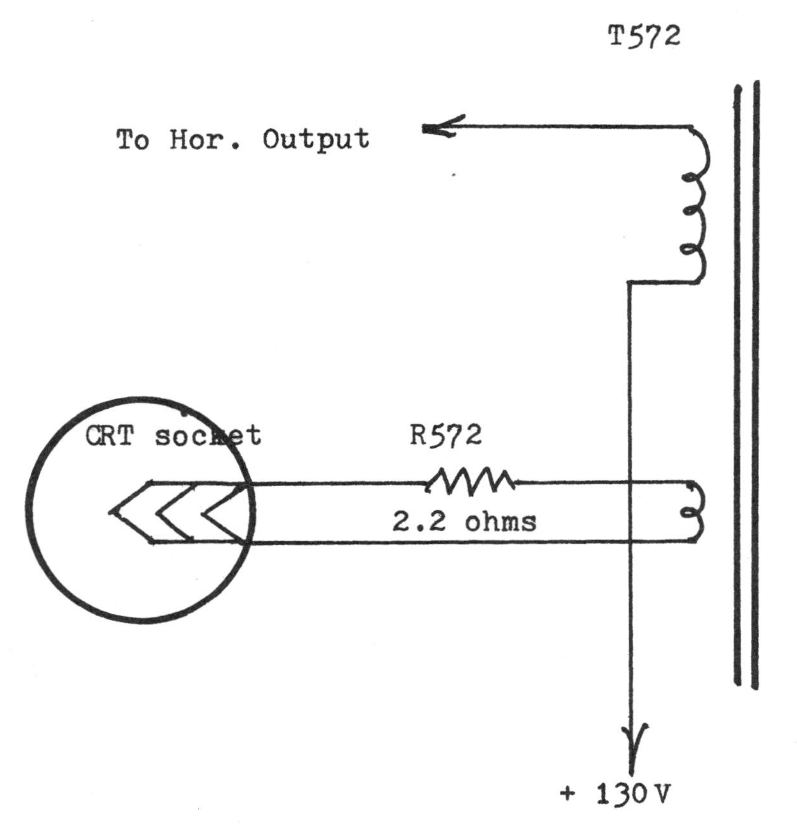

Fig. 5-8. A separate heater winding with low-ohm resistance is located on the flyback in series with heater or filaments of the CRT.

The no raster, normal high-voltage symptom can result from a bad heater pin connection. Poor or corroded CRT heater connections will not light up the picture tube. Check for bad heater terminals when the heater or raster slowly disappears. An open resistor (R572) can prevent heater voltage from the gun assembly of the CRT. Simply look toward the end of the gun assembly to see if the heater lights up.

Diminishing Raster – GE 13TVR01

In a GE 13TVR01 TV/VCR the raster would slowly fade away. A closer look showed that the filament or heater went out when the picture and raster disappeared. Soldering the heater terminal on the CRT socket terminals solved the fading-picture symptom.

High–Voltage Probe

Improper high voltage measured at the anode socket on the picture tube can determine if the horizontal output and high voltage circuits are functioning. A low high-voltage measurement indicates service problems in the horizontal or flyback circuits. Often, the sides are pulled in with low high-voltage at the anode socket. Low B+ voltage to the horizontal output transistor might produce low high-voltage reading. Also, improper drive voltage at the base terminal of the horizontal output transistor can cause low high-voltage. Intermittent and poor horizontal driver transformer terminals can also cause a low high-voltage symptom. Low or excessive high-voltage applied to the CRT might be caused by a defective safety capacitor.

A badly soldered joint on the flyback terminals can result in shrinkage at the sides of the raster. Breakdown of the capacitors within the horizontal yoke circuits can cause the sides to pull inward. The defective picture tube can cause a bright green raster, retrace lines, and shrinkage at the sides of picture.

To take a high-voltage measurement, slip the point of the high voltage probe underneath the anode rubber plug and socket. Be sure to ground the high-voltage probe to the ground strap or spring on the bell of the picture tube. Arcing will take place if the probe end doesn't touch the CRT high-voltage button. Sometimes, when the probe is inserted underneath the socket, the rubber plug and high-voltage lead will snap off. Be very careful when measuring picture tube high voltage. Compare the high-voltage measurement to that found on the service schematic.

Signal Tracing High–Voltage Circuits

A quick high-voltage measurement at the picture tube using the high-voltage probe can indicate if the flyback circuits are functioning properly. Critical focus voltage measurements at the CRT socket can indicate correct focus and high voltage. The average focus voltage is between 3 kV and 6.5 kV and in larger picture tubes, the focus voltage can vary from 6.5 kV to 9 kV.

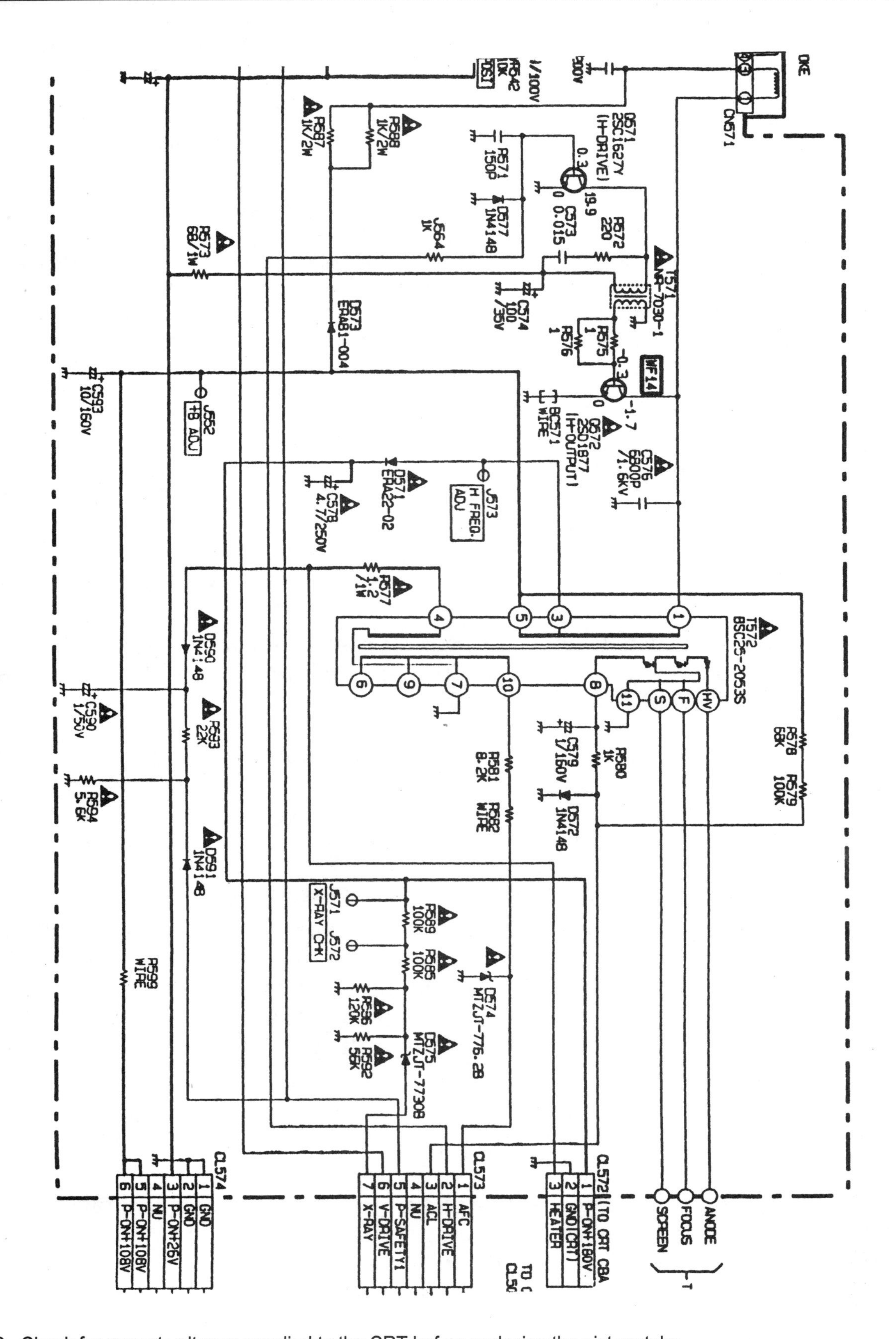

Fig. 5-9. Check for correct voltages supplied to the CRT before replacing the picture tube.

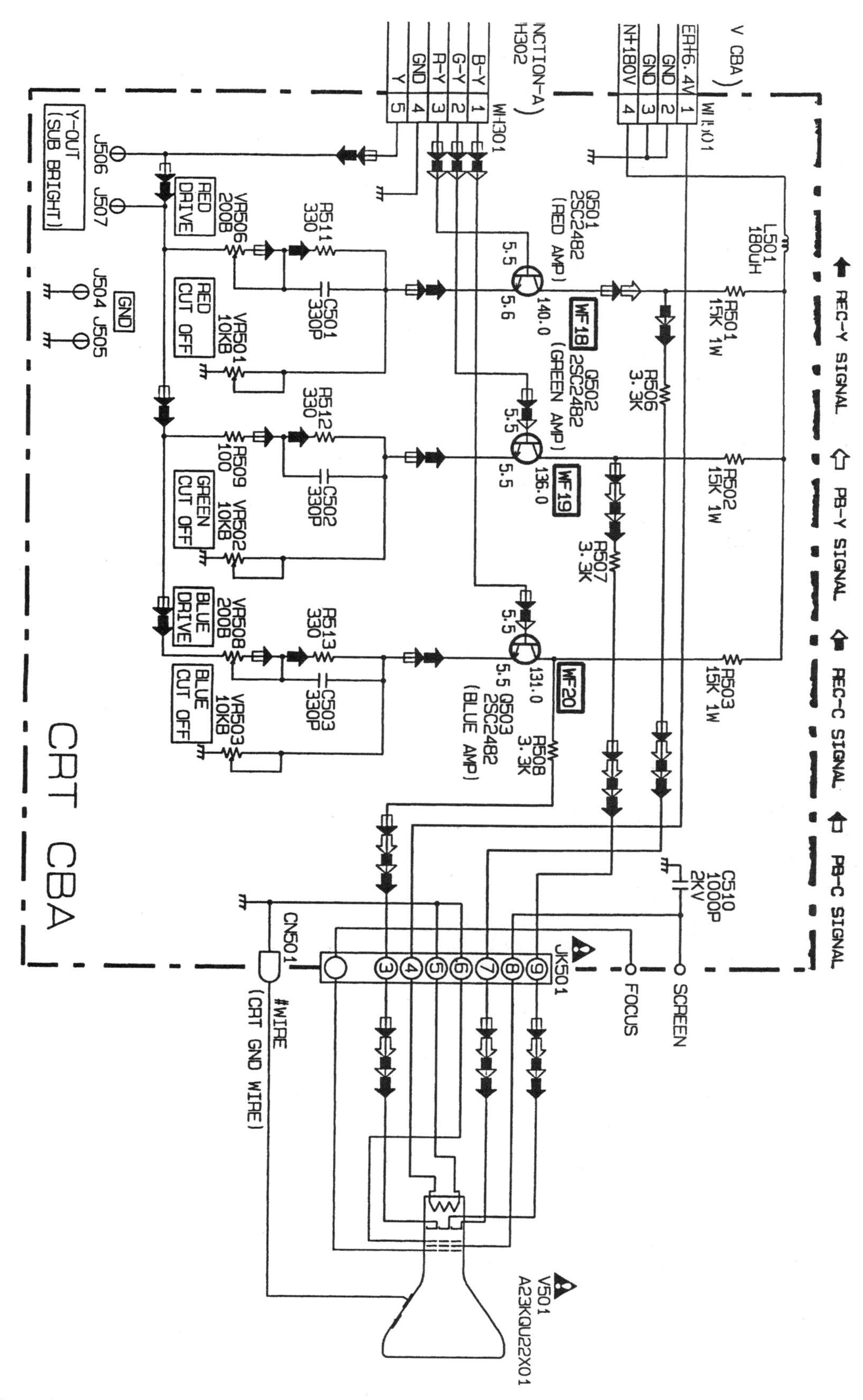

Fig. 5-10. The color amp circuits are mounted on the CRT PCB at the end of the picture tube.

Check the screen-grid terminal at the CRT socket for correct voltage and compare it to that found on the schematic. Most screen-grid voltages – in the smaller screens within the TV/VCR chassis' – are between +160V and 190V DC. When all voltages supplied by the flyback circuits are adequate and correct, proceed to the picture tube circuits for possible defects (*Fig. 5-9*).

CRT Circuits

Most CRT circuits are now found on the CRT board at the end of the picture tube socket (*Fig. 5-10*). In some color output circuits, the color preamp or driver transistors are with each color amp transistor, whereas in other CRT circuits, the red, green, and blue color amp transistors provide a signal to each color gun assembly. The color amps are powered by a +160V to 180V DC source with a heater voltage (6.4V) in a socket plug connector.

The B-Y, G-Y, and R-Y signals are fed through another socket plug connection and are sometimes soldered directly to the CRT board. The A-Y or sub-bright signal is fed to each emitter terminal of the color amps (*Fig. 5-11*). The R-Y signal is fed directly to the red color amp base circuit. Likewise, the G-Y signal is fed to the green color amp and the B-Y signal is fed to the blue color amp base terminal. Each color load resistor (R501, R502, R503) is fed with a +180V source. The color output of each color amp transistor is fed to the corresponding cathode color gun assembly.

Fig. 5-11. A B-Y, G-Y, and R-Y signal are each sent to a corresponding color amp transistor on the CRT neck.

The color picture tube has three different color gun assemblies mounted in the neck of the picture tube. When one of the heaters doesn't light up, the color will be missing on the screen of the CRT. The color picture tube consists of three gun assemblies that provide heater elements, which heat up the cathodes and emit electrons to the screen grid and anode circuits.

The voltage applied to the focus element sharpens the picture and raster and brings them into proper focus. By varying the focus control, the scanning lines can be adjusted for clear and clean lines. High voltage at the anode terminal speeds up and pulls the electron beam, striking it against the color beads at the front plate of the picture tube resulting in a raster or picture.

Symptoms of a Defective CRT

A bad heater connection or open heater element can be caused by a bad CRT socket or open heater-gun assembly. An intermittent picture or raster can result from poor heater connections or a bad socket. A weak gun assembly, a poor drive signal, or insufficient screen grid voltage can cause a weak or dim picture. When one color is missing on the screen, check for a defective gun assembly, color amp transistor, and color load resistor. The same problem can also occur when a poor brightness signal is applied to the color amp transistors.

Here are some other problems you may encounter with a defective CRT.

- ✔ A no raster with okay high-voltage can be caused by a bad CRT socket, no lighting of heaters, and poor socket connections.
- ✔ A bad solder joint on the heater plug or connector of the CRT filaments can cause a no raster symptom.
- ✔ Badly soldered joints or an open heater resistor can cause a no raster or no picture symptom.
- ✔ An open horizontal center control in some TVs may cause no raster, no CRT, and a lit heater.
- ✔ A no raster, no high-voltage symptom can be caused by improper high voltage or no high voltage from the flyback circuits.
- ✔ When the picture slowly fades away, check for badly soldered joints on the CRT board at the neck of the CRT.
- ✔ For intermittent loss of raster or filament, check for a badly soldered joint on Pins 9 and 10 of the flyback. Badly soldered joints on the flyback can cause an intermittent loss of raster (filaments), intermittent loss of vertical sweep, and an intermittently dead symptom.

- ✔ When the screen turns completely red, suspect a poor solder joint on the plug sockets to the CRT neck board.
- ✔ If the set has symptoms in which the raster is intermittent at startup and, once the raster is on, it won't go off but stays on, check for a badly soldered joint on connector CNO2 on the CRT neck board.
- ✔ A no raster, filament okay symptom can be caused by a defective picture tube, which may be rejuvenated.
- ✔ If the set has a bright screen with retrace lines and poor focus, this may look like either a bad CRT or a bad flyback, but the problem is actually caused by a bad CRT socket.

CRT Retrace Line Problems

If the set starts up, arcs over, and then shuts down, this can result from dust collected in the spark gap of the CRT. There are a variety of screen color changes that indicate CRT problems.

Bright Screen

- ✔ If the set has a bright screen and retrace lines, check small electrolytic capacitors (1µF to 4.7 µF).
- ✔ A bright picture with retrace lines, no visible video, and no on-screen display can be caused by a bad reference transistor or video amp found on the CRT neck board.
- ✔ In a set with a bright screen and retrace lines, check for a defective color or driver amp transistor.
- ✔ If there is excessive brightness, vertical collapse, and really bright retrace lines, check for open coils, inductance, or bad terminals.

Green Screen

- ✔ For a set with an intermittently bright green screen with retrace lines, check for burned or bad board connections on the load resistor.
- ✔ A really bright green screen at turn on followed by immediate shutdown can result from a clogged spark gap.
- ✔ A bright green raster with retrace lines can be caused by shorts in the CRT.

Red Screen

- ✔ When the set has a red screen and a retrace-line symptom that seems to be caused by a bad CRT, check the red spark gap for excessive dust. Blow out the excessive dust in the spark gaps.

Blue Screen

- ✔ A blue screen with retrace lines can result from a high-resistance leakage across the blue spark gap.

Color Amp Problems

- ✔ If there is no red in the picture, suspect a defective red amp.
- ✔ A purple picture with no green can be caused by a badly soldered joint on the green amp transistor.
- ✔ If there is a green raster and green retrace lines, check for a bad CRT lead wire.
- ✔ A no red symptom can be caused by a poor board connection at the red load resistor.
- ✔ When the picture turns purple and loses the color green, replace the defective green output transistor.
- ✔ A green picture with no blue can be caused by a bad foil or trace of B+ voltage to the blue collector load resistor.
- ✔ A poorly soldered joint on the collector and emitter terminals of the green color amp transistor can cause intermittent loss of green.
- ✔ For a blue screen problem, resolder all terminals of the color transistors on the CRT board (*Fig. 5-12*).
- ✔ An open coil to the red amp load resistor – located on the CRT board – can cause a red picture.
- ✔ To correct an intermittent loss of red, replace a defective red amp transistor.
- ✔ A badly soldered joint on the blue amp transistor – located on the CRT board – can cause a yellow raster.
- ✔ A no red symptom is caused by a defective red output transistor.
- ✔ Badly soldered joints on the collector terminals of all three different color amps – located on the CRT board – can cause the screen to turn all red or all blue intermittently.

Defective CRT

A defective picture tube can cause many problems, such as no raster, poor brightness, missing color, intermittent picture, poor focus, one-color screen, arcing in the gun assembly, bright screen with retrace lines, intermittently bright blue screen with retrace lines in the picture, dim picture, snapping and popping noises, black at the top and bottom, negative picture, and chassis shutdown.

Fig. 5-12. Resolder contacts on the color amp transistor terminals when the screen goes all green.

- ✔ An open CRT filament can cause a no raster, audio okay symptom.
- ✔ A heater-to-cathode short in the picture tube can produce an intermittent, bright blue or green screen with retrace lines.
- ✔ An intermittently bright screen can be caused by a defective CRT.
- ✔ If the relay clicks off and on and the set loads down, this can be caused by a defective picture tube.
- ✔ An internal short between G2 and the focus pin of the CRT can cause a raster okay, poor-focus symptom.
- ✔ A shorted gun assembly can cause a problem in which the screen turns black intermittently after a few seconds and the set shuts off and goes into shutdown mode.
- ✔ A red screen with black at the top and bottom is caused by a shorted CRT.
- ✔ If after turn-on, the raster dims then gets excessively bright in a short time, replace a leaky picture tube.
- ✔ A dim picture with no green can result from a bad green gun assembly in the CRT.
- ✔ If the picture is red only, then rejuvenate the CRT.
- ✔ A shorted picture tube can cause the screen to turn green intermittently with retrace lines. If the whole screen is green, then the problem is a defective CRT. Also, a heater-to-cathode short can cause a green picture.

- ✔ If the screen turns blue intermittently after a few seconds and the set shuts off, this is caused by a defective picture tube.
- ✔ A shorted CRT can cause the picture to have a pink cast and retrace lines.
- ✔ If by tapping the end of the picture tube, the picture sometimes clears up or goes negative, this indicates a flaking-off of cathode material lodged in the grid section of the picture tube.

Rejuvenating the Picture Tube

When the picture is dim, the picture tube might be restored by removing a short in the gun assembly with a combined CRT tester and rejuvenator test instrument. The weak gun assembly of each color gun assembly can strip off a collection of ions that have bombarded and collected on the cathode element. Rejuvenation of the picture tube can put a TV back into service and ready for resale. A rejuvenation process can restore brightness to a dim picture tube. Sometimes, a shorted CRT can be temporarily restored with an outside filament transformer.

Focus Blurry – Orion TVCR1320

In an Orion TVCR1320 TV/VCR, the focus was extremely blurry. Only 2 kV was registering on the focus terminal. The picture tube (V801) was defective causing the blurry focus problem.

A defective CRT sometimes can cause the horizontal output transistor to run hot, even after being replaced, and then it may blow again making a loud noise when it does. The arcing picture tube can cause damage to the high voltage regulator, horizontal output transistor, isolation resistor, and small electrolytic capacitors. A shorted and arcing picture tube can destroy the focus and screen controls. The shorted CRT can damage the beam-current resistor on the neck of the CRT board.

If there is severe arcing in the neck of the CRT, you should replace the CRT. Also replace the CRT when the picture becomes intermittently negative and cannot be rejuvenated. If when the screen control is turned up, the CRT begins to arcover, then replace the picture tube. After the set is on 10 minutes, if there are barely visible retrace lines, replace or try to rejuvenate the picture tube. If the retrace lines are always visible – when switching tuner channels – and there is a snapping and popping noise, replace the picture tube.

When constant arcing occurs in the gun assembly inside the yoke assembly – indicating a cracked or broken neck of the tube – you should replace the CRT. When the picture tube keeps arcing over in the spark gap assemblies, check for a defective CRT or excessive high voltage. Try to blow the dust from the spark gap before replacing the picture tube. The arcing flyback, rather than the bad CRT, can cause the arcing lines in the picture.

Improper focus of the picture – such as when the picture will not focus up or is way off to one side – is caused by low focus voltage. For a correct focus-voltage measurement, rotate the focus control with the high-voltage meter probe on the focus pin of the CRT socket. A dim or poorly focused picture might be restored with the rejuvenator CRT tester. If there is no control of focus and low brightness even with the brightness control fully on, you should replace the picture tube.

Other problems stemming from the CRT include:

- ✔ A bad focus or poor socket focus-pin connection resulting in a poorly focused picture.
- ✔ An internal short between the grid and focus element causing poor focus with an okay raster. This type of short can sometimes be blown away by applying a charged-up 10-μF, 250-volt electrolytic between G2 and the focus pin.

Chapter 6: Troubleshooting TV Video and Color Circuits

The video and chroma circuits of today are a lot easier to service than yesterday's video and color circuits were. Today, the video and color circuits are found inside the same IC as the deflection and IF circuits in many sets. Even the different sync circuits are located in the very same IC component. In the video circuits of a Sears 9-inch TV/VCR, IC301 contains the video/chroma/deflection/IF circuits.

The video signal is fed from Pin 51 of IC301 to a 4.5-MHz trap (CF302) to a buffer transistor (Q308) and out to the TV-video Input Select IC701. The Input Select IC701 feeds the video signal to Pin 44 of the Y/C process (IC401) (*Fig. 6-1*). A color video (C-video) signal is fed back from IC401 to the TV Micon (IC101), which controls the brightness, sharpness, contrast, color tint, and RGB outputs within IC301. The R-Y, G-Y, and B-Y, plus the Y signal, are fed from IC301 out of Terminals 22, 23, 24, and 25. The Y signal is amplified by buffer transistor Q301 and is connected to plug connectors CN301 and CN302.

The TV Micon (IC101) controls the volume, C-video input, S-CLK, S-Out, and S-In signals, plus Reset, SDA, SCL, V-Sync, H-Sync, and service-switching circuits. The brightness is controlled out of Pin 4, the sharpness is controlled out of Pin 3, the contrast out of Pin 6, the color out of Pin 5, and the tint out of Pin 7.

In an Emerson VT1920 TV/VCR, the video-amp signal is fed out of Pin 22 from IC201 to a buffer amp (Q205), to a video amp (Q701), and to the video-buffer transistor (Q702). The Video-In switch is coupled to the Video-Out signal on Pin 5 of the CP8007 connector. The Video-In signal from the VCR is fed to the video-input Terminal 25 of video/chroma/deflection IC401 (*Fig. 6-2*).

IC401 functions include the processing of the video/chroma/deflection/and Y-Out, with R-Y, G-Y, and B-Y fed to the color drive IC901. The brightness is controlled at Pin 22, the contrast is controlled at Pin 28, the color at Pin 26, the tint at Pin 34, and the sharpness at Pin 19. The Y output is sent out of Pin 4, with R-Y sent out of Pin 2, G-Y out of Pin 3, and B-Y out of Pin 1.

IC601 Micon controls the volume at Pin 15, the brightness at Pin 14, the contrast at Pin 13, the color at Pin 12, the tint at Pin 11, and the sharpness at Pin 10. A brightness

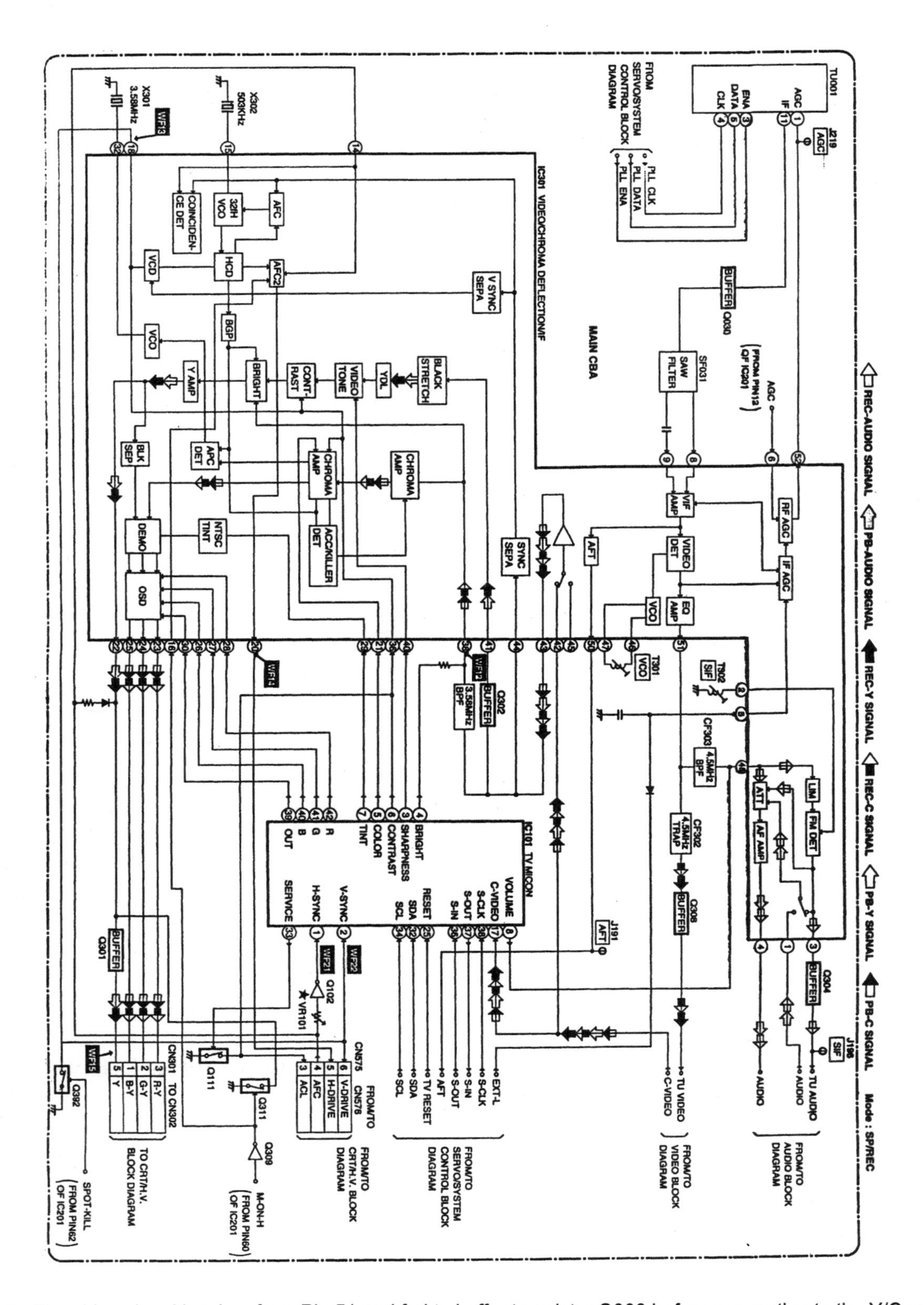

Fig. 6-1. The video signal is taken from Pin 51 and fed to buffer transistor Q308 before connecting to the Y/C process IC701 in a Sears 9-inch TV/VCR.

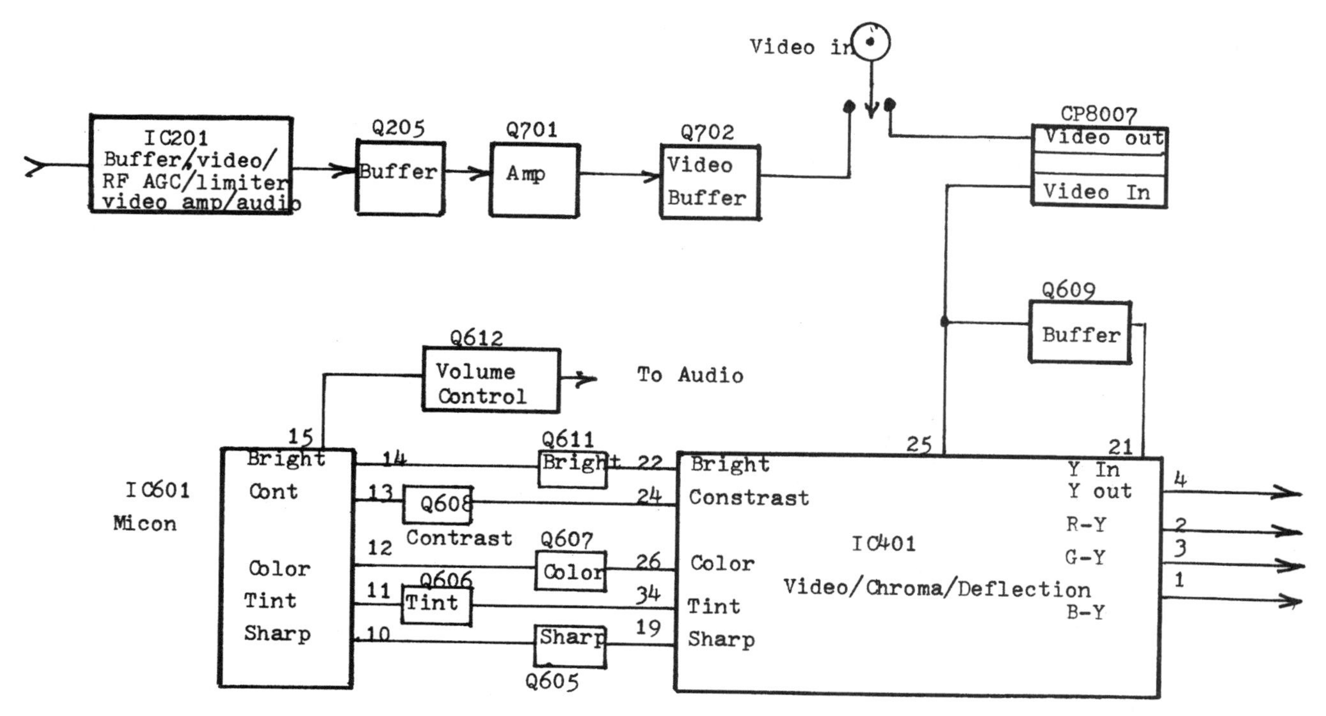

Fig. 6-2. Block diagram of the video circuits in an Emerson VT1920 TV/VCR.

control transistor is found between Pin 14 of IC601 and Pin 22 of IC401. The contrast-controlled signal has Q608 between Pin 13 and Pin 24 of IC401.

The controlled color at Pin 26 of IC401 has transistor Q607 in series with Pin 12 of IC601. A tint-controlled transistor (Q606) is controlled from Pin 11 of IC601 to Pin 34 of IC401. A sharpness transistor (Q605) is connected between Pin 10 of IC601 and Pin 19 of IC401. The sub color and tint controls are tied to respective color and tint amp circuits.

Video Circuit Problems

The video circuits control the picture, brightness, contrast, and sharpness of the picture. Picture problems can include:

- ✔ No picture.
- ✔ A negative picture.
- ✔ A smeared picture.
- ✔ A dark picture.
- ✔ An out-of-focus picture.
- ✔ A distorted picture.

- ✔ A snowy picture.
- ✔ A dim and washed-out picture.
- ✔ Lines in the picture.
- ✔ An intermittent picture.
- ✔ A picture that flutters.
- ✔ A noisy picture.
- ✔ No video or audio.

Problems in the luminance (Y) signal can include:

- ✔ A dim raster.
- ✔ No brightness or brightness control.
- ✔ White screen with bright retrace lines.
- ✔ Bright raster.
- ✔ Intermittent loss of raster.
- ✔ Dark raster.
- ✔ Darkness on the left half of the picture.

The Y signal from the video IC is also applied to the color-output transistor on the neck of the CRT board. The video-output IC might have a bright picture with poor contrast.

A picture with poor contrast might entail dim contrast, excessive contrast, contrast that is not adjustable, a really dark picture, a bright picture or washed-out picture, low control with retrace lines, a negative picture, an intermittent picture that goes dim or dark, weak contrast, a dim, washed-out picture, and brightness that cannot be controlled. Do not overlook a weak picture tube when there is low brightness and poor contrast. Very little contrast can result from a defective delay line. A defective Y/C-Process IC can cause a no raster symptom.

No Raster, High Voltage Okay

- ✔ If the TV has no raster, weak sound, and the channels change, but the audio does not, check for a bad voltage regulator.
- ✔ A no raster symptom with on-screen display may be caused by a defective diode and leaky capacitor in the video IC amp voltage source.
- ✔ If the set has no raster, no high voltage, and the CRT filament is lit, it may be the result of an open video/chroma IC.
- ✔ Advancing the screen control with a no raster symptom shows no vertical or improper vertical sweep indicates a defective component in the vertical circuits, causing video problems.
- ✔ A leaky or shorted vertical output IC might cause the low-voltage source to lower other voltage sources in the scan-derived flyback power supply.

- ✔ A no raster, no audio problem can be caused by a vertical output IC with open low-ohm resistors in the supply voltage circuit. Check all silicon diodes in the power supply that provides voltage to the video circuit.
- ✔ An open 1.2-ohm resistor on line from Pin 9 of the flyback can cause a no raster symptom.
- ✔ Symptoms such as no raster with intermittent high voltage and a dim picture can be caused by a 0.001-µF, 2-kV capacitor on the CRT board.
- ✔ A shorted zener diode can cause a no raster, no video, and no on-screen display symptom. The vertical output deflection IC with a 220-µF electrolytic tied to it also can cause a no raster symptom.
- ✔ A defective EEPROM IC can cause a symptom in which there is no raster, the high voltage is OK, and the filament is lit up.
- ✔ For a no raster symptom, suspect a defective service switch found in the CRT board.
- ✔ No raster or a very dark raster can be caused by an open, 1-kilom resistor on the picture-tube board.

No Video

- ✔ If the set has no video, a raster, and good audio, check for a bad crystal tied to the video/chroma IC.
- ✔ If after warm-up, there is a loss of video and sound and you cannot adjust the volume or channels, look for a bad microprocessor.
- ✔ No video with lines at the top can be caused by a badly soldered joint on the vertical output IC.
- ✔ For a no-video or audio-only symptom with snow, check for open resistance or an increase in resistance from the resistors off of the video/chroma IC (Fig. 6-3).

Fig. 6-3. A snowy screen in a set with no video.

- ✔ If the set has no video and normal on-screen display, check the memory IC.
- ✔ A defective video/chroma IC can cause a no video, audio okay symptom.
- ✔ A set with no video, but normal sound, can be the result of a shorted buffer amp transistor.
- ✔ An inoperative video symptom can result from water or other liquids spilling in back of the plastic cabinet, damaging resistors in the boost power supply.
- ✔ No video and a blank raster can be caused by a 12-volt IF regulator transistor.
- ✔ Check the 220-µF capacitor off of the vertical output IC for a no video, normal raster symptom.
- ✔ A defective automatic brightness limiter transistor can cause a no video symptom.
- ✔ If the set has only a white raster with normal audio, check for an open buffer transistor.
- ✔ The defective sync-kill transistor can cause a no video, audio okay symptom.
- ✔ If there is no video, but the raster and audio are okay, it can be the result of numerous poorly soldered connections where the signal board is mounted vertically on the main board.
- ✔ If the set experiences symptoms such as no video and no channel numbers displayed, with normal high voltage, raster, and audio, check the microprocessor.
- ✔ A defective delay line can cause a no video, color out-of-sync symptom.
- ✔ If a thin horizontal line appears (no vertical sweep) when the TV is turned off along with a no video or no audio, a leaky diode and burned isolation resistor in the voltage source from the flyback are the likely causes.

Excessive or Intermittent Brightness

- ✔ If the set has an intermittent bright screen and retrace lines, suspect a defective picture tube.
- ✔ When the intermittent picture goes dim or dark and there is no audio, check for badly soldered joints on the video/chroma IC.
- ✔ A defective screen control can cause intermittent brightness.
- ✔ Intermittent loss of video and audio can result from a badly soldered joint on the voltage regulator IC.
- ✔ If the brightness fluctuates when the TV is cold, the cause is a bad 1-µF electrolytic in the video/chroma section; the capacitor might test normal on a capacity tester but bad on an ESR meter.
- ✔ Badly soldered joints on the horizontal driver transformer can cause a loss of low tuner channels and a changing brightness symptom.
- ✔ To solve an intermittent video and brightness fluctuation problem, spray control cleaner fluid down inside the brightness and contrast controls.
- ✔ If the picture flashes in and out with poor audio and poor B+ regulation, a defective line-voltage regulator may be the cause.

Intermittent Video – Emerson VT1920

In an Emerson VT1920 the video would pop in and out after the TV/VCR operated for 30 minutes or so. The video signal was traced into the emitter terminal of video amp Q701. The video seemed to come in and out when transistor Q701 was moved or touched. At first, poor board contacts were suspected, but when Q701 was tested in the circuit, the transistor seemed to act up. The 25C945A transistor was replaced with a universal NTE85 transistor (*Fig. 6-4*).

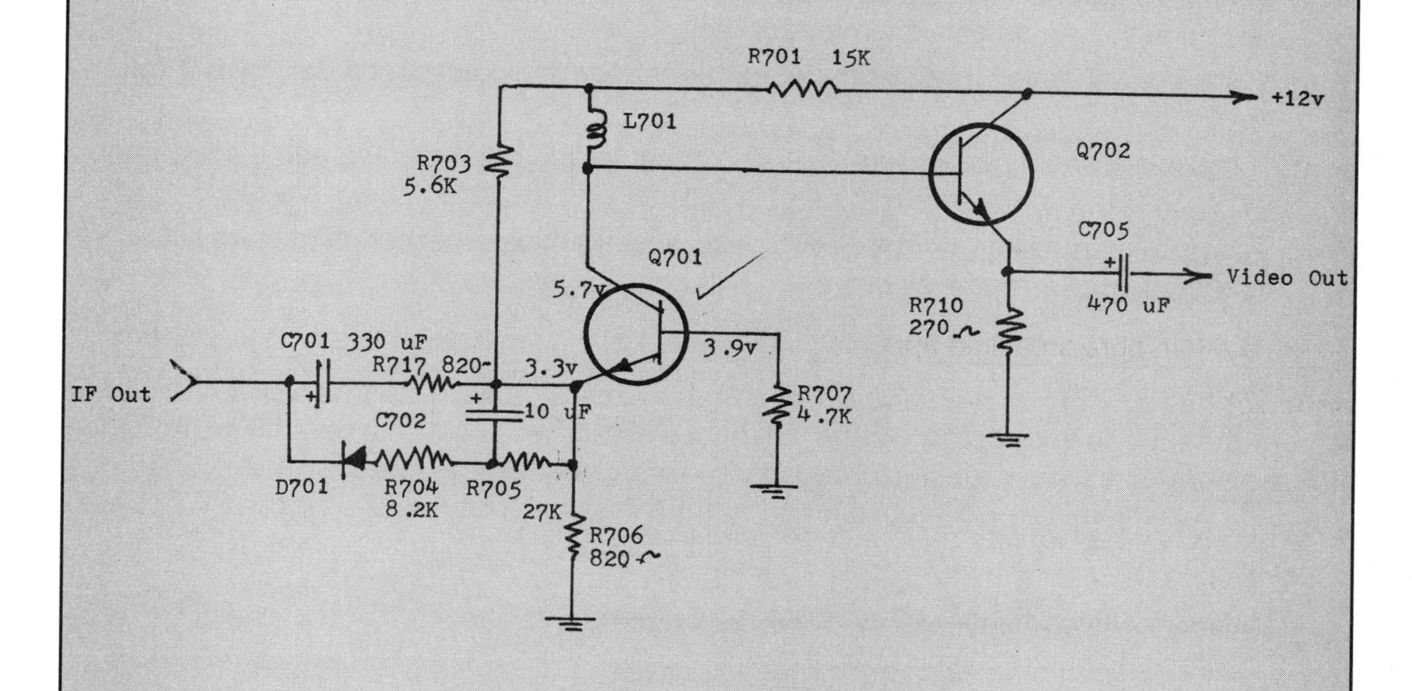

Fig. 6-4. Q701 caused intermittent video in an Emerson VT1920 TV/VCR.

- ✔ Intermittent loss of video, a dim picture, and a color okay symptom can result from a bad screen control in the flyback.
- ✔ When both contrast and brightness controls are turned up and there is a change in brightness and retrace lines, check for a defective drive reference regulator transistor.
- ✔ For intermittent brightness and a picture with no color, check for a defective PF SMD capacitor off of the video/chroma/deflection IC.
- ✔ If the TV has a bright picture, check for a dried-up or open electrolytic (30 µF, 250V) in the B+ line.
- ✔ Excessive brightness can result from burned or open resistors in the luminance circuits. Check for bad connection on coils found on the CRT board.
- ✔ A bright picture also might be caused by a leaky bypass capacitor on the CRT board. Also check for a defective luma transistor.
- ✔ For a set with a very bright picture and no retrace lines, suspect a defective 9-volt regulator.
- ✔ If there is a bright picture and the video is not visible but the on-screen display can be seen, replace the reference amp transistor and the video amp transistor.
- ✔ A defective transistor on the CRT board causes an extremely bright picture with screen control turned way down.

Bright Screen With Retrace Lines

If the set has a bright screen and retrace lines in the picture or an intermittently bright screen with retrace lines in the picture, suspect a defective CRT. Also a defective luminance buffer or buffer limiter transistor causes an extremely bright raster with retrace lines (*Fig. 6-5*). Check all three bias transistors in the color-output circuits on the CRT board.

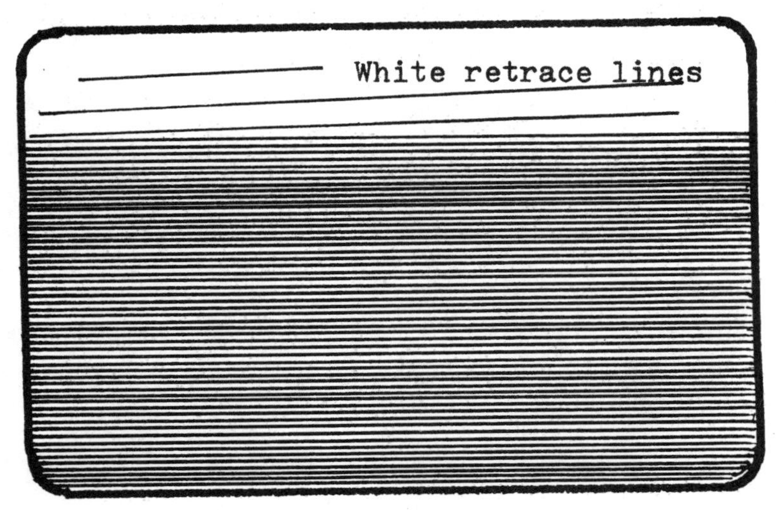

Fig. 6-5. A bright raster with retrace lines can indicate a defective video stage.

- ✔ A light pink picture with retrace lines can be caused by a defective red and blue bias transistor on the CRT board.
- ✔ For a bright green screen with retrace lines, check for a leaky green amplifier transistor.
- ✔ For a set with an intermittently blank raster with retrace lines, suspect an open 12-volt regulator transistor and a shorted silicon diode.
- ✔ For an extremely bright picture and retrace lines, check for low voltage at the secondary flyback winding with leaky PF capacitors in the +180V to 200V line. Also suspect open, low-ohm resistors (2.7 ohms) on the +215V line from Pin 2 of the flyback.
- ✔ If the set has no audio, no retrace lines, and a bright white picture, check the 9-volt regulator IC.
- ✔ If the set has a bright picture and retrace lines with very little video, replace both the reference and the video amp transistors.
- ✔ A problem with the video driver transistor may cause a bright picture and retrace lines.
- ✔ If there is no video and no raster with bright retrace lines when the screen control is turned up, the cause may be a problem with a luminance driver transistor.
- ✔ Excessive retrace lines can be caused by an emitter-to-collector short in the luma buffer (Y) signal transistor.

No Video/Retrace Lines – Sharp 13VTF40M

A Sharp 13VTF40M came in with symptoms such as no video, retrace lines in the raster, and a blue screen. Replacing IC201, which was defective, solved the video problem. Another TV/VCR of the same make and model had a no video problem. Replacing the 9.1V zener diode (D494) restored the video to the screen.

- ✔ If the set has a dim raster, no on-screen display, and retrace lines in the picture, suspect a character interface IC.
- ✔ For an excessively bright raster and retrace lines, check for badly soldered joint connections on the flyback.

Brightness Does Not Turn Down

- ✔ If the set has a bright picture, retrace lines, and the brightness cannot be turned down, check for defective low capacitors (30 µF, 200V) in the video voltage sources; test each electrolytic for ESR problems.
- ✔ If the set has extreme brightness, retrace lines, and the brightness cannot be turned down, suspect the resistors off the flyback.

- ✔ A loss of brightness control can result from a large resistor increasing in resistance within the base-voltage supply of the beam-limit transistor. Remove one end of the resistor and check for correct resistance. An open resistor in the base circuit of the beam-limit transistor can also result in an inability to lower the brightness.
- ✔ When the brightness cannot be turned down and there is extreme brightness and retrace lines, suspect a luminance buffer or luminance driver transistor.
- ✔ With a low brightness problem and an inability to control the focus, check for a bad picture tube.
- ✔ A shorted CRT can cause a really bright picture and an inability to control the brightness.

Dim or Dark Picture

- ✔ A set with a dim, washed-out picture even with the brightness and contrast controls turned completely up may appear to have a very weak video problem, but it is actually caused by a weak red gun assembly of the CRT.
- ✔ If the set has a dark picture, and rotating the brightness control makes no difference, check for an increase in resistance of resistors in the beam-current circuits.
- ✔ If there is no raster or a very dark picture, inspect the low-ohm (1 kilohm) resistors on the CRT board.
- ✔ When the TV is cold, if there is a dark picture and the brightness fluctuates, check for a small (1 µF to 4.7 µF) electrolytic off of the TV processor.
- ✔ When the picture darkens, make sure the CRT DAG ground wire is soldered to the common ground on the chassis.
- ✔ When the filament goes out and the picture becomes dark, check the filament resistor.
- ✔ A dark picture also can be caused by a defective electrolytic (100 µF) in the 12-volt source.
- ✔ An intermittently dim picture with normal high voltage can result from a 0.001-µF, 2-kV bypass capacitor on the CRT board.
- ✔ With a dark picture and an out-of-focus symptom, check for bad spark gaps.
- ✔ A dim raster with the brightness control fully on can be caused by a leaky luminance buffer transistor.
- ✔ A dim, washed-out picture can be caused by leaky bypass capacitors off of the video processor IC.

Black Lines, Dark Picture – Quasar VV8213A

In a Quasar VV8213A TV/VCR combo, several wavy black lines may appear with dark shading in the picture on both TV and VCR functions. To fix the problem, two 330 µF, 6.3V electrolytics in the shielded VCR power supply located on the TV board were replaced. These electrolytics might test okay with the ESR meter, but you should replace both capacitors to cure the problem.

- ✔ A dark picture with normal video can be caused by a leaky beam-limit transistor off of the video processor IC.
- ✔ For a dim raster with retrace lines, check the interface character IC within the on-screen display circuits.
- ✔ If there is a dark picture and you cannot adjust the picture or color level, the problem may be caused by a defective microprocessor.
- ✔ For a dim and dark picture or raster, suspect a defective picture tube.
- ✔ When the picture intermittently goes dim or dark, check for badly soldered joints on the luma/video/chroma/deflection IC (*Fig. 6-6*).

Fig. 6-6. Poorly soldered contacts on the luma/video/chroma/deflection IC can cause a dark picture.

Poor Contrast

- ✔ For poor contrast, check for an open or shorted contrast control. Use the ESR meter to test each low-value electrolytic off of the pin terminals of the luma/video/deflection IC.
- ✔ For a poor raster and washed-out picture, check for a defective normal-service switch on the CRT board.
- ✔ A green, faded picture can result from a defective EEPROM circuit.
- ✔ A defective delay line can cause very little contrast that cannot be varied with the contrast control.
- ✔ A weak gun assembly in the CRT can cause a dim and washed-out picture, even with the brightness and contrast controls wide open.
- ✔ A weak buffer or video amp transistor can cause a weak contrasting picture.
- ✔ A defective picture-in-a-picture transistor can cause a faint, distorted picture with no contrast.

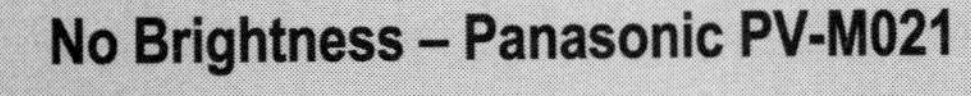

No Brightness – Panasonic PV-M021

In a Panasonic PV-M021 TV/VCR the screen had normal high voltage but no brightness. The signal was traced from Pin 8 on the luma/chroma/signal processor (IC301) to the base of the buffer transistor (Q309). No Y waveform was found on the emitter terminal. With in-circuit transistor tests, Q309 was found to be open with 12V at the emitter terminal. The 25B641 transistor was replaced with an NTE19 universal replacement (*Fig. 6-7*).

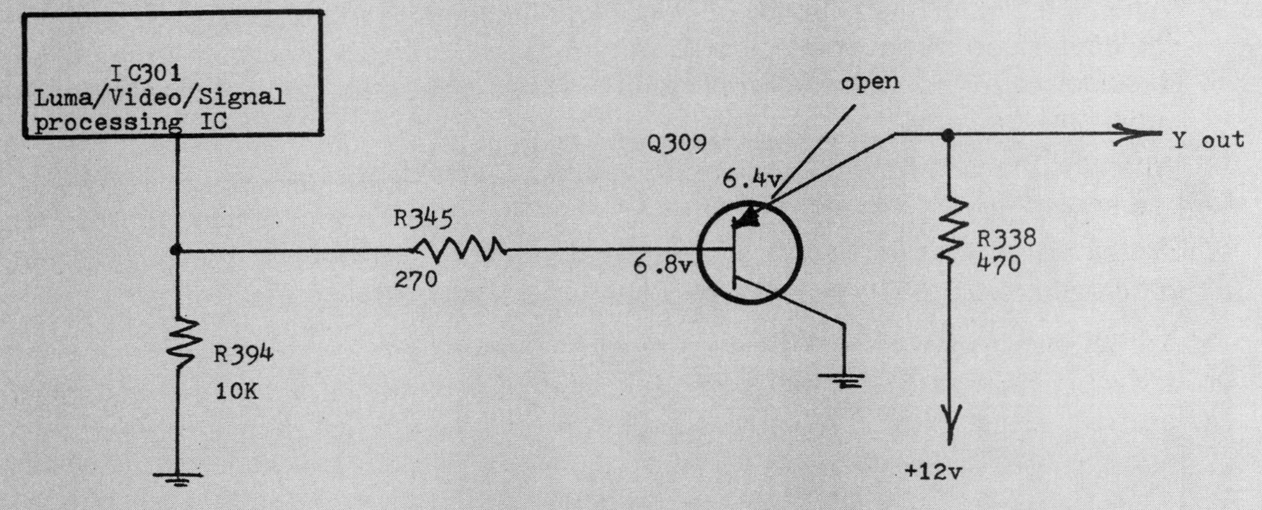

Fig. 6-7. No brightness in a Panasonic TV was caused by an open buffer transistor (Q309).

Snowy Picture

Although most snowy pictures are caused by a defective tuner, antenna system,and IF stages, you should suspect other areas, including a defective chroma/luma/video/IF/deflection IC, EEPROM, or even badly soldered joints.

- ✔ If there is a snowy picture and no video, this problem may look like a defective tuner, but it actually can be caused by an analog interface unit IC.
- ✔ A defective EEPROM can cause a snowy picture.
- ✔ When the picture goes snowy after warm-up, check for a defective 4-MHz crystal.
- ✔ A snowy picture on all channels can be caused by a defective Quad OP-Amp IC or a badly soldered joint on the IF module.
- ✔ An intermittently snowy picture might be caused by badly soldered connections of the luma/video/IF/deflection IC.

Intermittent Video

- ✔ The intermittent loss of video and audio can be caused by badly soldered joints on the IF transformer within the IF module.
- ✔ When the video and audio drift out and sometimes pop back in, and when tapping on the tuner restores audio and video, you should resolder all grounds in the tuner.
- ✔ When the video becomes intermittent and the brightness begins to fluctuate, clean the contrast control.
- ✔ Resoldering all grounds surrounding the IF shield can solve an intermittent video and audio problem.
- ✔ When the video is flashing and becomes dark, check for poor coil connections or a bad service switch.
- ✔ Symptoms such as very weak tuner action and barely visible video that pulls in and out can be caused by a defective luma/video/chroma/deflection IC.
- ✔ A defective delay line can cause intermittent video with audio okay.
- ✔ If there is intermittent loss of video and sound after warm-up, and if you can't adjust the volume or channels, the problem may be caused by a bad microprocessor.
- ✔ Intermittent video can be caused by a defective video amp and video buffer transistor.
- ✔ A poorly soldered connection on the video amp terminals can cause intermittent video.
- ✔ With intermittent video symptoms, suspect an intermittently functioning T-chip.
- ✔ When the picture flashes in and out with poor audio and B+ regulation, check the line voltage regulator.

Smeared Video

For smeared video, check the small capacitors (1 μF, 4.7 μF, and 10 μF) off the video/luma/color/deflection IC with the ESR meter. A smeared video with color can be caused by a defective coil and open or poor board terminal connections of the peaking coils found in the video circuits.

Resolder all suspected connections in the area around the coils found in the video circuits. For a smeared picture with ghosts, which may look like an alignment problem, resolder all suspected components in the area of the delay line. With a smeared video problem that seems like it might be a bad CRT, do not overlook a defective luma/color/video/IF/deflection IC.

Some other video problems you may encounter include:

- ✔ A defective CRT can produce distorted, blurry, and dark video.
- ✔ A defective flyback can cause a blurry, distorted picture.
- ✔ A defective comb filter can create heavy jail bars in the picture.
- ✔ A dim, washed-out picture can be caused by a defective bypass capacitor on the luma/video/color/deflection IC.
- ✔ No raster and a washed-out, negative picture – especially when the screen control is turned up – can be caused by a defective normal service switch found on the CRT board.
- ✔ One or more weak gun assemblies inside the picture tube can cause a dim, washed-out picture, with brightness and contrast controls wide open.
- ✔ Intermittently washed-out video with a negative picture can be caused by a bad solder joint on the video board. Flex the board and pay particular attention to the solder joints near the connectors.
- ✔ For a negative picture, check for open coils in the video amp circuits, as well as an open coil off of the comb filter.
- ✔ A weak CRT in all gun assemblies can cause a very dull, negative, washed-out picture that is largely unaffected by the tint control.
- ✔ Sometimes, by tapping on the end of the CRT, the negative picture will come and go; when this occurs, either rejuvenate the picture tube or replace it.

Delay Line

- ✔ The delay line may be defective if the picture has very little contrast and the contrast control cannot be varied.
- ✔ Intermittent video with normal audio may be defective delay line.
- ✔ If there is no video and the color is out of sync, the delay line may be defective.
- ✔ A bad solder joint in the delay line can produce intermittent, dark horizontal lines over the entire screen.
- ✔ If the normal picture goes dark with a reddish raster and no video after the set has operated for several hours, it can be the result of a defective delay line (*Fig. 6-8*).

Loss of Picture – Emerson VT1322

An Emerson VT1322 TV/VCR had no picture. The video signal was traced with the scope. When the scope probe touched a small SMD resistor, the picture popped in. A badly soldered joint on SMD resistor R410 (1 kilohm) brought back the picture in the TV/VCR combo.

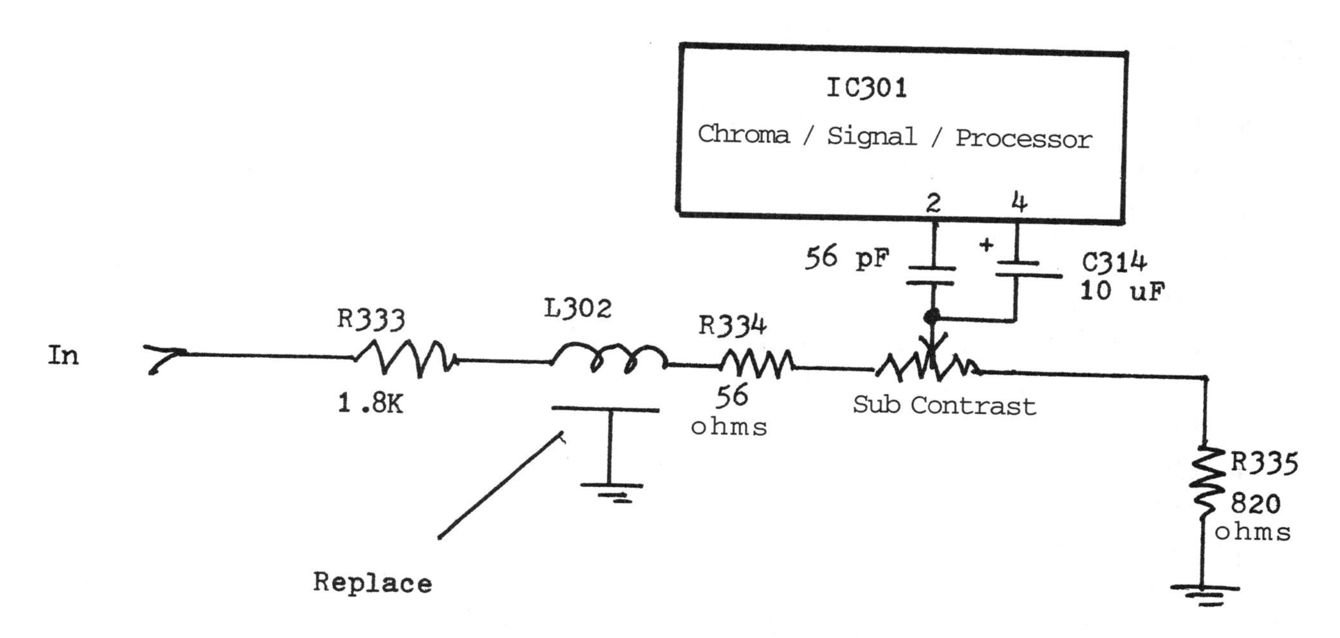

Fig. 6-8. A defective delay line (L302) can cause many video problems with the picture.

In addition to problems resulting from a defective delay line, the delay line can cause other problems as well including:

- ✔ Poor video caused by a delay line that is shorted to ground.
- ✔ A ghostly, smeared picture caused by poor solder joints on delay line terminals.
- ✔ Intermittent, dark horizontal lines over the entire screen caused by a badly soldered joint on the delay line.
- ✔ Intermittent loss of video caused by bad solder joints on the delay line.
- ✔ Poor contrast as a result of a leaky delay line.
- ✔ A washed-out picture with retrace lines can be caused by an open delay line.

Lines in the Picture

Besides retrace lines in the picture, the TV might have different video problems like dark lines, dark areas, and shaded areas that can be seen in the picture.

- ✔ Noisy lines in the picture or noise on all channels can be caused by a defective SAW filter component.
- ✔ A wavy band of vertical dark lines in the picture can be caused by a defective regulator IC and decoupling electrolytic capacitors in the video and chroma circuits.
- ✔ A defective 330 μF, 35V capacitor in the video-supply voltage will cause symptoms such as stretched vertical linearity and a 1-inch section of black horizontal lines at the top one-third of the screen, which, as the TV warms up, move up the screen. Use the ESR meter to check all electrolytics.

- ✔ To solve the problem of a line at the top of the screen and a no-video symptom, resolder all terminal pins on the vertical output IC.
- ✔ Intermittent white bands on the screen or a loss of raster can be caused by a badly soldered joint at the base terminal of the output color transistor on the CRT board.
- ✔ A bar on the left side of the screen can result from a bad filter in the video voltage sources. Also, video bars can result from an open resistor across the vertical yoke winding.

On-Screen Display Problems

The on-screen display signal is controlled by an interface or control IC. It is fed from an OSD amp and OSD driver to the green color output circuit. In the Sears 9-inch TV/VCR, the OSD is inside the video/chroma/deflection IC301 and controlled by Micon IC101. If the set has symptoms in which there is a bright picture and retrace lines, the video is not visible but the on-screen display is, then check for a bad reference and video amp transistor.

Some on-screen display problems include:

- ✔ On-screen channel numbers with extra lines coming from the digits when the TV is cold and a complete loss of all control functions when the set warms up caused by a defective control microprocessor.
- ✔ On-screen display shifting to one side caused by a defective EEPROM.
- ✔ When the on-screen display shakes severely and a wavy picture results from a remote-control selection, replace the main filter capacitor.
- ✔ If the picture turns green with retrace lines after the TV is on for 10 minutes, check for a leaky picofarad (pF) capacitor on the base of the OSD-amp transistor.
- ✔ If there is no raster except for the on-screen display, check for a leaky 220-µF, 35-volt electrolytic tied to the vertical output IC.
- ✔ A defective video buffer transistor can cause loss of on-screen display(*Fig. 6-9*).
- ✔ A defective analog interface IC can cause problems such as a snowy picture, no auto programming, no on-screen display, and no volume control.
- ✔ If there is no on-screen display, look for a leaky OSD sync transistor.

Troubleshooting the Video Circuits

To service the video circuits, connect a color dot-bar generator to the TV antenna posts. Start at the first IF amp transistor or IC and signal trace the generator signal with a demodulator probe and scope. If a dot-bar generator is not handy, then scope the IF and video circuits with a strong TV signal tuned in on the TV/VCR. Of course, the line-signal waveforms will be different than with the color dot-bar generator. The waveforms found in Sams Technical Publishing's PHOTOFACT® schematics are taken with a color dot-bar generator (*Fig. 6-10*).

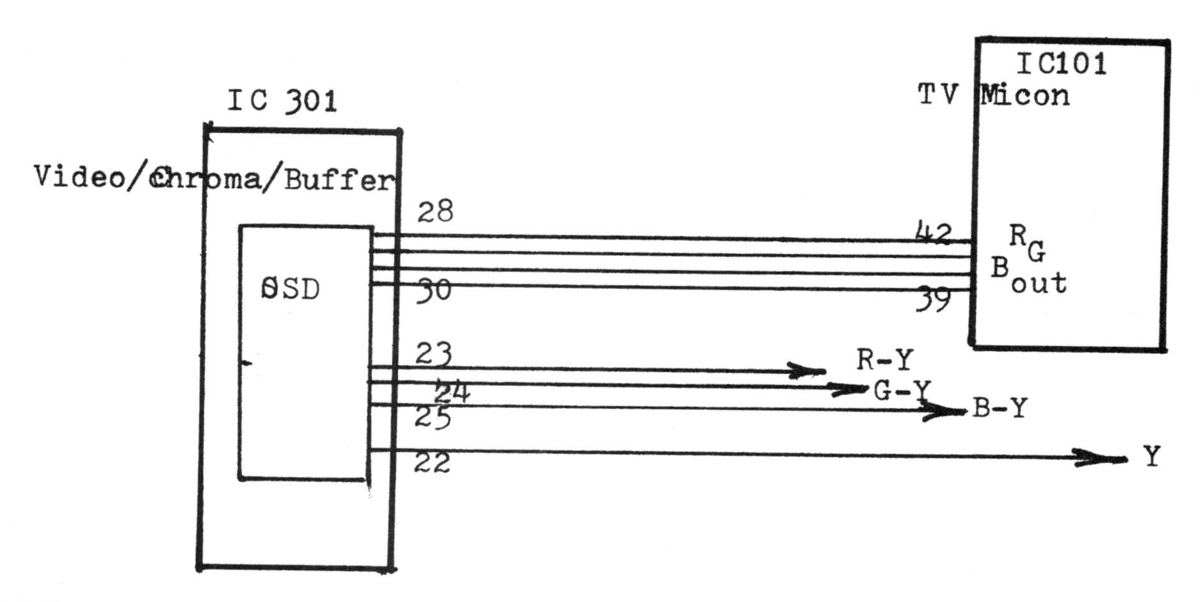

Fig. 6-9. Within some TVs, on-screen display problems can be caused by the sync and video buffer IC301.

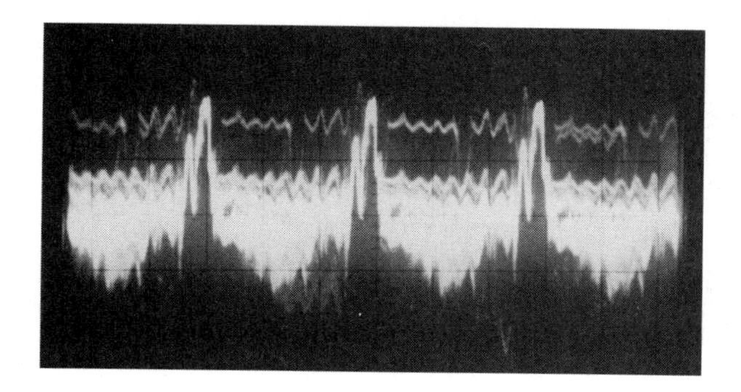

Fig. 6-10. The video waveform signal with the set tuned into channel 5.

Check the video waveform on the base of the video amp and emitter terminal for a good waveform. If the video waveform is normal at the video output pin, proceed to the video/chroma IC circuits. A video signal can be injected at the base of the video amp or input terminal of the video/chroma IC. Scope each stage for a loss or no signal.

For example, in a 9-inch Sears TV/VCR, scope the video signal out of Pin 51 of IC301 (video/chroma deflection/IF). Signal trace the video signal through the buffer transistor (Q308). When the video signal stops, troubleshoot that circuit by taking critical voltages and resistance measurements. For instance, if the signal went into the base circuit of Q508 and didn't come out the emitter terminal, check the transistor with in-circuit tests and take a critical voltage test on each terminal.

The video circuits can be signal traced if you inject a video signal into each video circuit and trace the signal with an oscilloscope. When the video signal is traced into the input

Pin 44 of the Y/C processor (IC401) and not out at Pin 41, look for a defective IC, components tied to IC pins, and an improper voltage source (*Fig. 6-11*). Make sure that the IC is defective before removing and replacing it.

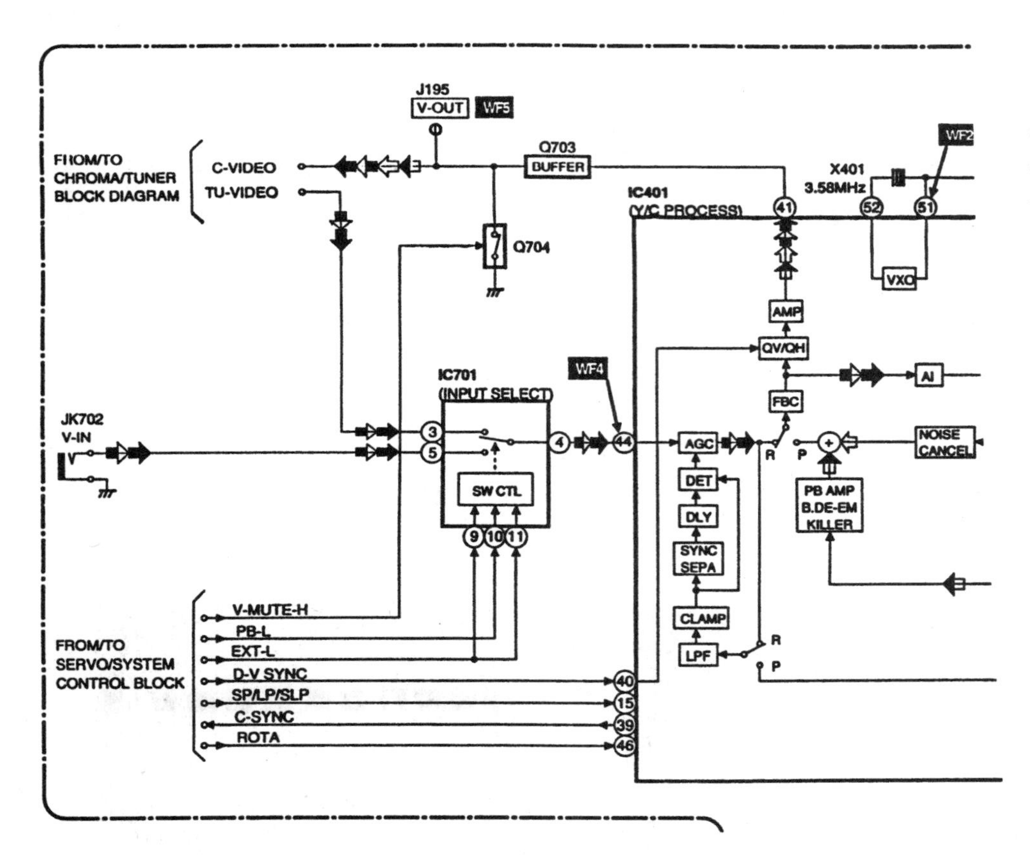

Fig. 6-11. The video signal comes in on Pin 44 and goes out of Pin 41 of the Y/C process IC401 in a 9-inch Sears TV/VCR.

TV Chroma Circuits

When presenting the chroma circuits here, a Sears 9-inch TV/VCR combination unit is used as the example in parenthesis. The chroma circuits are included with the video (Y) circuits inside the Y/C processor (IC401). A color signal is sent back to the video/chroma (IC301), where a 3.58 MHz color oscillator is located. The color signal is developed, amplified, and demodulated with the three color signals out of Pin 23 (R-Y), Pin 24 (G-Y), and Pin 25 (B-Y). These color signals are sent to the color output transistor mounted on the CRT board. The luma or brightness signal (Y) is found at Pin 22 and fed to the emitter-terminal circuits of the color-output transistors. Each color transistor amplifies the separate color and is fed to drive the color gun found on the CRT board.

No Video – Zenith SLV-1940S

In a Zenith SLV-1940S TV/VCR there was no video from either the TV tuner or VCR. The sound and raster were normal. Replacing IC02 on the VCR board solved the no video problem.

The most important color waveforms are taken on Pin 32 of IC301 for the color 3.58 MHz VCO waveform and the output pins of the R-Y, G-Y, and B-Y of IC301. If the 3.58 MHz color waveform is missing, then no color is found in the picture (*Fig. 6-12*). The color-output waveforms of all three demodulator color signals are quite similar. All three demodulator color signals can be signal-traced to each color output transistor found on the CRT board (*Fig. 6-13*).

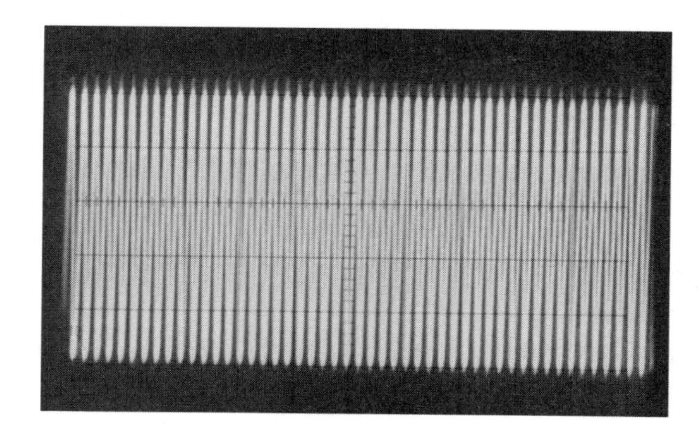

Fig. 6-12. The color waveform of the 3.8-MHz crystal VCO circuit.

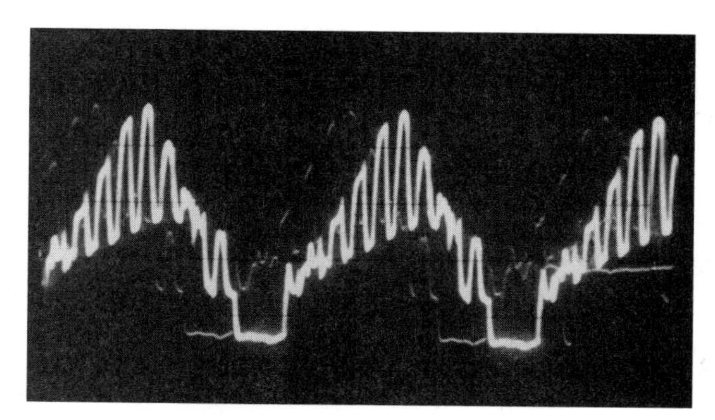

Fig. 6-13. One of the demodulators output waveforms as fed to the CRT board.

Color Circuit Problems

No Color or Intermittent Color

- ✔ For a no color or intermittent color problem, check the 3.58 MHz crystal.
- ✔ A badly soldered joint on the 3.58 MHz crystal can cause a picture with no or intermittent color.
- ✔ No color in the picture can be caused by a defective luma/color/deflection IC.
- ✔ Measure the supply voltage applied to the color circuits; a low-voltage source can produce a no-color symptom.
- ✔ Check all IC and transistor color components for poorly soldered terminals when there is no color.
- ✔ A leaky luminance-driver transistor can cause an inoperative color symptom.

- ✔ A defective trimmer capacitor in the color circuit can cause a no color symptom.
- ✔ A defective ceramic filter can cause either no color or intermittent color in the picture.
- ✔ A defective SMD picofarad (pF) capacitor on the luma/chroma/deflection IC can cause intermittent or total loss of color.
- ✔ The no-color symptom can result from leaky bypass capacitors off of the Y/C IC terminals to the common ground, a missing pulse from the flyback, an open peaking coil, or open coils on the comb-filter network.
- ✔ Where there is virtually no color or only traces of color in the picture after warm-up, check for a bad center post on the color gain control.
- ✔ When there is no color and no audio, check for a burned low-ohm isolation resistor from the flyback voltage source.
- ✔ No color, luminance okay, check for a bad solder joint on the +11-volt supply line of the flyback.
- ✔ No color, no video, dim raster, can be caused by a shorted buffer transistor between emitter and collector terminals (*Fig. 6-14*).

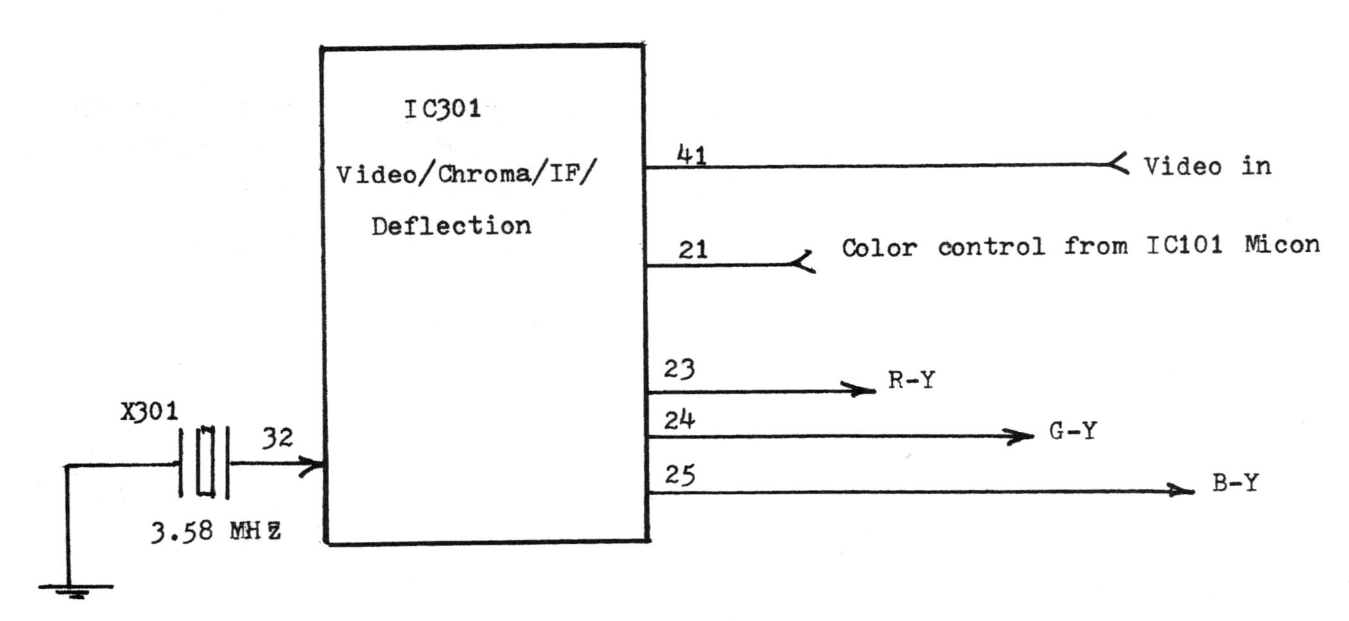

Fig. 6-14. The various color components that can cause problems in the color circuits.

- ✔ When the color is intermittent, it may be caused by poor socket connections on a PIP module.
- ✔ An intermittent picture, a weak picture, and no color can result from a defective SMD-picofarad (pF) capacitor off of the luma/color/deflection IC terminals.
- ✔ If there is a loss of color when changing channels, check the 3.58 MHz signal waveform.
- ✔ Badly soldered joints on the color and tint controls can lead to a loss of color control.

Weak Color

- ✔ A badly soldered connection or cracked rings of PC wiring can cause weak color or changing color in the picture.
- ✔ For weak color symptoms check for open resistance or an increase in resistance in the resistors on the color amp transistors or IC components.
- ✔ If there is a color change after the TV is on for a few minutes, look for leaky zener diodes in color voltage sources.
- ✔ A defective CRT can cause poor or weak color.
- ✔ Check for a weak color amp transistor or open luma/color/IF/deflection IC.
- ✔ An open trace on the voltage supply line feeding both tint and contrast control can cause no tint or a change in contrast.

Color Sync Problems

Color sync problems can be caused by a defective delay line, an increase in resistance off of the chroma/luma/IF/deflection IC, a bad solder connection around the chroma IC, or a badly soldered joint on the voltage regulator in the voltage source.

Other problems associated with color sync include:

- ✔ When the color goes out of sync and diagonal color bars appear, replace the color 3.58 MHz crystal. You should also check for poor board connections or component connections within the color sync circuits.
- ✔ A defective color microprocessor can cause color bars to appear.
- ✔ Improper adjustment of the trimmer capacitor in the chroma circuits can cause a color-out-of-lock symptom with a one-inch band of color across the screen.
- ✔ A poor ground on the flyback can cause a color-out-of-sync symptom.

One Color Missing

- ✔ For missing colors in the raster or picture, suspect color problems within the demodulator output IC circuits, bias or drive transistors, color output transistors, and color guns in the CRT.
- ✔ When the screen turns all red or all blue, check the red and blue color output transistors of the CRT board on the neck of the picture tube.
- ✔ A poor connection on the red output transistor collector load resistor can cause a loss of red in the picture. Scope the red demodulator IC outputs for a red waveform (*Fig. 6-15*).
- ✔ No red in the picture can be caused by a defective luma/color/deflection IC or a defective red output transistor. Also check for an open coil in the red output circuit.
- ✔ For a red screen problem, inspect the CRT board for open foil or trace.
- ✔ An intermittent red screen can result from excessive dust in the CRT spark gap. Intermittent red in the picture or raster can be caused by an intermittent red driver or output transistor, a badly soldered connection on the red output transistor, or a defective red gun assembly in the picture tube.

Intermittent Blue Screen – Sylvania VT190

The screen of a Sylvania VT190 TV/VCR had an intermittent blue screen. Extremely low voltage was found at the 5V regulator Q701. The 5-volt regulator transistor was replaced with a 25D205 transistor, thus restoring the color raster.

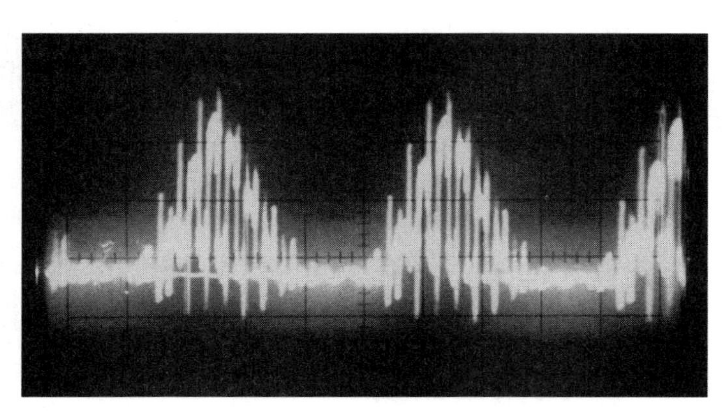

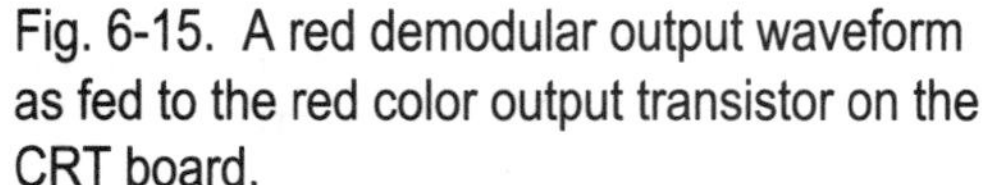

Fig. 6-15. A red demodular output waveform as fed to the red color output transistor on the CRT board.

Green Color Problems

When the color changes the appearance of the raster, check for defective components in the color output and CRT circuits. A defective green output transistor can cause an all green raster, whereas a defective green gun in the picture tube can change the color of the raster. A defective luma/chroma/deflection/IC and demodulator circuits can result in a loss of green color in the picture. When a normal picture turns into a green picture, suspect a bad picture tube. For a loss of colors, resolder all griplets or traces in the color output circuits.

When the picture turns purple, suspect a defective green output transistor. A picture with all green and no blue can be caused by a bad trace between the blue and green output transistors on the CRT board. A defective green amplifier can cause a bright green screen with retrace lines in the picture. A bright green raster with retrace lines may be the result of a shorted picture tube and can cause the demodulator IC to fail. A faded green picture can be caused by a defective EEPROM IC.

Blue Color Problems

If the screen is all blue or if green is missing from the picture, check the CRT and blue output transistor. A leaky T-chip IC can cause a no blue and no green symptom or a no color symptom. An all blue screen can result from broken or cracked traces within the blue color output transistor on the CRT board. If a bright screen becomes excessively blue and the chassis shuts down, check for excessive dust in the blue spark gap on the picture tube circuits. An open blue output transistor on the CRT board can cause a loss of blue in the picture. With bad solder joints on the collector terminals of the color-output transistors, the screen intermittently turns all red or all blue. A defective EEPROM IC can cause symptoms such as no audio, no customer menu, only 1 inch of vertical sweep, and an all blue screen.

Blue Screen, Okay On-screen Display – Magnavox CCR095AT04

In a Magnavox CCR095AT04 TV/VCR the screen was all blue with an okay on-screen display. The VCR would play tapes and the colors were normal. A bad SF03 SAW filter caused the all blue screen symptom in the TV circuits.

Chapter 7: Servicing Control Functions, AGC, Sync, and Audio Circuits

The TV system control or interface unit might control the TV circuits with a mechanism or servo-control IC operating the VCR functions. IC601, a control Micon, controls the volume, brightness, contrast, color, tint, and sharpness of the Emerson TV/VCR circuits (*Fig. 7-1*). The volume-control signal is fed from Terminal 15 to a volume-control transistor (Q612) and fed to Pin 2 of IC201. The brightness is fed from Pin 14 of IC601 to a bright-control transistor (Q611) and fed to Pin 22 of video-chroma IC401. A contrast signal is controlled from Pin 13 to a contrast transistor (Q608) to Pin 28 of IC401. The color-controlled signal at Pin 26 of the video/chroma/deflection IC appears from a color-control transistor (Q607) and is fed from Pin 12 of IC601. The tint and sharpness-control circuits are controlled in the same manner.

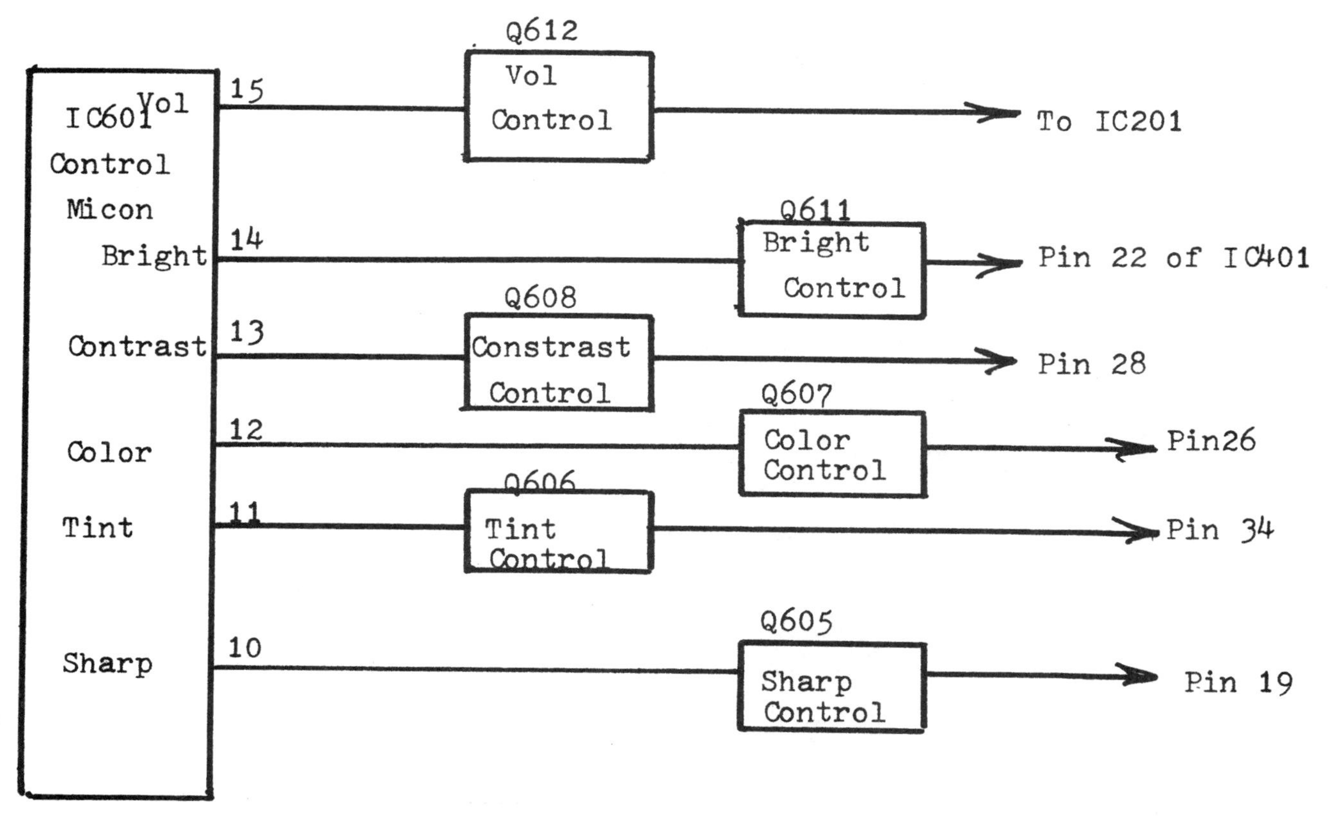

Fig. 7-1. The Micon control IC601 in an Emerson VT1920 TV/VCR controls the volume, brightness, contrast, color, tint, and sharpness.

In a Sears 13-inch color TV/VCR combo, the different control circuits are found in a digital-to-analog (d/a) converter (IC101). The brightness control is fed to the bright-control transistor (Q152) from Pin 45 of the video/chroma/deflection IC (IC301). The sharpness control operates from Pin 11 of IC101 to a sharpness-control transistor (Q155), into Pin 10, and then to the video circuits (*Fig. 7-2*). The contrast is controlled from Pin 46 to the contrast-control transistor (Q151) and to Pin 36 of IC301.

Likewise, the color and tint controls are operated through respective transistors (Q153 and Q154) to the color circuits in the chroma section of IC301. Interface or combination-system control circuits might control both TV and VCR operations.

Check the system, interface, or d/a converter IC when one or more control functions are not operating. Determine if either the remote or keyboard controls are functioning. Check both remote and keyboard operations for one or more functions that don't operate. When one function does not operate, check the signal from the control IC and the control transistor. Don't forget to check the supply sources of each suspected control IC.

When the tint display or closed captions are shifted far to the right or appear larger or smaller in size, replace the microprocessor. When the TV keeps changing channels by itself and won't stop, check the channel pushbutton for low resistance across the switch contacts (*Fig. 7-3*). When characters are seen in rows and columns, check for a defective character IC control. A bad microprocessor or control IC might cause problems such as inoperative auto programming, no screen display, no volume control, and snowy reception. With intermittent loss of functions such as volume up and down and channel up and down, check for a bad microprocessor.

Check each control transistor for open or leaky conditions between the control IC and the controlled circuit. When the set is cold and experiences a complete loss of all

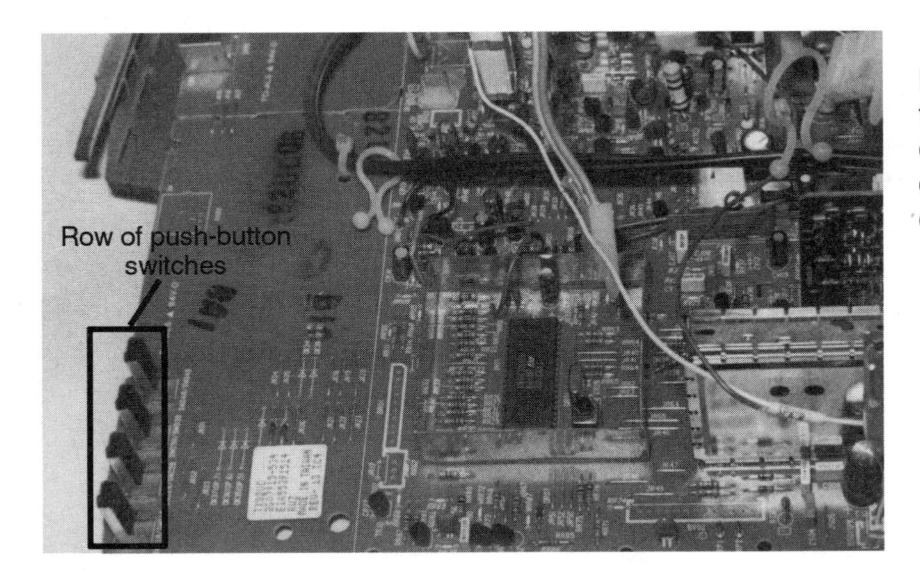

Fig. 7-3. Separate pushbutton-type switches can operate the channel up/down, volume up/down, power on/off, and contrast on the front panel.

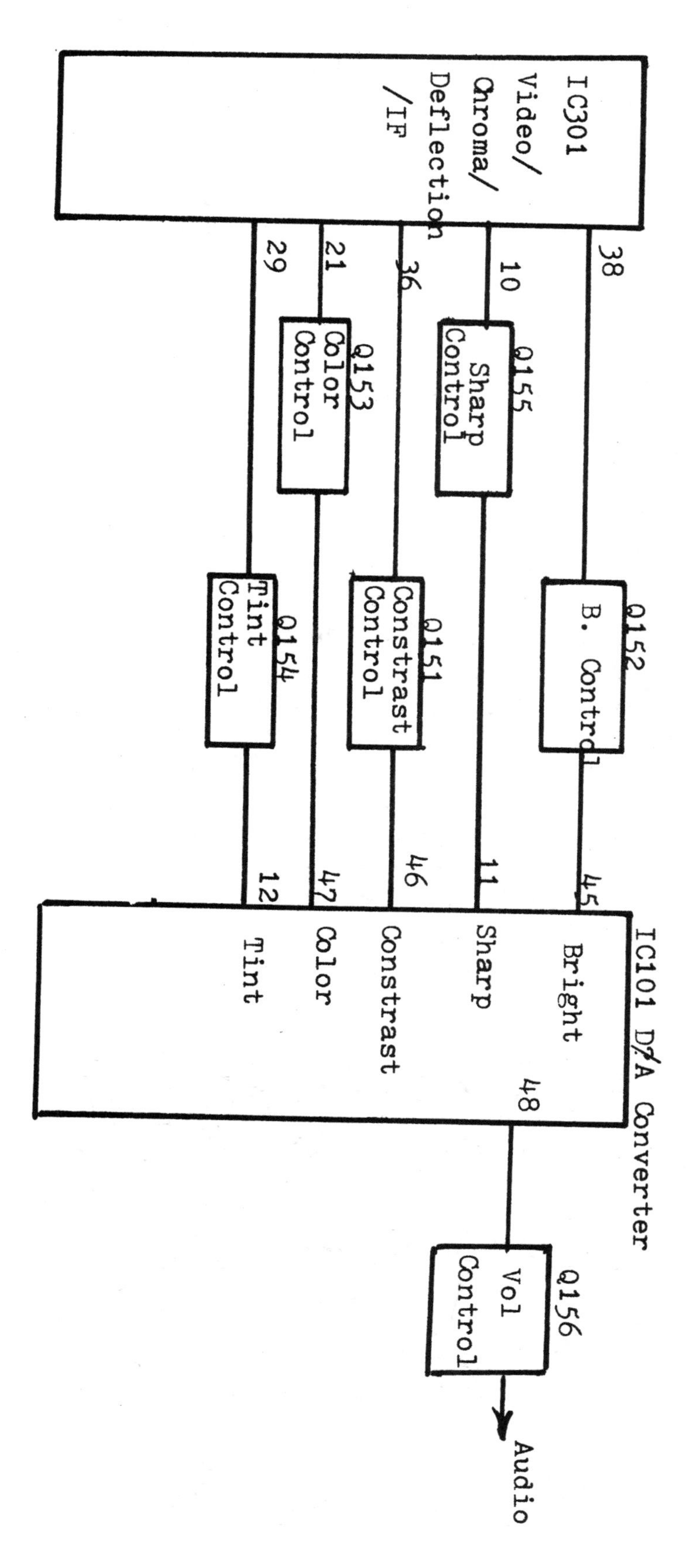

Fig. 7-2. In a Sears 13-inch TV/VCR, the d/a converter (IC101) controls the brightness, volume, sharpness, contrast, color, and tint functions.

functions such as channel up and down and volume up and down, suspect a faulty microprocessor. A defective microprocessor or control IC can also cause improper adjustment of picture level, color level, time set, and so on. The character interface IC can cause a dim raster with retrace lines and no display. The VCR system control functions are covered in the Part II of this book.

AGC (Automatic Gain Control) and Sync Problems

Servicing the AGC and sync circuits in a modern TV chassis is much easier than in early TV chassis'. The AGC circuits are controlled from the IF/video DET/AFT/AGC/ limiter IC. In the Emerson VT1920 chassis, the RF-AGC voltage is fed to the varactor tuner, and the IF-AGC is controlled inside the same IC (*Fig. 7-4*).

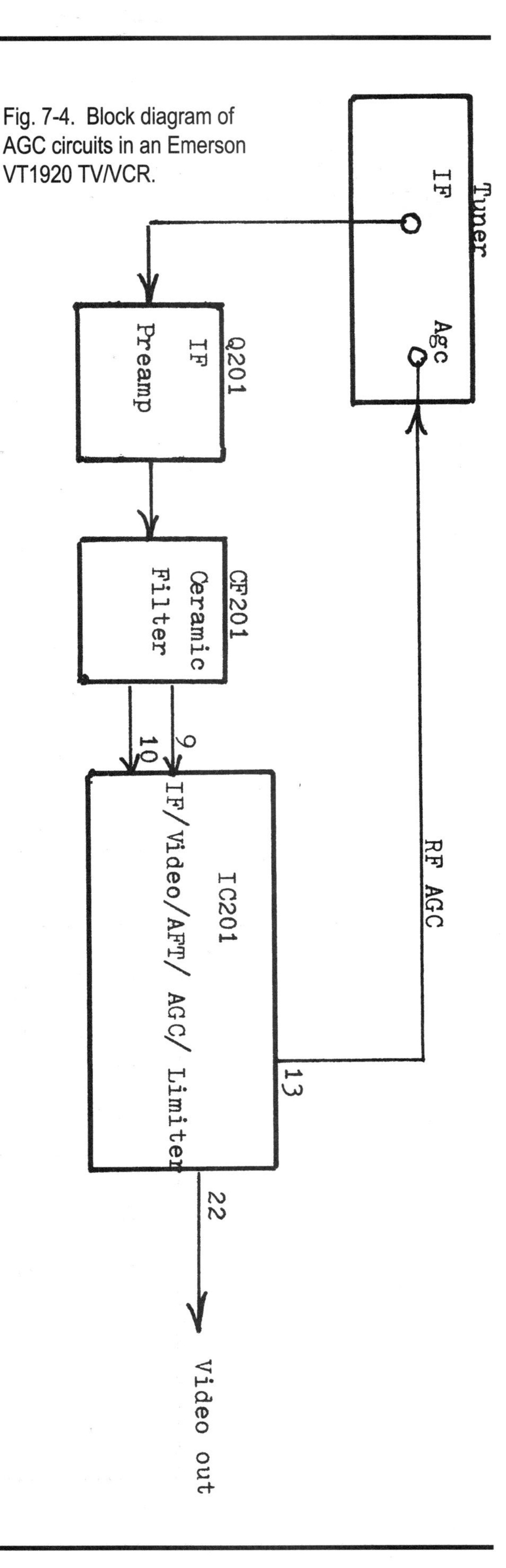

Fig. 7-4. Block diagram of AGC circuits in an Emerson VT1920 TV/VCR.

The RF-AGC develops within the TV demodulator circuits of a Panasonic PV-M2021A TV/VCR, appears at Pin 19 of IC701, and is fed to the RF stages of the VHF-UHF tuner. The IF-AGC circuits are fed to the IF amp inside IC701 and can be checked at test-point TP705 on Pins 17 and 18 of IC701. An IF-AGC voltage from the IF-AGC stage inside the TV demodulator IC, is fed directly to the VIF-amp stage. The input-IF signal is fed from a SAW filter (FL701) to Pins 22 and 23 of IC701. Check the RF and IF-AGC voltages applied to the tuner (Pin 19) and the IF-AGC voltage at TP705 (*Fig. 7-5*).

When improper voltages are found on the RF and IF-AGC terminals, check for a defective IC. Double-check the RF voltage at the varactor tuner and the IF-AGC at test points on the IC. If there is no signal from the timer,

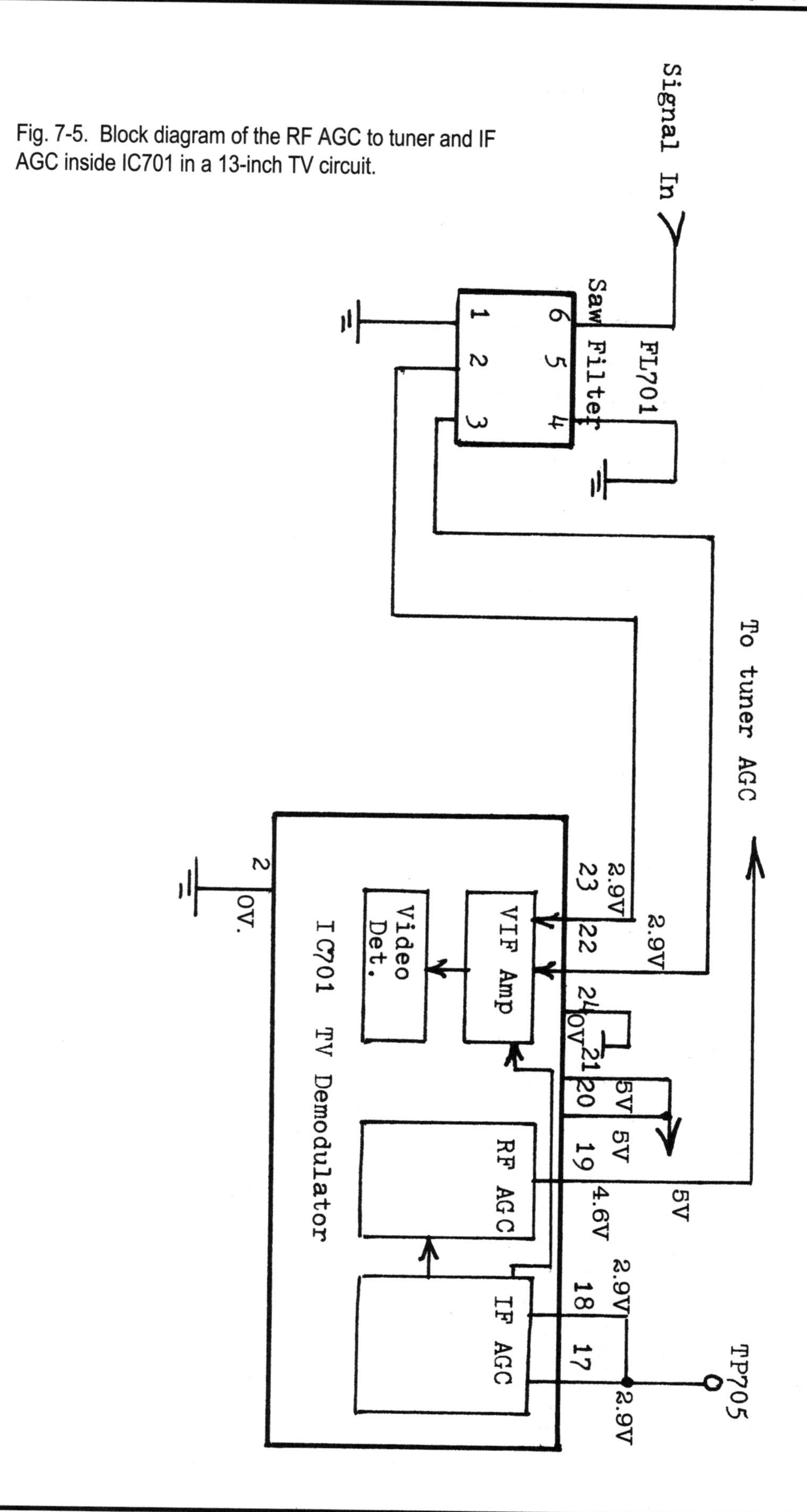

Fig. 7-5. Block diagram of the RF AGC to tuner and IF AGC inside IC701 in a 13-inch TV circuit.

check for normal RF-AGC voltage at the tuner. Inject an internal RF voltage at the tuner and notice if the picture returns. Notice that the AGC voltage will change when a different channel is tuned in. Often, the AGC circuits are functioning when a change of voltage is found on the AGC-tuner terminal.

AGC problems consist of a very dark or light and unstable or snowy picture. Problems and possible causes may include:

- ✔ A white raster as the result of a defective AGC or video circuit.
- ✔ A snowy picture from excessive positive voltage at the varactor-tuner AGC terminal.
- ✔ A white screen from excessive positive voltage applied to the IF stages.
- ✔ Waving motion in the picture and noise in the audio from poor AGC action.
- ✔ Excessive snow in the picture caused by a defective AGC circuit.
- ✔ Improper AGC voltages, which in turn, can cause erratic pulling of the picture and poor contrast from a defective IC or tuner.

Double-check the varactor tuner and IC circuits when AGC problems occur.

The AGC circuits can be checked with correct scope waveforms and voltage measurements on the tuner IC and tuner. Check for a video waveform at the output of the IF stage of an IF/video DET/AFT/AGC/limiter IC. A defective IC can cause no video output waveform and a white screen. A changing dark picture may result in video waveform at the IC terminal of a defective IC.

To make sure that the correct voltage is applied to the tuner, check the voltage at the tuner and RF-AGC terminal on the IC. Replace the defective IC when improper AGC voltage is found at the RF-AGC output and there is no video at the video output terminal of the IC.

For an unusual picture, check for defective electrolytic capacitors in the power supply source of the AGC-IC. Shunt the decoupling capacitor at the voltage source and notice if the picture returns to normal. Clip the electrolytic into the circuit after pulling the AC power plug.

- ✔ For excessive pulling and tearing of the picture, double-check the small electrolytics in the AGC circuits.
- ✔ For equivalent series resistance problems, check all electrolytics within the AGC circuits with the ESR meter.
- ✔ Pulling of the picture on the right side of screen may be the result of defective small electrolytic capacitors in the AGC and video circuits.
- ✔ A smeared and waving picture can be caused by a defective AGC-IC.

- ✔ When the picture is dark and has lines in it, and when the audio has a buzzing sound, solder all pins on the AGC-IC.
- ✔ When the picture shifts and becomes snowy, it may appear to be an AGC problem, but it actually stems from badly soldered joints under the IF shields.
- ✔ Check RF or AGC voltage terminals that have very low voltage applied to the tuner or IF section for leaky bypass capacitors.
- ✔ A scrambled picture that appears to be an AGC problem may be caused by poorly soldered IF IC pins with a cold solder joint feeding the Vcc voltage source on the IC.
- ✔ Check the large electrolytics in the AGC circuits if there is a stationary hum bar on all channels, which may appear to be an AGC problem, and if there is a grainy picture and a buzzing sound in the audio.

Sync Problems

The vertical and horizontal sync circuits are found in the luminance/chroma signal-process IC. The sync signal is taken from the video input signal and fed to a sync separator that then feeds both vertical and horizontal sync circuits. The vertical sync signal is fed to a countdown vertical circuit. A horizontal sync signal is fed to the AFC and to the horizontal oscillator circuit. Both the vertical and horizontal sync are tied to each circuit inside the large luma/chroma-processing IC (*Fig. 7-6*).

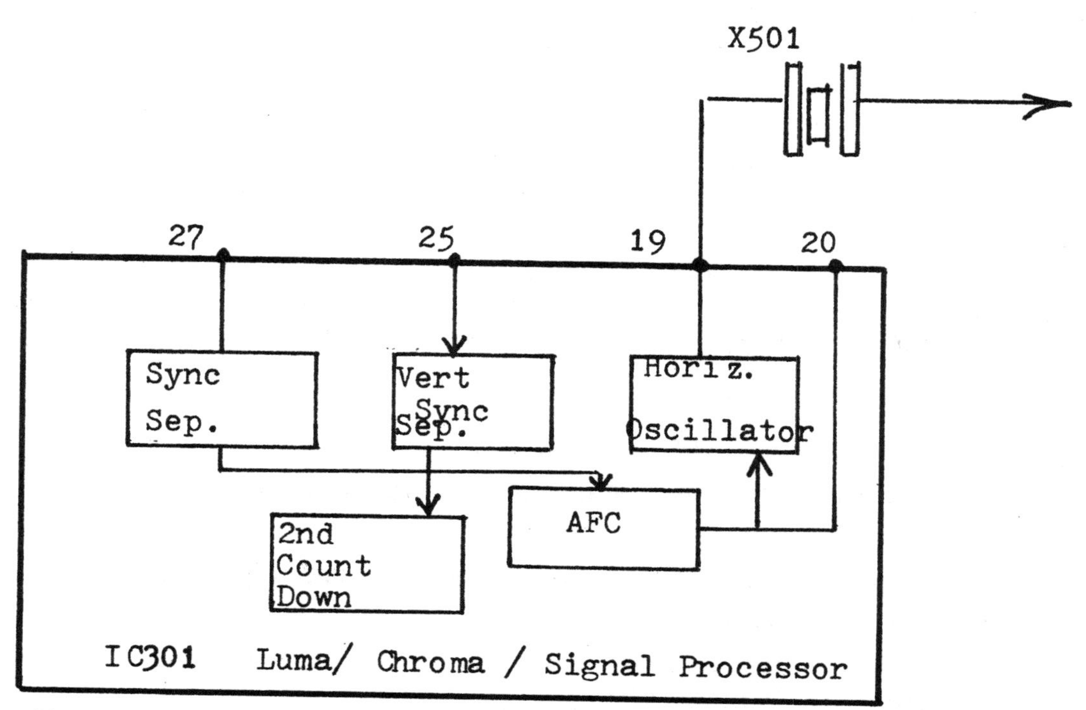

Fig. 7-6. the video and sync pulses are separated and applied to the vertical and horizontal circuits inside the luma/chroma/signal processor IC301.

- ✔ Loss of horizontal and vertical sync can be caused by a defective luma/chroma process IC.
- ✔ Badly soldered joints on the grounds of the IF pins, inside the tuner, can cause a loss of video, audio, and horizontal/vertical sync.
- ✔ No sync can result from poorly soldered connections on the resistors or capacitors tied to the video input of the luma/chroma IC.
- ✔ A loss of both horizontal and vertical sync with horizontal tearing can be caused by open electrolytics in the voltage source of the IC.
- ✔ A defective sync-amp driver transistor also can cause a loss of horizontal or vertical sync.
- ✔ Open bypass capacitors on the luma/chroma-processing IC can cause a loss of horizontal sync and tearing at the bottom of the picture.
- ✔ No horizontal sync can be caused by a defective sync-processor IC.
- ✔ Bad solder joints on the pins of the flyback can cause poor horizontal sync and a loss of color.
- ✔ Problems with the large filter capacitor (220 µF, 200V) can cause a loss of horizontal sync and horizontal tearing.
- ✔ A badly soldered joint on the horizontal hold control can cause the loss of horizontal sync.
- ✔ When the horizontal will not hold and the high voltage will not adjust, suspect a defective hold-down safety transistor. Also resolder all the pins on the horizontal driver transformer and flyback.
- ✔ Check for defective diodes in the sync circuits. These diodes might test okay, but if in doubt, replace them.
- ✔ A defective sync-kill-switch transistor can cause a sync problem.
- ✔ Symptoms such as no vertical lock and picture jitters can be caused by a defective chroma/luma-process IC or a defective sync-processor IC.
- ✔ To correct vertical roll throughout the picture from bottom to top, replace the main filter electrolytic.
- ✔ A bad service switch can cause vertical roll or a loss of height.
- ✔ For poor vertical sync and intermittent vertical sweep, check for leakage of the service switch to ground.
- ✔ When the picture rolls down all the time, suspect a T-chip.

Troubleshooting Sync Circuits

Scope the video signal coming into the luma/chroma IC. Take a vertical sync waveform at Pin 25 of IC301. If the video is coming in and there is a vertical sync waveform at Pin 25, check for a defective IC301. Take a critical voltage measurement at the voltage supply pin and compare to the schematic. If the supply voltage is low and there is no sync waveform, replace the luma/chroma IC. Remember that the luma and chroma circuits might be functioning with defective internal sync circuits and a rolling or horizontal tearing of the picture.

Audio Circuits

In a Sears 9-inch TV/VCR, the sound circuits are taken from IC301 (luma/chroma/IF/AF amp), out of Pin 51, and fed to the audio output IC801. The sound appears at Pin 48 and is fed internally to the attenuator and AF-amplifier stage. An FM-detector coil at Pin 2 is also fed to the attenuator stage. The AF-amp stage amplifies the TV audio and appears at output Pin 4 of IC301. The audio output is fed out of IC301 to Pin 1 of the input terminal of audio-output IC801 (*Fig. 7-7*).

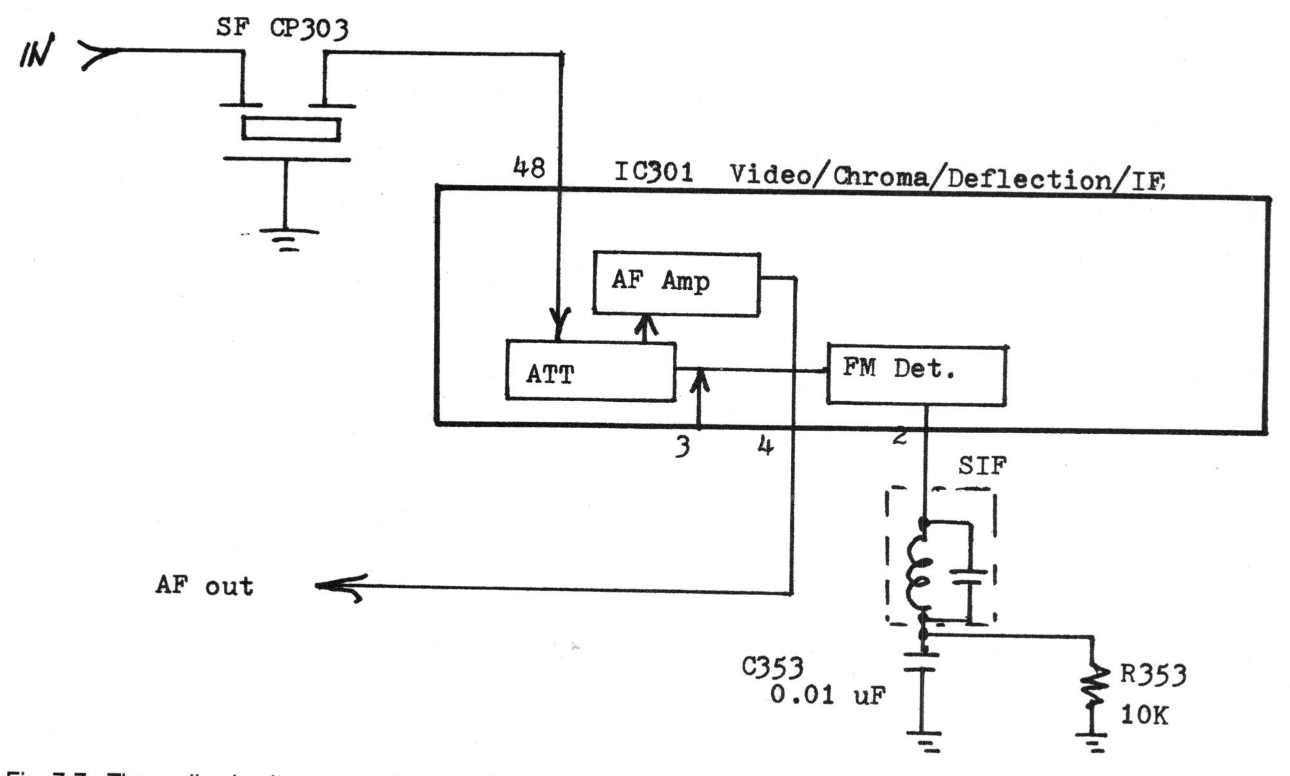

Fig. 7-7. The audio circuits are developed inside the video/chroma/deflection/IF IC301.

In a Sears 13-inch TV/VCR, the TV audio and video signal appears out of Pin 51 of IC301, where the video and sound are separated. The sound goes to a 4.5 MHz bandpass filter and into Pin 48 of a limiter stage. The sound goes through the FM detector and out of Pin 3. An audio-buffer transistor (Q210) is found between IC301, and the TV-audio output is fed to an IC-output circuit (*Fig. 7-8*).

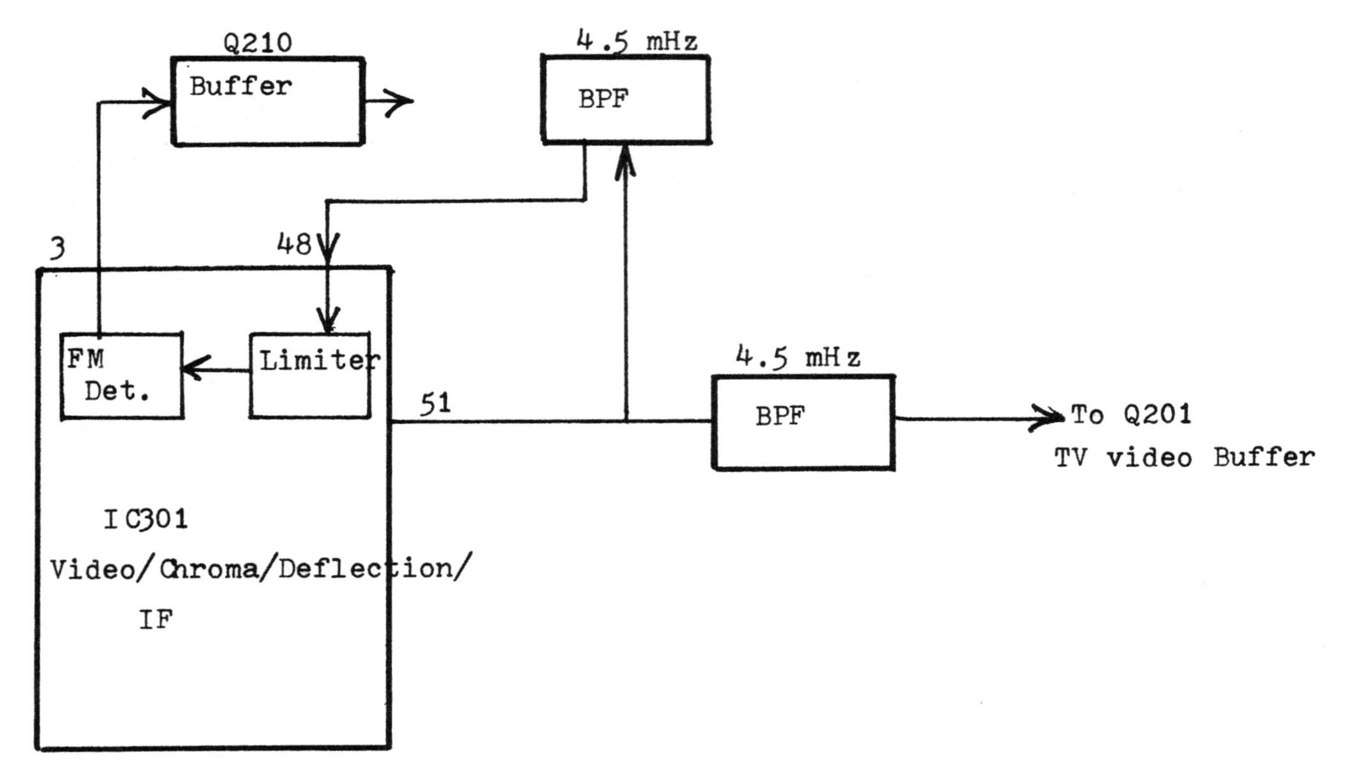

Fig. 7-8. The sound circuits in a Sears 13-inch TV/VCR are separated from the video circuit and fed into the audio circuits inside IC301.

Most of the sound problems found in the audio input circuits are loss of sound, weak sound, and intermittent sound. Weak and intermittent sound can result from a defective IC. Garbled or intermittent audio can result from a bad IC or improper grounds in the sound IF circuits. Check for poor trace or foil connections in the IF circuits and use the ESR meter to check all capacitors in the audio input circuit. A badly soldered joint on the silicon diode from the flyback winding that provides voltage to the audio input IC can cause intermittent sound problems.

- ✔ If there is a motorboat engine-type noise in the sound and picture, replace a defective audio input IC.
- ✔ Badly soldered terminals on the audio input IC or chroma/luma/IF IC can produce intermittent sound and video.
- ✔ A defective 4.5 MHz crystal filter can cause intermittent audio.

- ✔ A badly soldered joint on the shield covering the bottom side of the audio-IC circuits can cause intermittent audio.
- ✔ Check the buffer transistor after the input IC for poor contacts resulting in erratic sound.
- ✔ If the volume goes up and down as the picture is changed with poor source voltage regulation, check for a defective line-voltage regulator.

Defective Microprocessor

A defective microprocessor, Micon, or system control IC, can cause many different problems. If there is no audio after the TV warms up, check for a thermal problem with the microprocessor IC. A leaky capacitor off of the microprocessor can create a problem in which the volume is all the way up when the TV is turned on; the volume can be lowered, but comes right back up again. These are some problems that can result from a defective microprocessor:

- ✔ The volume goes up to full as the set is turned on.
- ✔ Sound intermittently jumps up to a very high level.
- ✔ Loss of audio control after the TV warms up.
- ✔ The audio becomes loud randomly and stations cannot be programmed.
- ✔ All functions lock up and there is no volume up or down and no channel up or down.
- ✔ Low sound at turn on.
- ✔ Raising the volume control results in no sound.
- ✔ Intermittent loss of volume up and down control.
- ✔ There is no audio.
- ✔ There is a loss of sound and video.
- ✔ An inability to control the volume on the different channels.

Audio Output Circuits

Often, audio output circuits in the TV/VCR consist only of an IC audio output component (*Fig. 7-9*). In the audio output circuit (IC801) of a 9-inch Sears TV/VCR, the audio input is fed into Terminal 1 and the output is fed into Terminal 7. The audio output is capacity-coupled to an earphone jack and a PM loud speaker. Transistor Q801 provides muting, controlled by the system control (IC201) on pin Terminal 3 of the output IC.

In a Sears 13-inch color TV/VCR, the sound-output circuit is coupled through electrolytic C4203 to Pin 1 of IC4201. The amplified audio signal appears on Pin 7 and is capacity-coupled with C4271 to a socket, and then on to the earphone and speaker. A mute signal is applied to the audio-mute transistor (Q4201) and to Pin 3 of the audio output (IC4201). The +12V supply voltage for the audio-output IC is found on Pin 8 (*Fig. 7-10*).

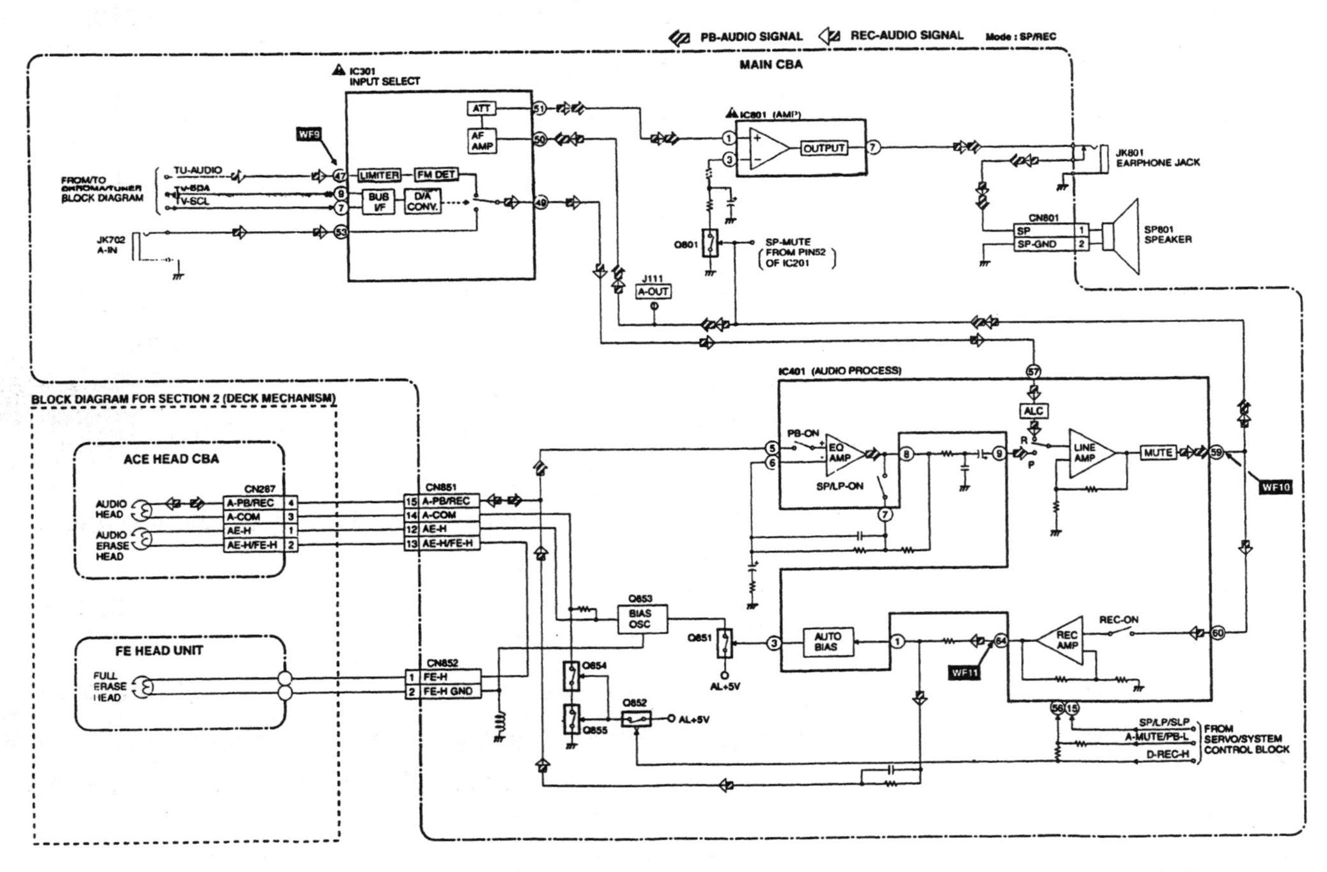

Fig. 7-9. The audio output is fed from Pin 51 of IC301 to audio output IC801 in a Sears 9-inch portable TV/VCR.

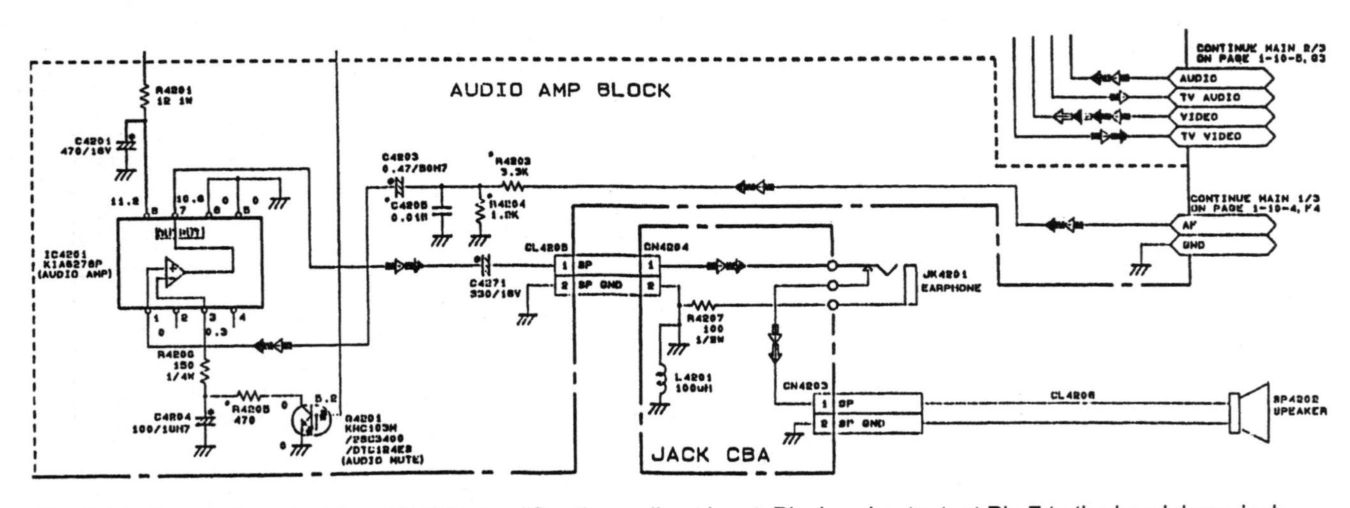

Fig. 7-10. The audio output Amp IC4201 amplifies the audio at input Pin 1 and output at Pin 7 to the headphone jack and speaker.

Sound problems caused by the audio output IC include no audio, weak audio, distorted audio, and intermittent audio. Causes include:

- ✔ A defective output IC can cause intermittent problems with poorly soldered connections on the IC terminals.
- ✔ A defective coupling-output electrolytic can cause intermittent audio.
- ✔ A distorted and motorboating sound can result from a defective output IC.
- ✔ A no-audio problem can be caused by an improper voltage supply to the output IC, with poorly soldered joints on the low-ohm isolation resistor in the power source.
- ✔ Weak and distorted sound can result from a defective output IC.
- ✔ No audio can be caused by a defective diode and an open, low-ohm fusible resistor in the 12V power source.
- ✔ Intermittent audio can be caused by a defective audio-output IC, electrolytic-coupling capacitor, and intermittent voice coil in the speaker.
- ✔ Open resistance or an increase of resistance in the resistors tied to the output IC can cause garbled audio.
- ✔ A defective output IC can cause the no-audio symptom with a loud noise at turn-on.
- ✔ A dropped or frozen cone can cause poor and distorted audio in the speaker.
- ✔ No audio can be caused by an open or damaged voice coil in the speaker. No doubt, the volume was turned up to full and blasted the speaker voice coil.

Other audio-related problems and possible causes include:

- ✔ No audio and no vertical sweep due to a defective voltage regulator in the voltage source.
- ✔ No audio and no video due to an open, low-ohm fusible resistor.
- ✔ No audio and distorted audio as a result of a defective speaker.
- ✔ No audio and a bright white picture caused by a defective power-line regulator.

No Audio - Quasar VV1220A

After the audio-output stage in a Quasar VV1220A TV/VCR was hit by lightning, considerable parts damage was suspected. Besides burn marks on the antenna connection and AC input cord, two transistors and AN5265 IC were damaged with a 1K, 1/2-watt resistor. The AN5265 IC was replaced with a universal linear TV sound, IF sound preamp, and AF output universal replacement (NTE 1789). Both damaged transistors were replaced with ECG289 universal replacements.

If the screen control is turned up and shows no vertical sweep and no sound, check for a leaky vertical output IC, burned isolation resistor, and silicon diode in the power source. A no-sound, no-video symptom can be caused by a defective EEPROM IC. No audio and an excessive popping noise in the speaker can result from a defective audio output IC.

For no audio and no color, check for an open, low-ohm resistor off of the pin terminal on the flyback of the 12V line. The no-audio symptom can be caused by a defective transistor or no-command signal from the system-control IC with a muted audio-output circuit. When there is no audio or only faint sound, it may be caused by a poorly soldered joint on an SMD capacitor in the mute circuits. If there is no audio, no vertical sweep, and the channels can't be changed, the cause may be an open 1.5-ohm resistor at Pin 5 of the flyback; no voltage from the flyback can cause a loss of four different voltage sources.

When the sound is always muted or there is no sound, suspect the system-control IC. Also, a defective mute transistor can cause inoperative sound. For intermittent or muted sound, don't overlook poor mute-transistor terminal connections.

Intermittent Sound

Intermittent audio can be caused by:

- ✔ A defective audio-output IC.
- ✔ A defective electrolytic-coupling capacitor between the speaker and output IC.
- ✔ A defective microprocessor.
- ✔ Poorly soldered speaker connections or voice-coil connections.

With the sound turned up, push up and down on the speaker cone with both thumbs; if the sound cuts in and out, replace the PM speaker. Intermittent volume also can result from poor tuner-board connections. Intermittent volume and greatly decreasing volume can be caused by intermittent closing of the down switch. Poorly soldered joints or traces with low-ohm resistors in the audio-output circuit can also cause intermittent audio.

Poorly soldered joints on the 12-volt regulator IC can cause intermittent loss of audio and video. A badly soldered joint on the shields covering the bottom side of the audio IC can cause intermittent audio. Intermittent or erratic sound can result from poorly soldered stakes on the tuner module.

Intermittent Audio and Shutdown - Symphonic

A Symphonic TVCR901 TV/VCR would shut off after about 15 minutes with intermittently functioning audio. Transistor Q1012 became leaky when hot and caused the intermittent shutdown. After replacing the leaky 2SA1318 transistor, the sound also returned to normal.

Intermittent Sound - RCA

The sound was intermittent in an RCA 13TVR60 TV/VCR. At first, the sound-output IC (IA01) was suspected. When the sound was signal-traced from output Pin 2 of IA01 IC, the sound seemed normal. The sound at the speaker was intermittent and cut up and down. Electrolytic-coupling capacitor CA07 (470 µF) was found to be intermittent and was replaced (*Fig. 7-11*).

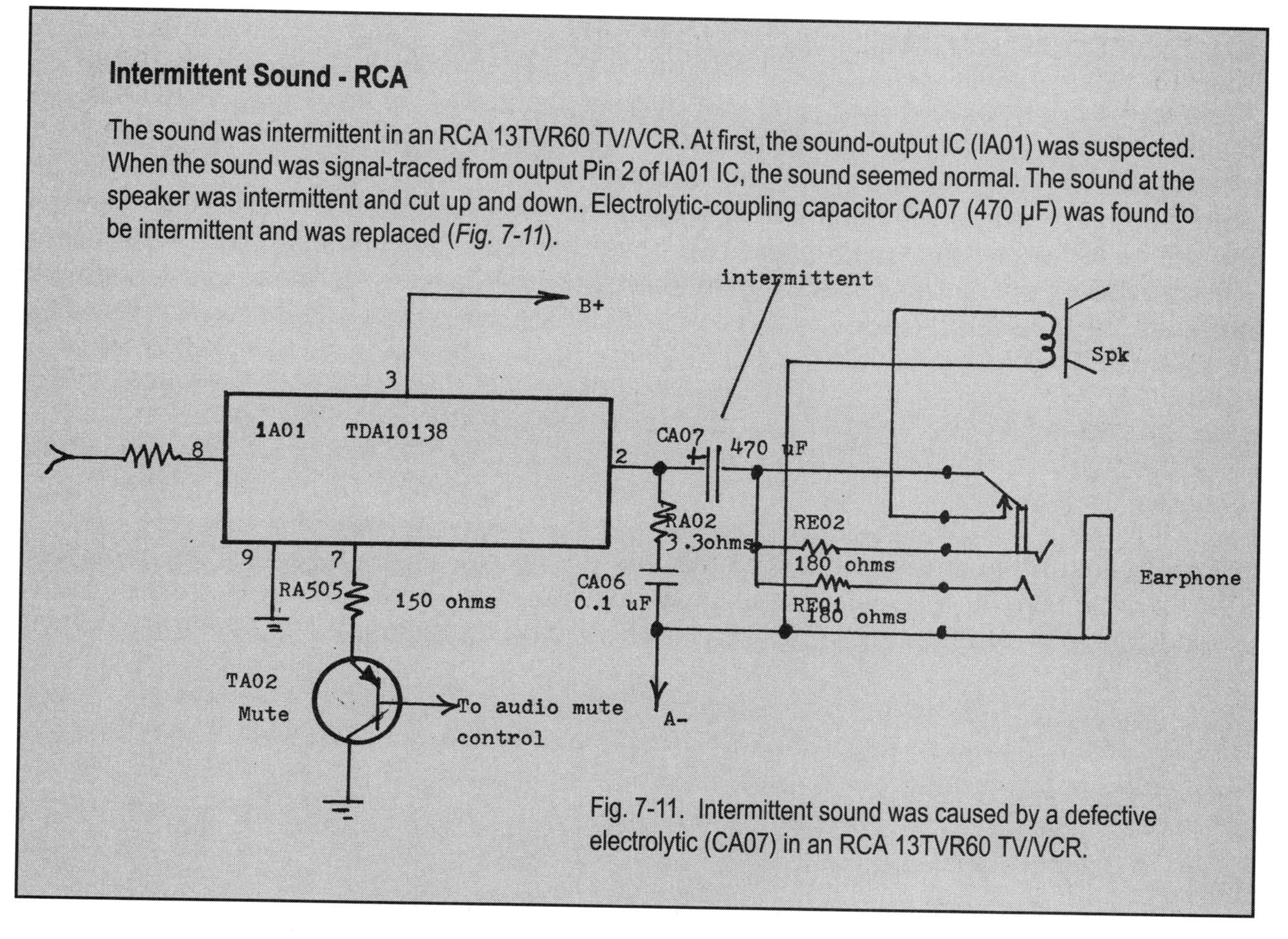

Fig. 7-11. Intermittent sound was caused by a defective electrolytic (CA07) in an RCA 13TVR60 TV/VCR.

Weak or Low Sound

Problems and possible causes:

- ✔ Very low audio and distorted sound may require readjustment of the sound IF transformer.
- ✔ For weak and distorted audio, check the SMD transistors in the sound circuit.
- ✔ If there is low sound at turn-on and raising the volume results in no sound, check for a faulty microprocessor.
- ✔ Weak or low audio can be caused by liquid spilled inside the back of the TV/VCR.
- ✔ Loss of sound may be caused by a defective MPX decoder.
- ✔ If the volume runs up and down by itself and the raster intermittently fades out and returns, check the digital-to-analog IC.
- ✔ For a weak sound system, check for an open input or output coupling electrolytic. Use the ESR meter to check the suspected capacitor in the circuit.
- ✔ When intermittent loss of audio and noisy sound occur, check for badly soldered joints on the input IC or the output IC terminals.

Distorted Sound

A defective audio-output IC can cause distorted sound problems (*Fig. 7-12*), as can readjustment of the audio coil or a change of capacitors in the same circuit. Adjustment of the discriminator or quadrature coil can solve some audio-distortion problems. Simply adjust the core slug until the sound is clear. A defective input-sound IC can cause weak and distorted audio in the speaker. Also check for leaky SMD transistors in the sound circuits when the audio is distorted. A defective output IC and speaker can cause distorted audio at low volumes.

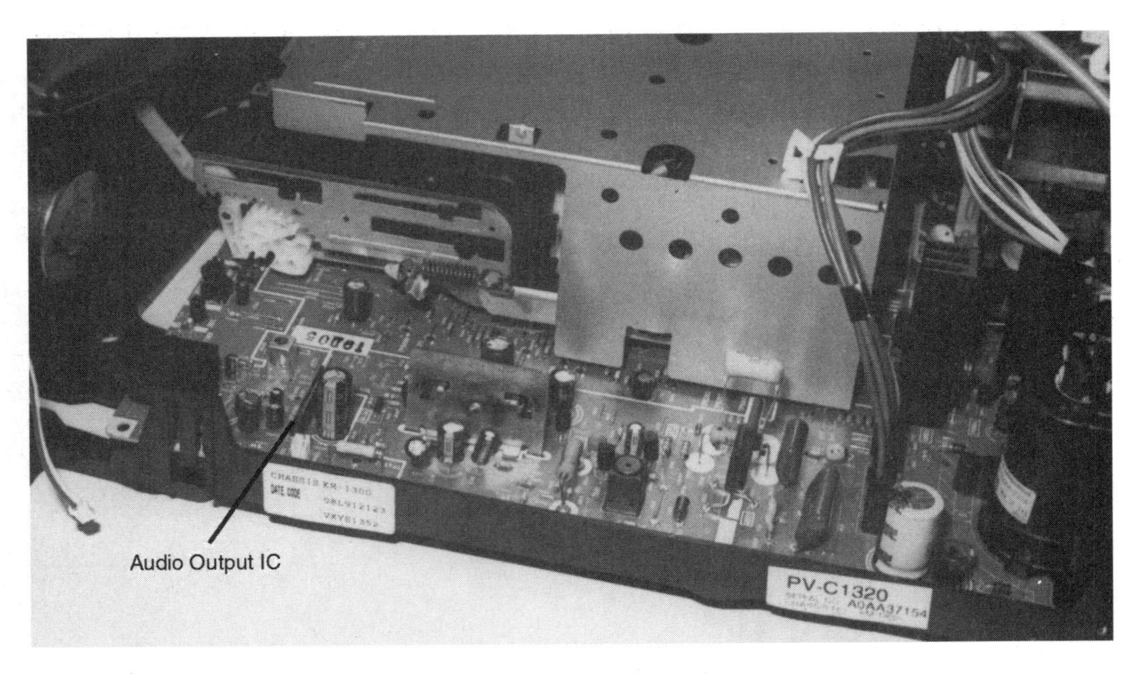

Fig. 7-12. The audio output IC is located quite close to the small PM speaker.

Improper Noises in the Sound

Problem and possible causes:

- ✔ Erratic noises in the speaker can be caused by poor solder joints on tuner stakes.
- ✔ Severe hum in the sound may be a defective line-voltage regulator IC.
- ✔ A severe buzz in the audio can be caused by leaky zener diodes and electrolytic-decoupling capacitors in the voltage source of the sound-output circuits.
- ✔ Distorted hum in the sound can be the result of a defective main filter capacitor.
- ✔ Severe hum in the audio and a weaving picture can be due to a dried-up filter capacitor. Check all electrolytic capacitors with the ESR meter.
- ✔ If there is noise in the speaker that sounds like popping corn after the set warms up for 2 to 10 minutes, check the audio-output IC.

- ✔ Very poor audio and a hissing sound in the speaker like a rushing waterfall can be caused by a defective tuner.
- ✔ If there is a motorboat engine-type sound in the speaker, the sound-input IC may need replaced and the IF sound coil and RF AGC control may need adjusted. It may also be caused by a defective audio output IC.
- ✔ Severe buzzing and intermittent audio can be caused by poorly soldered joints on the IF sound-transformer circuit.
- ✔ For symptoms that seem like an AGC problem with a grainy picture (buzzing in the audio, hum bars on all channels), check the 220-μF, 16V electrolytic.
- ✔ If the speaker is noisy, check the small bypass capacitors directly off of the system-control IC, volume, or sound line.
- ✔ A pulsating sound in the audio with intermittently blank raster can be the result of a leaky diode in the low-voltage regulator circuits.
- ✔ Squealing noises when the volume control is adjusted can result from defective small (4.7 μF to 10 μF) electrolytic capacitors off of the audio-output IC terminals.

Distorted Audio - Sears TVSR090252

In a Sears TV/VCR the audio was normal on Pin 1 of the sound-output IC801 and distorted at output Pin 7. The voltage on supply Pin 8 was only 11.7V, which was off 0.80V. R802 was found to be burned and was replaced. The voltage on Pin 3 was zero. The auto-mute transistor (Q801) tested normal in the circuit. Replacing IC801 and the 150-ohm resistor solved the distorted audio symptom (*Fig. 7-13*).

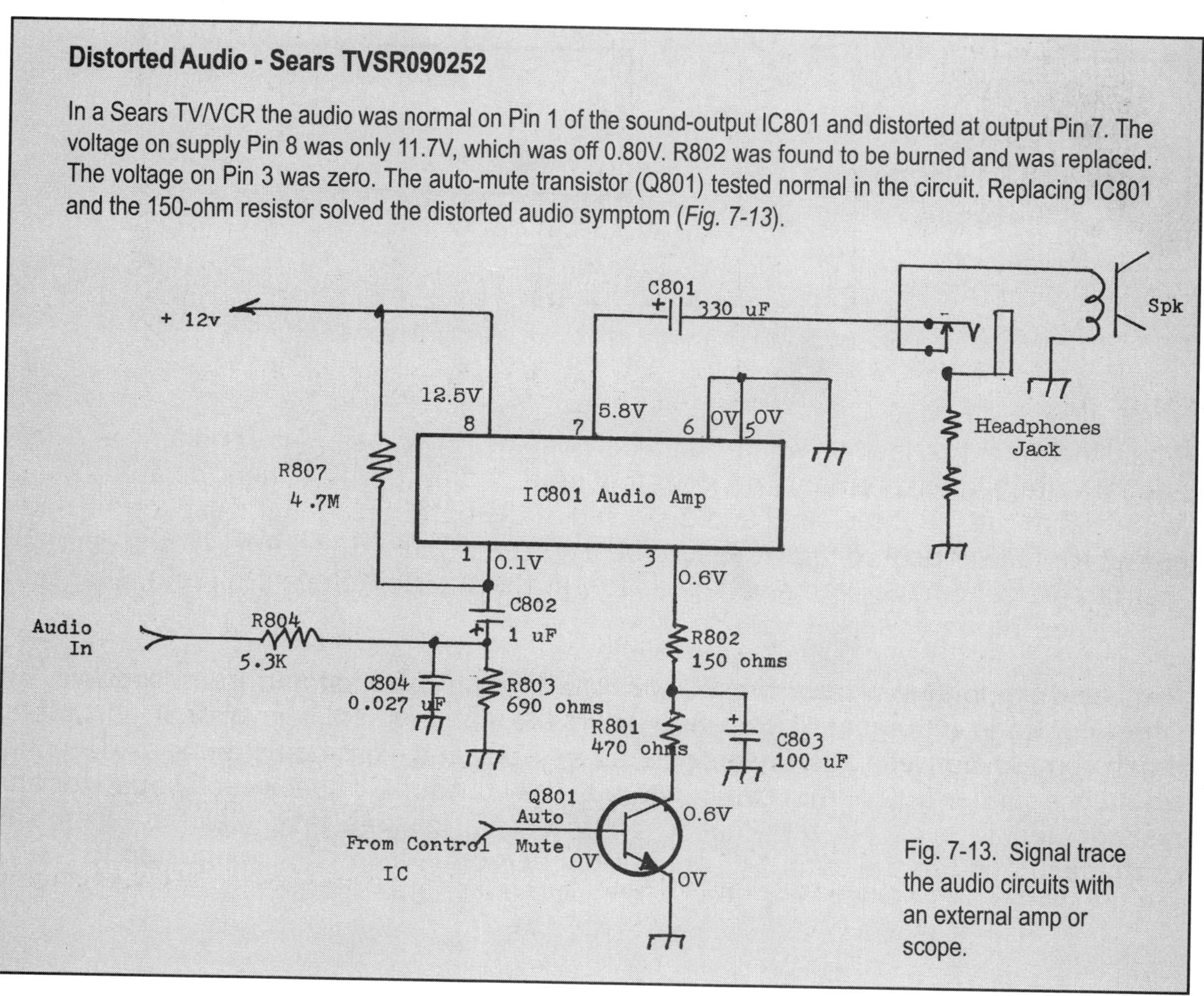

Fig. 7-13. Signal trace the audio circuits with an external amp or scope.

Troubleshooting Sound Circuits

Sound circuits can be signal-traced by using a clear TV station and by checking the waveform after the audio detector. Signal trace the audio clear through all audio circuits to the speaker. For a dead, distorted, or weak audio stage, a sine or square wave can be injected at the audio-input circuits and traced to the speaker with the scope. Another method is to use an external audio amplifier and signal trace the sound circuits up to the speaker (*Fig. 7-14*).

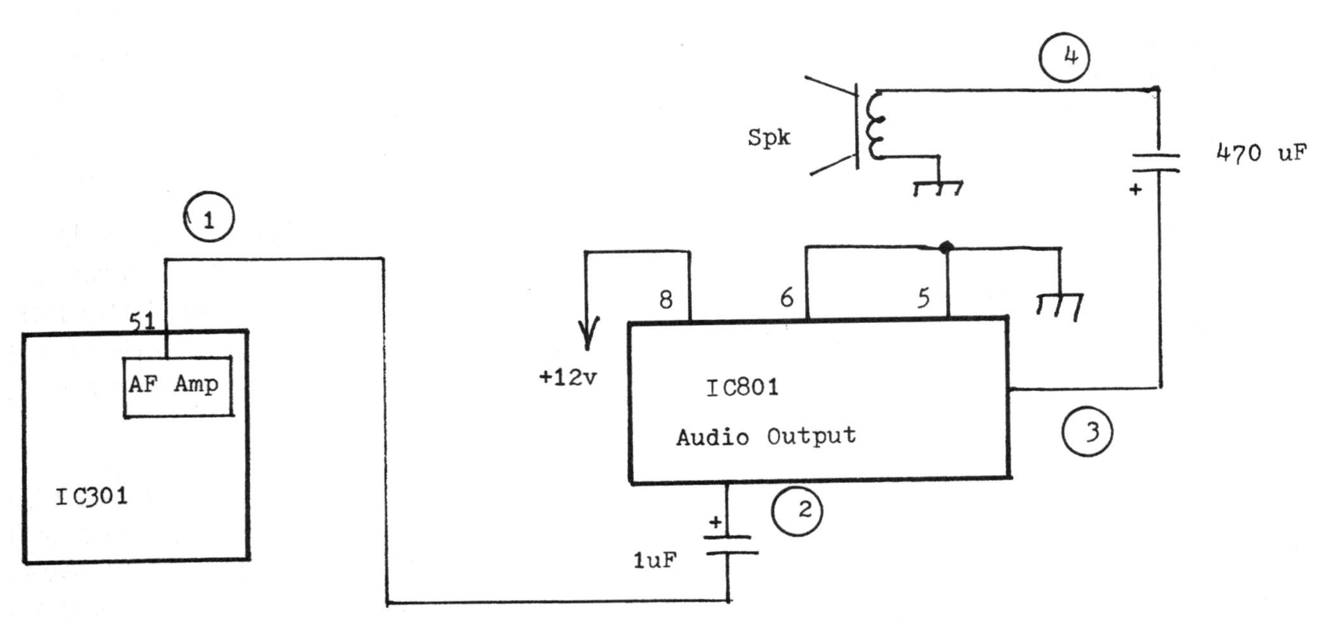

Fig. 7-14. Signal trace the audio circuits with an external amp or scope by the numbers.

When the signal is lost or when distortion begins, check each component and take critical voltage measurements on both the input and output IC. When the audio signal is normal at the input terminal and distorted at the output terminal, suspect the audio output IC. Take critical voltage measurements on each pin of the output IC and compare them to the schematic. If the voltage is low at the supply-voltage pin (Vcc), suspect a leaky IC or improper supply voltage.

Remove the supply pin from the PC wiring with a solder wick and iron. Notice if the supply voltage increases. Now shut down the chassis. Take a critical ohmmeter measurement between the unsoldered supply pin and common ground. A low-ohm resistance (under 500 ohms) can indicate a leaky output IC. Check each capacitor and resistor tied to each pin terminal to ground for a leakage test. Make sure that all components tied to the output IC are normal before replacing the suspected IC.

Speaker Problems

A dead speaker can have an open voice coil or broken wires from the voice coil to the speaker terminals. A bad speaker-wiring connection can cause a loss of sound or intermittent sound. Often the speaker voice coil can be blown open from too much volume. With the volume turned full on, the solid-state output IC might be damaged with an open speaker. Check the speaker-cone resistance for an open voice coil. Usually, the total resistance with the DMM on the low-ohm scale should be a 0.5 to 1 ohm difference from the speaker impedance. For instance, an 8-ohm speaker might measure 7.5 ohms on the DMM.

Buzz in Audio - Panasonic PVM-2021

In a Panasonic PVM-2021 with its TV audio OK, a buzzing noise was heard in the playback (PB) mode. When signal-traced with a scope, IC4001 was suspected. IC4001 was replaced with the original replacement LA7282M. This solved the noisy audio problem.

A damaged speaker cone can make an intermittent noise in the audio, whereas a loose cone can make a blatting noise. The speaker cone can come loose from the framework and produce a vibrating noise. The dropped or frozen speaker cone on the center pole piece can cause a tinny and distorted sound. Place both thumbs on each side of the speaker cone and press down on the cone. If the cone rubs against the center magnet, then replace the speaker.

Check the speaker cone for holes or tears. Small holes punched into the cone can be repaired with speaker or contact cement. Make sure the voice-coil spider is glued and not vibrating when the speaker moves in and out. A warped speaker cone might cause the voice coil to drag against the center pole piece. Remember that the speaker should be connected at all times to the amplifier-output IC or to a load resistor in order to prevent further damage to the output IC. Substitute another PM speaker in order to check the defective speaker in the TV/VCR combo.

Part II:

Repairing VCR Circuits and Mechanical Movements

Chapter 8: Servicing VCR Speed Problems

Speed problems generally consist of no speed, slow speed, erratic or intermittent speed, and fast speed rotation. Usually slow or no speed problems are caused by a broken motor belt, loose belt, or oil on the belt surface. Speed problems are also caused by the various motors (loading, capstan, reel, mode, drum, and cylinder). Today's TV/VCR might have two or more motor belts, several plastic gears, and might have a loading and capstan motor belt as well.

To service the VCR section and get at critical parts, the VCR chassis must be removed from the TV chassis. The VCR mechanism is mounted on top of the TV chassis (*Fig. 8-1*). Some VCR units are mounted at the top of the TV cabinet. Pull or lift out the TV/VCR

Fig. 8-1. The VCR mechanism is located under a metal shield, shown here in a Panasonic 13" TV/VCR.

Fig. 8-2. Remove metal screws from each corner of the metal cover after removing TV/VCR chassis from picture tube assembly.

chassis from the front picture tube plastic cabinet. Remove all cables and plugs that connect the TV/VCR chassis to the picture tube and speaker. Move the TV/VCR combo chassis away from the picture tube cabinet. Lay the CRT and front cabinet face down on a padded blanket or pad to protect the front of the CRT.

Removing the VCR Chassis

The metal cover over the VCR deck must be removed before you can look at the VCR mechanism. Generally, removing four metal Phillips-type screws frees the cover (*Fig. 8-2 and 8-3*). A magnetized Phillips screwdriver is handy because the screws can easily be lost inside the mechanism. The screws can be replaced with the same screwdriver and inserted into the correct mounting holes. Keep the magnetized screwdriver away from the tape head assemblies.

Now you can service any component on top of the chassis and make a few adjustments before removing the whole chassis. The video and audio control head, tape guides, and pressure roller can be cleaned from here if no other service is needed. The VCR chassis must be removed to get at the components underneath the mechanism and to the VCR PC board.

To do this, remove three or four screws in the corners holding the tape deck to the main chassis. Remove all cords and plugs from connectors on the main CBA. Obtain the correct harness to connect the main CBA to the VCR deck when doing warranty service or repairing a TV/VCR that you sold (*Fig. 8-4*). The VCR deck can now be connected to the TV chassis for the servicing of the tape deck.

Fig. 8-3. The top of the VCR unit is visible after removing the metal cover.

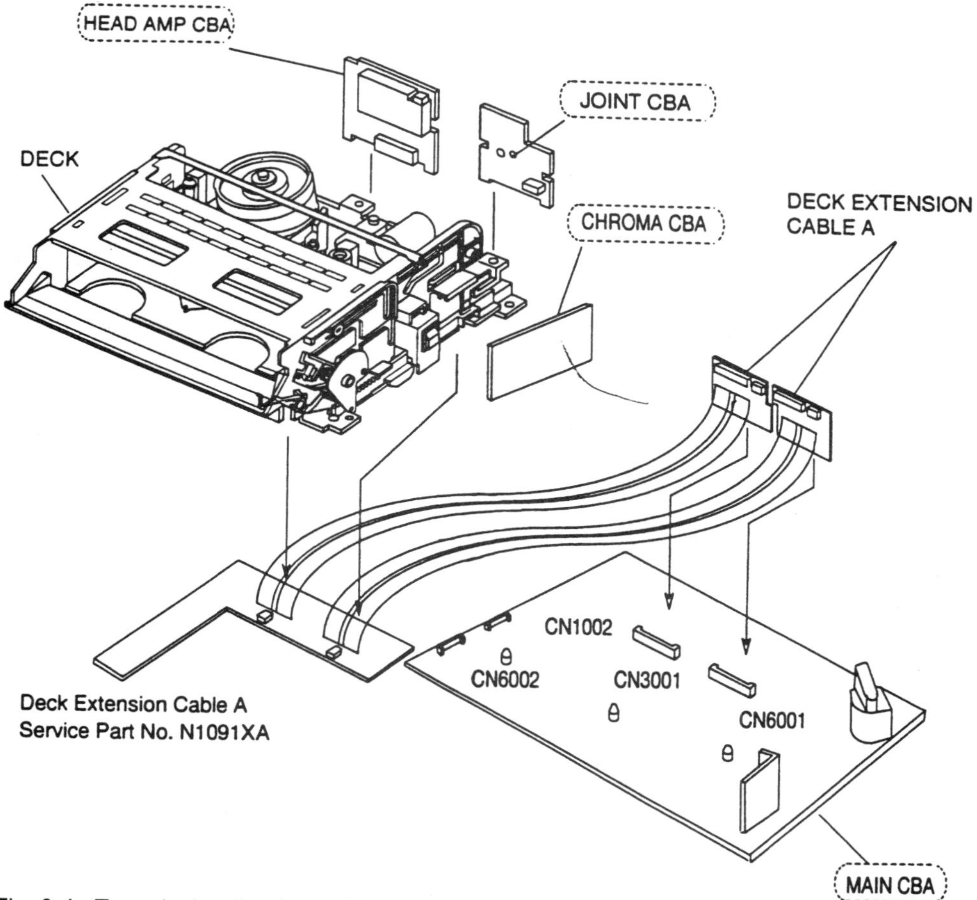

Fig. 8-4. Tape deck extension cables connect the VCR so the it can be connected to the main chassis for servicing (Sears 9" TV/VCR diagram shown).

The Different Speeds

Most TV/VCR decks employ VHS or VHS-C cassettes and operate at three different speeds: standard (SP or SL), long play (LP), and extended play (EP or SLP). Using standard play (SP) or SL speed on a T120 cassette allows a two-hour use. Using long play (LP), the same tape can be used for up to four hours. The extended play (EP) or SLP (standard long play) allows six to eight hours of usable time on a T120 cassette. A selector switch sets the correct speed for both recording and playback operations.

Test Equipment and Hand Tools

The same test equipment used to service the TV section can be employed to repair the VCR circuits. Most of the required test equipment is already found on the service bench. For electrical adjustment the required test equipment is listed below.

Required Test Equipment

1. NTSC pattern generator
2. AC voltmeter
3. Alignment tape
4. Blank tape
5. DC voltmeter
6. Oscilloscope - dual-trace with 10-1 probe 40 MHz frequency range
7. Frequency Counter

Required Hand Tools

These tools are needed for electrical and mechanical adjustments. Again, most are already found on the service bench.

1. Long and short Phillips-type screwdriver
2. Several sizes of flat blade screwdrivers
3. Lock screw wrench or X-nut adjustment screwdriver
4. Reel table height fixture
5. Post adjustment screwdriver
6. VHS alignment tape
7. Back tension meter
8. Torque gauge

VCR Tape Path

The VHS tape moves out of the supply reel assembly at the tension lever and tension pole, moving toward the number 3 guide. The tape moves past the flying erase head (FE). The flying erase head is mounted before the video head in the drum assembly to erase any previously recorded material on the tape. The flying erase head provides clean, glitch-proof recording.

The tape then moves past a tape guide spindle or S-guide roller and S-slant guide to the cylinder or drum assembly.

The cylinder or drum assembly contains the tape heads that record or playback any recorded material on the tape. These slant guides pull the tape out of the cartridge and wrap the tape around the drum. In record mode, the tape head applies the music or recording to the tape as it rolls around the tape head. Likewise any sound recordings are picked up off of the tape and applied to the head and amplifier circuits in playback mode.

Then the tape moves past the audio control head (ACE). In record mode, the upper half of the ACE head records the audio information on the upper edge of the tape. When in play mode, the upper half of the ACE head reads the audio information on the tape (*Fig. 8-5*).

When the tape moves past the ACE head in record mode, the lower portion of the ACE head – specifically called the control head - records coded information on the lower edge of the videotape. This coded information keeps the different motors operating at the correct playback and recording speeds. When dirt or residue collects on the ACE head, distorted or garbled sound and improper recording and playback occur.

The number 8 guide is usually used as the reference height when adjusting the tape path. The videotape is pressed between a rubber pinch roller and capstan drive shaft that pulls the tape past the tape heads. The capstan motor rotates the capstan drive via a motor drive belt. The speed of the capstan motor determines the rate at which the tape is pulled through the tape path.

To prevent the tape from pulling or the VCR from eating the tape, excess tape is threaded past the number 9 guide to the cartridge. Often the number 9 guide is pulled out at the reverse mode. The take-up reel moves the excess tape, feeding it back into the VHS cartridge.

Cleaning

To ensure a proper, working tape path for recording and playback operations, a thorough cleaning of the tape heads and guide assemblies is necessary. A good cleaning can

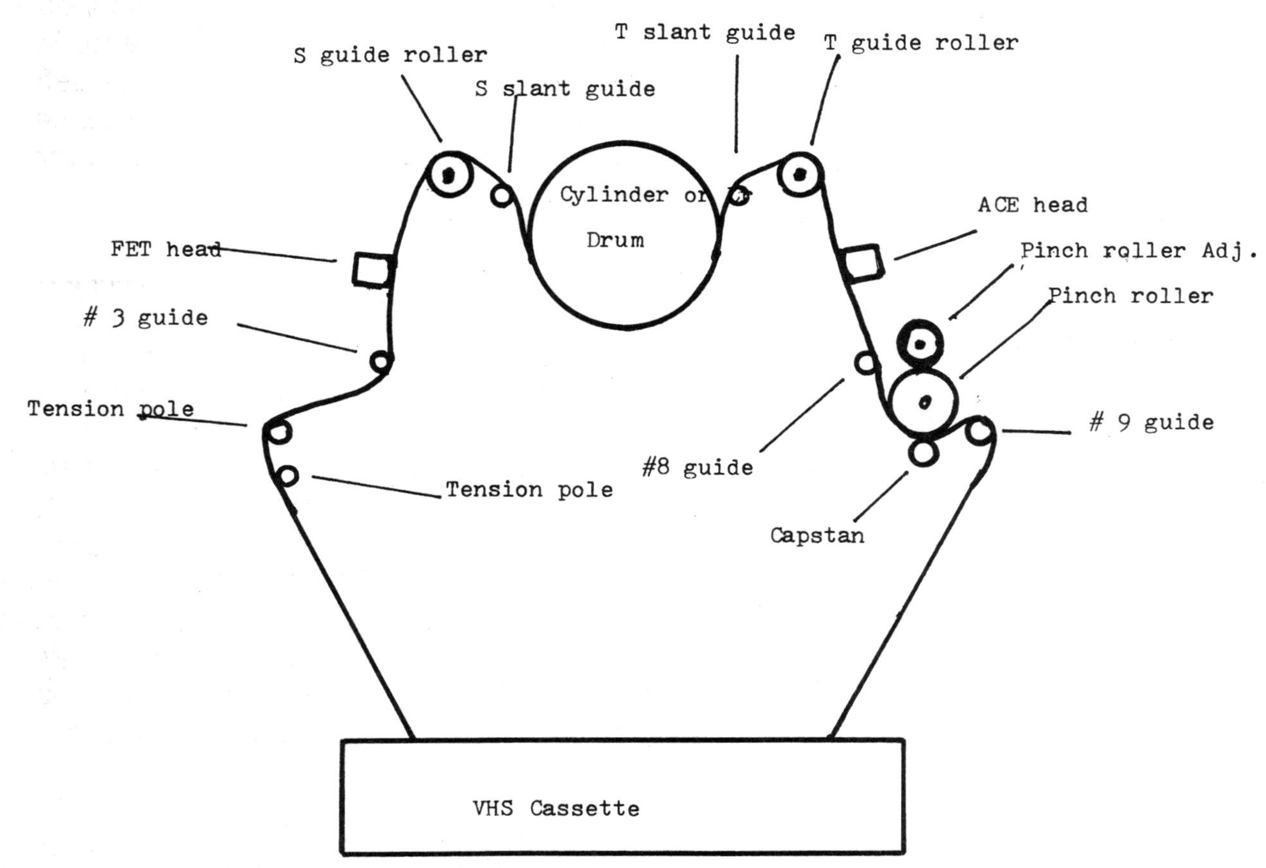

Fig. 8-5. The tape path when loaded around the various guides, cylinder, and pulleys.

also cure many service problems. Clean all parts of the tape transport (upper drum with video heads/pinch roller/ACE head/FE head) using 90% Isopropyl alcohol.

Each time a TV/VCR chassis appears on the service bench, clean up the VCR components immediately. Inspect and replace any worn or defective rubber motor drive belts after cleanup. The video head surface should be cleaned with a cleaning or chamois skin stick. Although a cleaning cartridge can be used to remove excess oxide dirt from the various tape path components, a good hand cleaning does a much better and more complete job. Be careful when applying any cleaning spray to the head surfaces though. It can blow and spray out onto other parts and drip down inside the VCR chassis. Always allow ample time for excess liquids to dry before starting up the VCR.

Remove the top shield of the VCR to get at the cylinder or drum area. Put on a rubber glove to avoid touching the upper and lower drum with your bare hands. If thin rubber gloves are not available, hold the drum gently by the top rim to prevent fingerprints on the tape area. Place a few drops of 90% Isopropyl alcohol on the head of a cleaning stick or chamois skin and, by slightly pressing it against the head tip, turn the upper drum to the right and to the left (*Fig. 8-6*).

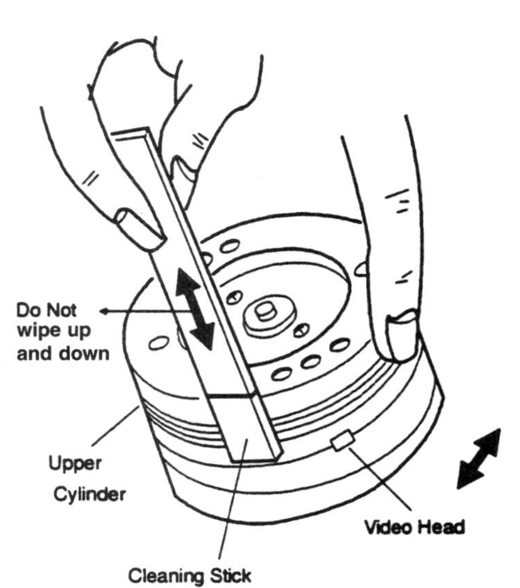

Fig. 8-6. Hold the top of the cylinder/drum assembly gently with fingertips. Clean tape head horizontally with cleaning stick.

Do not move the cleaning stick up and down vertically or you can damage the tape head. Wait for the cleaned parts to dry thoroughly before operating the unit. Do not reuse a stained head cleaning stick or chamois skin for cleanup; throw it away.

Cleaning the Audio Control

Clean the audio control head with a cotton swab dipped in 90% Isopropyl alcohol. Be careful not to damage the upper drum and other tape running parts. Do not clean the control head surface vertically (*Fig. 8-7*). Move the cotton swab back and forth horizontally to clean off oxide dust and dirt from the control head. Wait for the control head to dry thoroughly before operating the VCR or damage may occur. Wipe off all tape guides and rollers with a swab dipped in alcohol to clean up the tape running path.

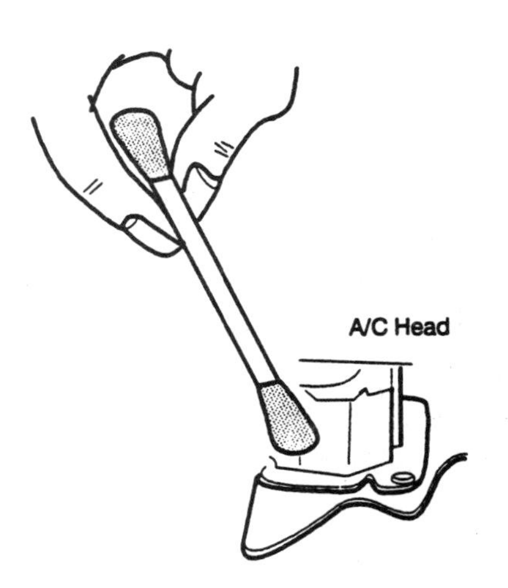

Fig. 8-7. Clean the A/C head with a cotton swab and 90% isopropyl alcohol.

Periodic Check and Lubrication

Like an automobile, the VCR transport must be lubricated and checked after so many hours of operation. The cylinder assembly should be checked every 1000 hours of operation. A change of the pinch roller assembly, pulley assembly, loading belt, band brake assembly, main brake S assembly, main brake T assembly, T brake arm assembly, capstan motor, capstan belt, F brake assembly, clutch assembly, and arm idler assembly should be changed every 2000 hours.

The loading motor assembly, AC head assembly, reel base assembly, ground brush assembly, and FE head should be changed every 3000 hours of operation. All worn components should be changed or replaced when the VCR transport is in for a service problem.

Belt Replacement

A worn, loose, or stretched loading belt can cause intermittent or erratic loading problems. A cracked or broken loading belt can result in the VCR cartridge not loading. A loading belt with grease or oil in the belt area might cause erratic loading. Sometimes the loading belt can come off and be pulled into the gear area, which keeps the loading motor from having any movement, thus the cassette does not load. Inspect all belts when the VCR is brought in for service. Always replace all belts with the original part number when possible. Do not replace a flat belt with a square belt.

- ✔ A worn, loose, or stretched capstan belt may be the cause of garbled audio in EP mode (*Fig. 8-8*).
- ✔ A worn or slick capstan motor drive belt that slips may be the result of a dry or frozen capstan bearing.
- ✔ A new capstan belt that continually comes off the motor pulley needs realignment. Realign the belt and roller with the capstan motor.
- ✔ Loss of playback and recording functions can be the result of a cracked or broken capstan motor belt.
- ✔ When capstan speed is slow, drags, and produces worn and fluttery tape speeds, clean old, hardened grease or oil off the capstan motor.

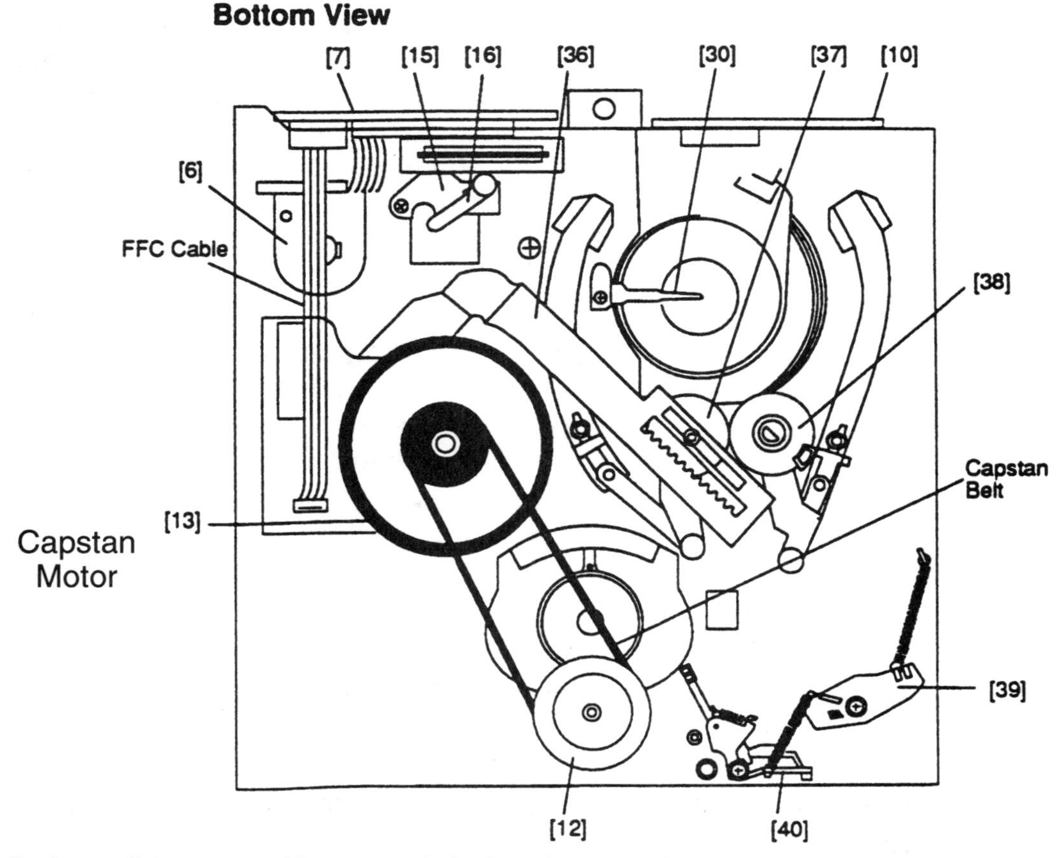

Fig. 8-8. Replace a dirty, worn, or shiny capstan belt when slow or erratic tape movement occurs.

Intermittent Tape Loading – Panasonic PV-M2021

No tape loading or intermittent tape loading can result from a defective system control IC, loading motor driver IC, or a defective loading motor. A different polarity voltage is fed into pins 1 and 9 of the loading motor IC6003 to replace or make the loading motor go forward. Critical voltage measurements on pins 2 and 8, and 7 and 3 pointed out a defective loading motor (see *Fig. 8-9*).

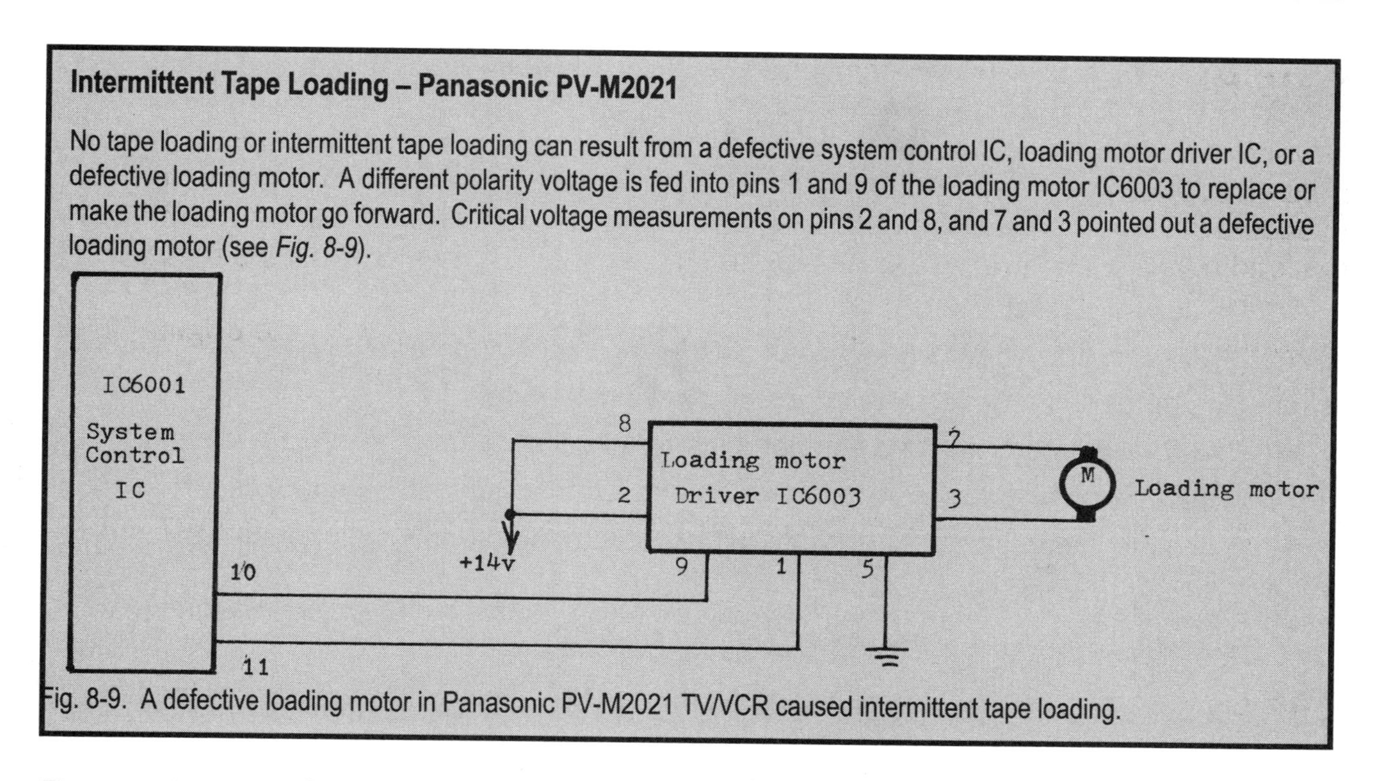

Fig. 8-9. A defective loading motor in Panasonic PV-M2021 TV/VCR caused intermittent tape loading.

Poor Board Connections

Poor board connections of various components soldered to the PC wiring can cause a variety of intermittent and erratic problems in the VCR. Bad connections of a transistor element or an IC component can be very difficult to locate. Broken sections of PC board wiring, traces, tie-through, and connections can produce intermittent operation. Although the low-ohm range of the VOM or DMM can be used to trace out the broken or cracked traces, the ESR meter is ideal for locating poor connections and broken traces or foil wiring.

The ESR meter sounds off with an interrupted beep and the meter hand flickers when a poor connection is discovered from trace to a corresponding component. An open trace between components will not show a reading on the meter. The poor component connection on a longer PC wire can show some sign of a resistance measurement on the meter. Be sure to discharge all electrolytic capacitors in the circuits to be tested and remove the power cord. As with an ohmmeter, you can damage the ESR meter if a voltage occurs across the meter probes.

For poorly soldered connections on the board terminal of one IC or microprocessor, push the meter probe against the top pin terminal and the other probe into the soldered connection of PC wiring. The meter probes are quite sharp and can be placed on different terminals without shorting out any two terminals. The sharp probe bites through spray or rosin residue for a proper connection on a soldered joint. Badly soldered joints will show some resistance on the ESR meter, which is indicated by an erratic beeping sound.

Speed Problems

Too Slow

Possible Causes:

- ✔ Defective capstan motor
- ✔ Brake arm that erratically releases causing the capstan to rotate slowly
- ✔ Old, hardened grease on the capstan or flywheel bearing
- ✔ Bad IC regulator feeding the capstan motor circuits
- ✔ Defective low-voltage IC regulator to the capstan motor voltage supply source
- ✔ Defective electrolytic capacitors in the capstan motor supply voltage source
- ✔ Cracked or loose capstan belt
- ✔ Stretched, loose, or shiny and smooth loading belt

Too Fast

Possible Causes:

- ✔ Poorly soldered joint on the motor terminals
- ✔ Defective capacitors off the servo PCB (in play mode or cannot select proper record speed)

Fast Speed - Symphonic TVCR1381

The tape speed was too fast in a Symphonic TVCR1381 TV/VCR.

Check these areas:

- ✔ Defective electrolytic in the main power supply
- ✔ Badly soldered joints on the drive motor assembly and PCB
- ✔ Leaky or shorted zener diode in the power supply

Specific to the Capstan Motor:

- ✔ Capstan motor driver IC
- ✔ Defective servo IC
- ✔ Defective capstan motor

Specific to the Cylinder Motor:

- ✔ Defective cylinder motor (check for poorly soldered connections on the motor)
- ✔ Badly soldered joints on the cylinder drum stators on the PCB
- ✔ 4.19 in MHz crystal (also, the cassette does not load and the LED display does not light up)
- ✔ Cylinder motor PCB for defective diodes (intermittent shutdown or drum motor PG signal)

Specific to Capstan Motor:

- ✔ Defective capstan or drum motor
- ✔ Loose magnet inside the capstan motor
- ✔ Defective capacitors off the servo PCB (in play mode or cannot select proper record speed)
- ✔ Defective PG head (resolder connector terminals)
- ✔ 1µF 60-volt electrolytic off the capstan FG line (in play mode - check the gap from flywheel to FG sensor when capstan is running at full speed)
- ✔ Defective capstan motor (in record or play mode with no FG pulse)
- ✔ Defective capstan motor IC (high pitched squeal from audio playback with capstan running at full speed)
- ✔ Poorly soldered connection on the capstan drive IC or a defective capstan IC
- ✔ Capstan control transistor (rewind and fast-forward shut down)

Specific to Sound

- ✔ Pressure roller with a roller spring out of place.

Erratic Speed

Possible Causes:

- ✔ Voltage regulator transistor or voltage source IC to the capstan motor
- ✔ Defective electrolytic capacitor in the voltage source
- ✔ Defective drum CBA
- ✔ Defective servo IC
- ✔ Voltage regulator IC (drum motor does not work and capstan motor runs fast) - spray the regulator IC with freeze spray or coolant to make it act up

Loading Motor Problems

No Loading

Possible Causes:

- ✔ Defective loading motor and bracket(no play either – may load tape slowly then quit)
- ✔ Motor driver IC in the Syscon circuit
- ✔ Open 2.7- or 3.9-ohm resistor in the B+ supply voltage (motor does not run)
- ✔ Bad housing motor driver IC
- ✔ Open fuse (overheats loading motor)

Slow Loading

Possible Causes:

✔ Defective loading motor or binding load assembly

No Loading or Improper Loading

Possible Causes:

✔ Defective loading motor (head may spin a few seconds and shut off)

✔ Defective loading motor IC (with a normal mode switch) (see *Fig. 8-10*)

✔ Defective motor block assembly (loading motor and mode switch – begins loading then shuts off)

✔ Slipping pulley belt or loose loading belt (replace)

Recording Problems

Sometimes the actual tape is the problem. If a cassette will not record, check for a missing knock out at the back of the tape. By removing the plastic tab, the cassette or recording can be saved. If it is missing, place a piece of tape over the opening to record over the existing recording. Double-check the recording and play it back again before erasing the recording.

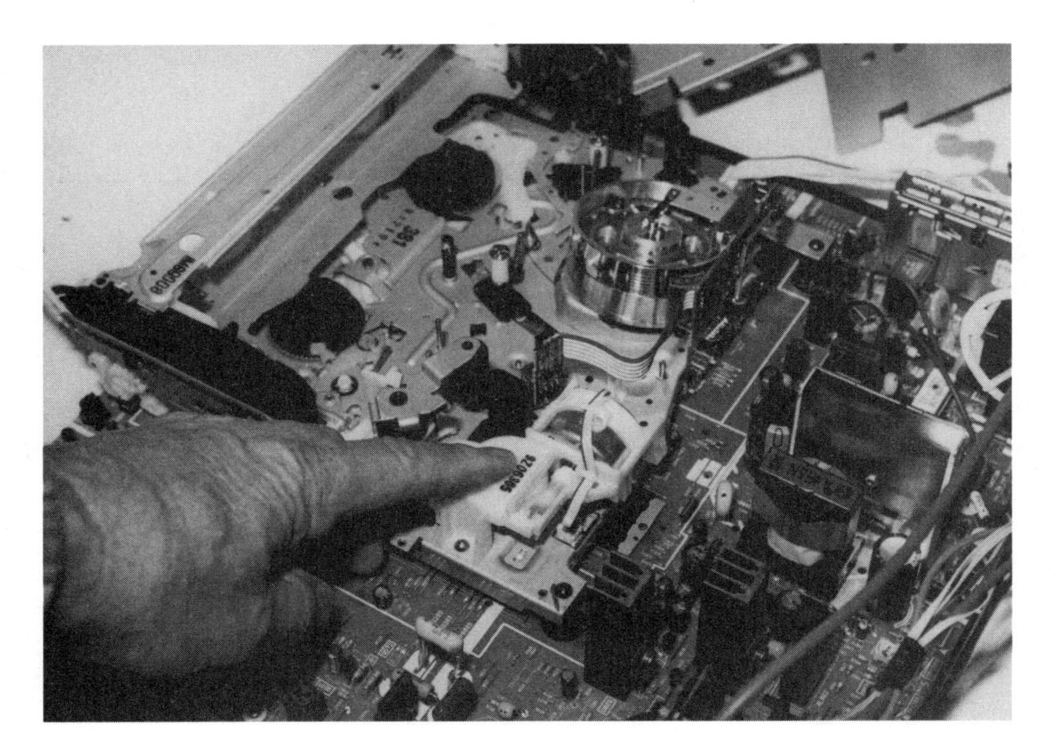

Fig. 8-10. The loading motor drives a worm gear to load the cassette into the tape transport.

Slow Loading - Zenith TVSA1320

The 13 inch TV/VCR loading motor was suspected as the cause of no loading and slow loading of cassettes. The input and output voltages were monitored at the loading motor driver IC204. Sometimes the loading motor voltage was present at the loading motor terminals and other times no voltage was found on terminals 3 and 7 of IC204 (see *Fig. 8-11*). Replacing the loading motor driver IC204 solved the problem.

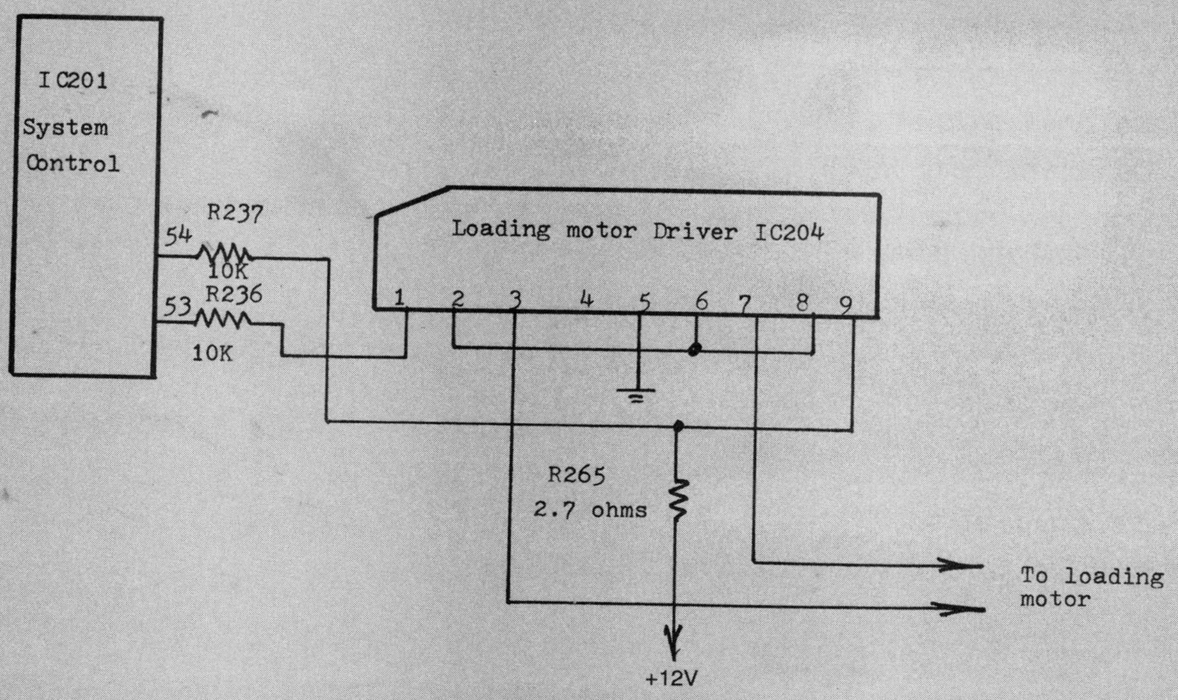

Fig. 8-11. The defective loading motor IC204 cause slow loading in a Zenith 13" TV/VCR.

Loads Empty Tray

Possible Causes:

- ✔ Defective loading motor IC (can appear to be a sensor problem)

Loads/Unloads and Immediately Unloads/Loads

- ✔ Defective capacitor on the cylinder motor drive PCB
- ✔ Defective loading motor

No Eject

- ✔ Cracked motor assembly (fast forward assembly may also be out of line)

RCA Capstan Motor - 13TVR60

A tape was stuck inside of a RCA 13TVR60 TV/VCR combo. After removing the excess tape, the transport was checked with a blank cassette and it was determined that the loading motor would run and then stop. Replacing the loading motor solved this problem.

Cassette is Jammed

Possible Causes:

- ✔ Defective loading motor drive IC
- ✔ Open low-ohm resistor in the 12-volt source to the loading motor IC (also power shuts off after a few seconds and display blinks)
- ✔ Worn clutch assembly (loading motor still runs)

Capstan Operation Problems

Tape Stuck

Possible Causes:

- ✔ Bad capstan motor or a frozen capstan bearing

No Operation

Possible Causes:

- ✔ Flat armature or a dead spot on capstan motor (motor may start when the pulley is rotated)
- ✔ Open low-ohm SMD resistor in the voltage power source
- ✔ Improper or low voltage at the driver IC
- ✔ The 4.7 to 10 µF electrolytic capacitors in the servo and driver IC circuits (they may shut down after loading and ejecting tape - check all electrolytic capacitors with the ESR meter.)

Squeal - Emerson VT1920

The capstan motor squealed in play mode and the speaker produced a warbling in an Emerson VT1920 TV/VCR. The capstan motor was removed and the bearings cleaned out and lubricated. Noticeably, the bottom capstan bearing was dry, which created the squeaking noise when the capstan motor began to operate and when it quit running.

Improper Operation

Possible Causes:

- ✔ Defective capstan motor driver IC (unit may stop after two or three hours of operation in play mode)
- ✔ Poorly soldered connections on capstan motor driver IC terminals
- ✔ Defective capstan motor driver IC (movement shuts down after five seconds)

Dead Unit with No Cylinder/Drum Rotation

No Rotation

Possible Causes:

- ✔ Motor supply voltage
- ✔ Defective voltage regulator IC
- ✔ Drum motor driver IC
- ✔ Poor connection between drum motor and servo IC
- ✔ Bad tape-end lamp
- ✔ Cylinder servo IC (replace)
- ✔ Lower drum PCB connections
- ✔ Bad contacts
- ✔ Bad solder joints on the cylinder CBA
- ✔ Open low-ohm resistor in the voltage source of the cylinder motor

Tape Loads and Motor Shuts Down

Possible Causes:

- ✔ A defective drum motor (*Fig. 8-12*) (inspect the trace lines with the ESR meter for cracked wiring or poorly soldered components)
- ✔ Bad tape-end sensors

Motor Runs then Shuts Down

Possible Causes:

- ✔ Cylinder motor assembly (replace)
- ✔ Cylinder servo IC (replace)

No Spin - Symphonic TVCR9F1

The tape head drum would not spin in a Symphonic TVCR9F1 TV/VCR. Additionally, the tuner experienced snow on high and low channels. After taking a critical voltage test on the drum motor and the 6-volt regulator IC, it was determined that IC651 should be replaced.

Fig. 8-12. A defective cylinder/drum can cause cylinder movement to stop.

Fast Forward and Rewind Problems

With the VCR in play-fast-forward mode, the tape travels from the left (supply reel) to the right (take-up reel). The VCR can eat tapes in play or fast-forward mode.

VCR Eats Tape

Possible Causes:

- ✔ Defective reel motor IC (take-up reel frozen, no fast-forward, rewind okay)
- ✔ F/R bracket-broken shifter (in eject mode)
- ✔ Defective diode in the 5-volt supply line (no fast-forward or rewind)

No Fast–Forward Only

Possible Causes:

- ✔ Tape-end sensor

No Fast–Forward or Rewind

Possible Causes:

- ✔ Brake or brake solenoid assembly (replace brake solenoid switch)
- ✔ Bad mode switch (check brake and brake solenoid assembly first)
- ✔ Worn rubber damper (on the top right side of the mechanism)
- ✔ Broken mode switch (reburnish or replace when VCR goes to record from fast-forward and rewind)
- ✔ Broken arm gear on mechanism underside
- ✔ F/R bracket-broken shifter
- ✔ Bad spot (flat armature) on capstan motor (see *Fig. 8-13*)
- ✔ Defective photo-interrupter under the take-up reel (does not play either)
- ✔ Broken spring from loading motor assembly (all other functions work)
- ✔ Dirty cam switch (clean or replace)
- ✔ Voltage regulator IC (does not play either)
- ✔ Defective switching regulator (no fast-forward/rewind in play mode)
- ✔ Servo/system control IC (no eject or play either)

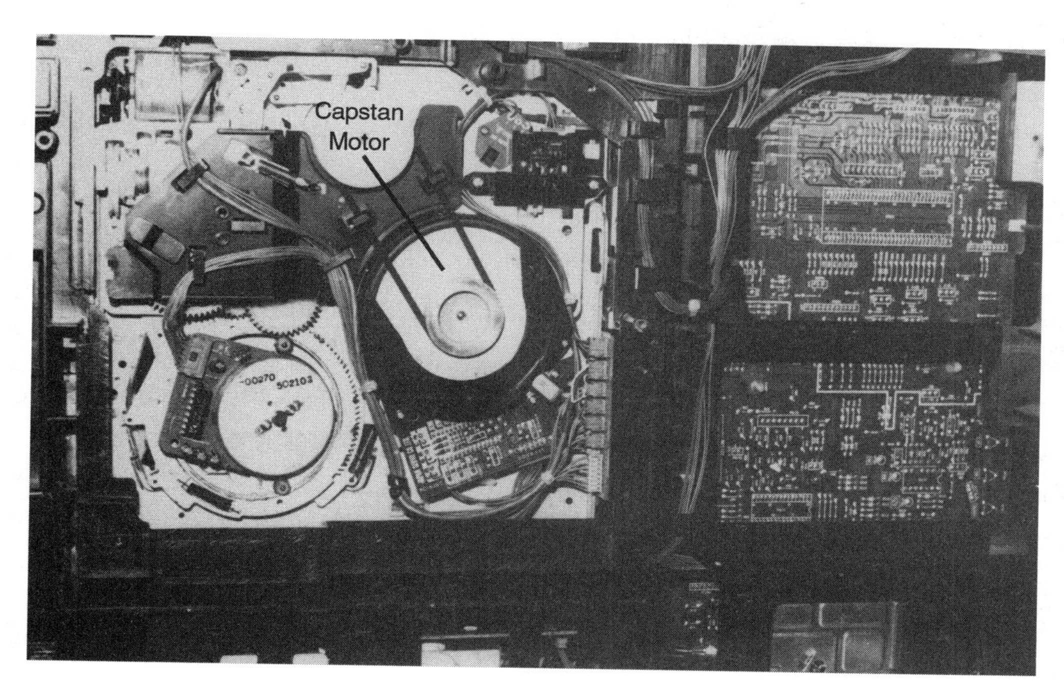

Fig. 8-13. Suspect a defective capstan motor when fast-forward and rewind do not work.

Rewind Mode Operation Problems

When the VCR is in rewind mode, the tape moves through the unit from right to left. Always replace belts, the drive tire, and the plastic idler (gum and idler gear) when a no-rewind symptom occurs. Take critical voltage measurements on all terminals of a defective voltage regulator when investigating a no-rewind operation.

No Rewind or Fast-Forward

Possible Causes:

- ✔ Gear AY assembly
- ✔ Dirty cam switch (VCR may shut off and play back too fast – clean or replace)
- ✔ Bad capstan motor
- ✔ Bad plate assembly (rewind and fast-forward can also slow – brakes do not shut off)
- ✔ Main lever latch under the main cam gear
- ✔ Broken mode switch
- ✔ Damaged teeth on the cam gear

No Rewind Only

Possible Causes:

- ✔ Bad idler
- ✔ Defective capstan motor IC in the servo circuit
- ✔ Capstan motor assembly

Intermittent Rewind and Fast-Forward

Possible Causes:

- ✔ Reel motor assembly (rewind and fast-forward stop after two or three seconds and review and fast-forward in play mode are fine)
- ✔ Defective take-up reel sensors (rewind starts then stops after seven to ten seconds)
- ✔ Dirty IP leaf switch (switches to rewind three to five seconds after power-up and does not accept a tape)
- ✔ Syscon IC (rewind and fast-forward shut off after one to three minutes, the reel and end sensors test okay)
- ✔ Capstan control transistor (if the Syscon IC tested okay)

Slow Rewind and Fast-Forward

Possible Causes:

- ✔ Defective supply reel sensor (also no memory functions)
- ✔ Bad plate assembly (rewind and fast-forward can also stop – brakes do not shut off)

Rewind When Tape Inserted

Possible Causes:

- ✔ Tape-end sensor

Switches to Rewind and Other Functions Randomly

Possible Causes:

- ✔ Replace or reburnish the mode switch (switches to record, review, and fast-forward randomly)
- ✔ Defective tape end sensor (switches from play mode to rewind mode)
- ✔ Shorted take-up LED sensor (switches to stop from play and reverse)
- ✔ Bad start sensor (after a tape is inserted, rewind and fast-forward stop and the tape ejects)
- ✔ Bad contact on the supply sensor PWB (switches from rewind to stop and then the tape ejects)
- ✔ Defective cassette housing end sensor (stops or switches to rewind and fast-forward randomly).
- ✔ Ratchet lever assembly (fast-forward switches to stop, the take-up reel stops, and the supply reel spills out tape)
- ✔ Badly soldered connection of the driver IC (starts, stops, and rewinds randomly)

Rewind Does Not Stop

- ✔ Possible Causes:
- ✔ Left tape-end sensor (also causes tape to tear)

Blows Fuse

- ✔ Defective capstan motor (occurs during play, rewind, or fast-forward)

Take-Up Reel Problems

The take-up reel and supply reel should rotate freely when the VCR is operating. The idler wheel shifts to the take-up reel in play mode. In some VCRs, the take-up reel motor operates the take-up pulley to prevent the tape from spilling out or being eaten. When the take-up reel does not rotate or turns erratically, excess tape is spilled out inside the VCR mechanism. Additionally, a defective reel motor, defective sensors, reel drive IC, idler reel, and reel drive assemblies can cause tape problems. In lower priced TV/VCRs, the take-up reel might operate from a belt or idler pulley (see *Fig. 8-14*).

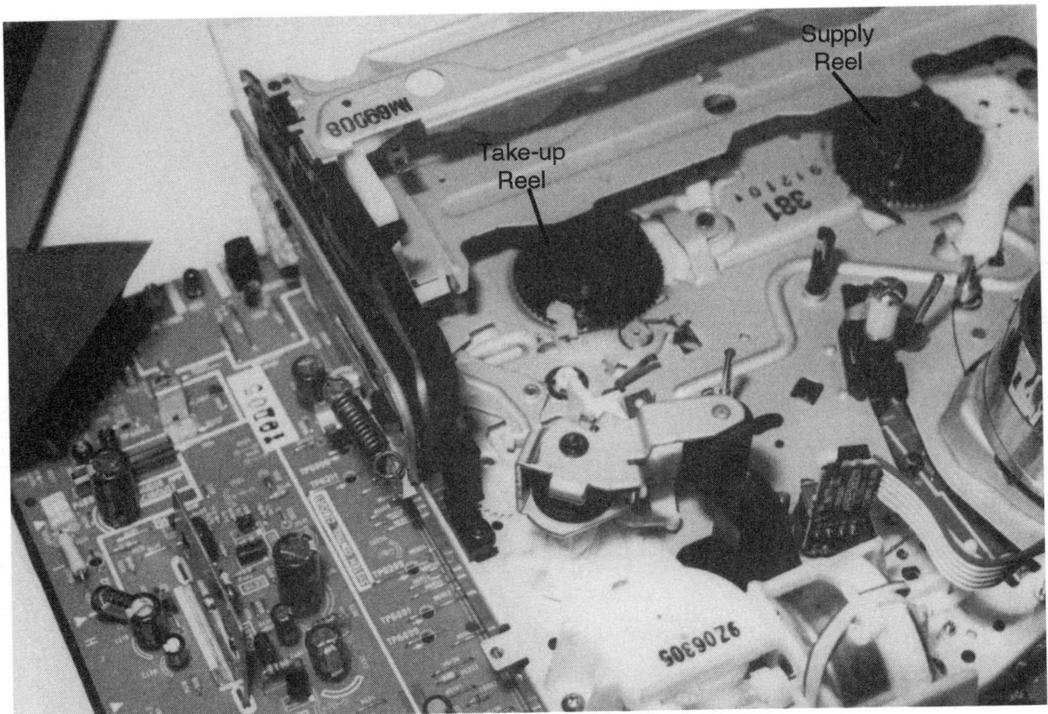

Fig. 8-14. A worn tire, belt, or pulley in an inexpensive TV/VCR combination can cause the take-up reel to slow down, operate erratically, and spill excess tape.

VCR Shuts Down from Play, Fast-Forward, or Rewind

Possible Causes:

- ✔ Bad supply reel sensor (play mode)
- ✔ Take-up reel sensor (play, fast-forward, and rewind modes)
- ✔ Reel sensor under the take-up reel (shuts off after 10 to 20 seconds in play, rewind, and fast-forward modes)
- ✔ Defective capstan reel drive IC (play mode)
- ✔ Reel motor assembly and driver IC (rewinds or fast-forwards for two or three seconds and stops)
- ✔ Reel photo sensor (stops after anywhere from few minutes up to an hour or two in play mode)
- ✔ Reel sensor assembly (the unit plays a few seconds, stops and shuts off)
- ✔ Photo sensor under the take-up reel (shuts down in play or record)
- ✔ Take-up end sensor (jumps to rewind from play randomly)
- ✔ Reel clutch assembly (play mode switches to stop sporadically)
- ✔ Loading motor IC (shuts down after a few seconds, eats tape, and reel does not turn)

No Playback - Sears 9-Inch TV/VCR

No playback action was found in a 9-inch Sears TV/VCR with a misaligned pinch roller. The riding pin on the pinch roller assembly was bent out of line and did not follow in the groove of the cam gear. The pin of the pinch roller arm assembly is positioned in the end of the groove of the upper cam. Try to repair or replace the defective pinch roller assembly (*Fig. 13-10*).

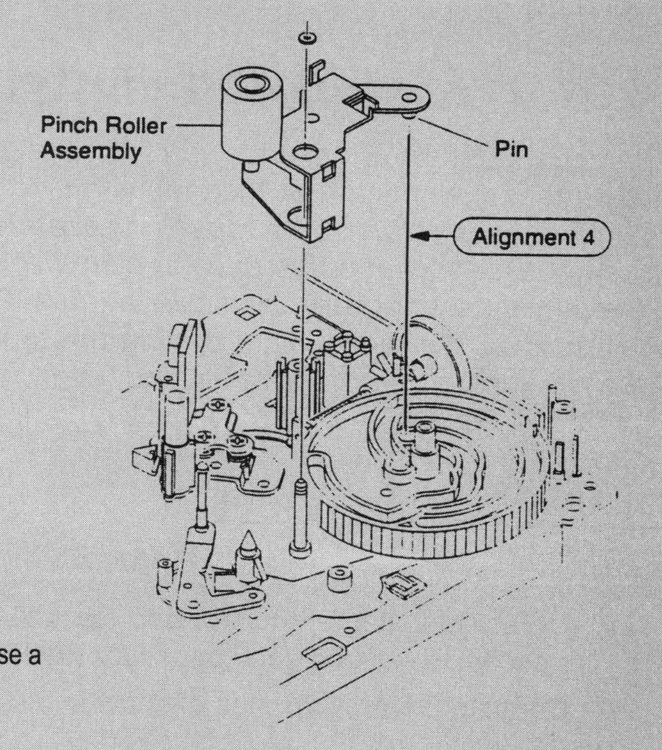

Fig. 8-15. A bent pinch roller assembly can cause a no playback action in a 9-inch Sears TV/VCR.

No Take-Up Operation

Possible Causes:

- ✔ Idler reel (play mode)
- ✔ Erratic Take-Up
- ✔ Dragging take-up reel (also causes audio quiver from speakers)

Reel Motor Spins and VCR Does Not Stop

Possible Causes:

- ✔ Regulator transistor in the power supply (no other working functions)

No Play

Possible Causes:

- ✔ Reel drive unit

Reel Motor has Low Torque

Possible Causes:

- ✔ Open resistors on the IC driver IC
- ✔ Clutch assembly (replace).

Wow (low–end) and Flutter (high–end)

The capstan shaft is driven by a roller, belt, or DC motor. The speed of the capstan determines the speed that the tape is pulled through the transport. A rubber pressure roller is applied against the tape and capstan to pull the tape past the heads and guide assemblies. Wow and flutter can occur when the speed of the tape movement varies. Generally speaking, they are slow variations in the pitch of audio due to variations in the speed of the mechanism driving the tape. Both audio and video can be affected. Always replace the drive belts and idler tires for slow, erratic, or wow and flutter conditions.

Display is Intermittent

Possible Causes:

- ✔ Defective voltage regulator and electrolytic capacitor (47 µF) in the power supply (reel spins - check each capacitor with the ESR meter)
- ✔ Capstan driver IC (bands of snow in display, garbled audio)

Poor or Garbled Audio

Possible Causes:

- ✔ Worn upper bronze bearing in capstan motor assembly (motor makes vibrating noise)
- ✔ Worn or not completely round pressure roller
- ✔ Defective clutch gear assembly (also causes slack in take-up)
- ✔ Dirty capstan motor (also speed is slow or drags – clean and re-lubricate)
- ✔ Capstan driver IC (bands of snow in display, garbled audio)
- ✔ Hardened felt on the clutch assembly (audio playback and record had flutter at slower speeds)
- ✔ Clutch assembly too tight (idler operates slower and take-up reel quivers creating wow and flutter, worse in record mode)

Squeal from Capstan Motor

Possible Causes:

- ✔ Bottom bearing of capstan (squeal in play mode)

Chapter 9: Solving Loading and Unloading Problems

Loading problems can include not ejecting a tape, chewing up or eating tape, no loading, slow loading, or the machine loads the tape then shuts down. Also, improper tape loading and unloading can occur in one or two different motions: the holder will not take the tape at all or it pauses and immediately ejects the tape. Sometimes the tape gets jammed in the machine and will not eject. This chapter covers the symptoms a VCR might display when having loading and unloading problems and the possible causes for each.

A mode switch is common to most VCRs and can be either a gear or slide-bar type. A defective or dirty mode switch can cause several different problems in the VCR (*Fig. 9-1*).

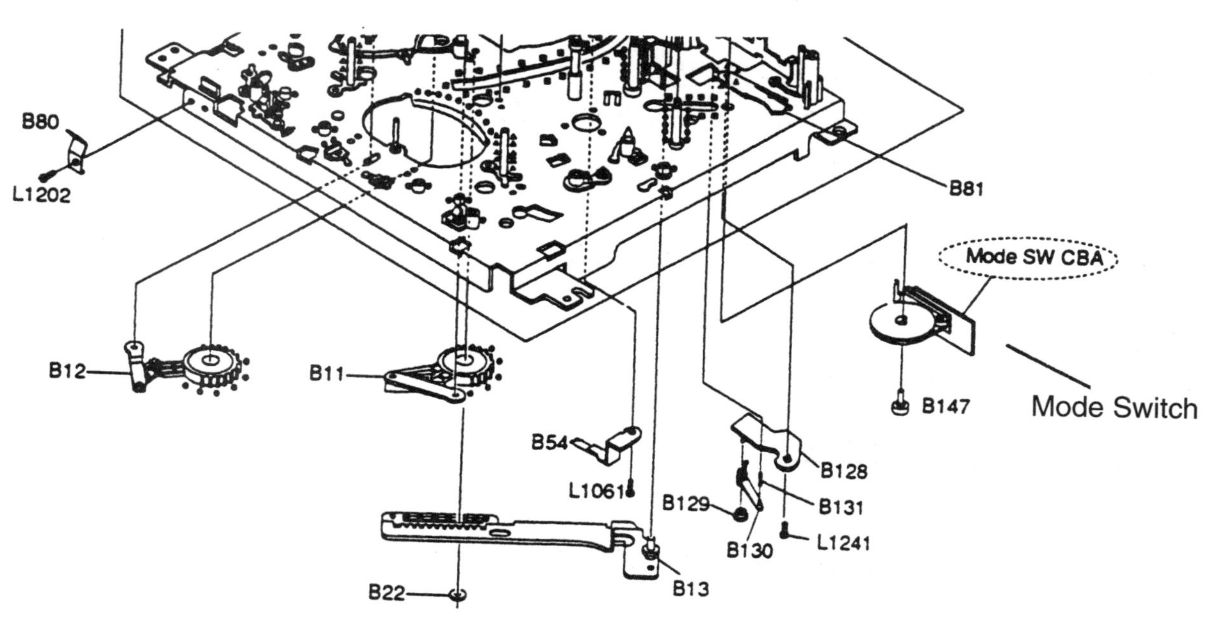

Fig. 9-1. Replace or reburnish the mode switch when the tape is stuck inside transport and the VCR powers off.

Cassette Bracket Does Not Stay Down

Possible Causes:

- ✔ Defective cassette holder roller

Cassette Does Not Accept or Load Tape

Possible Causes:

- ✔ Worn wheel, gear assembly, or rotary switch
- ✔ Defective linking gear cam in a broken cassette assembly
- ✔ Loose cassette housing
- ✔ Shorted zener diode and fuse in the loading motor circuits (power shuts off)
- ✔ Poor ground connections to the front load switches (cassette might not load - right side carriage gears may also need to be replaced)
- ✔ Broken cassette housing loading gear
- ✔ Large electrolytic (470 μF) capacitors in the 12-volt source
- ✔ Cam assembly and slant base track (even with new loading motor belt – clean and lubricate)
- ✔ Start sensor and cassette switch
- ✔ Eject drive gear (replace)
- ✔ Defective loading motor
- ✔ Rotary switch (replace - erratic loading, no rewind)
- ✔ Worn cassette basket (change entire cassette basket assembly)
- ✔ Transistors in the power supply (power shuts off and appears to work only intermittently under load)
- ✔ Switch and start sensor or carriage assembly
- ✔ Mode switch (several attempts – replace or reburnish)
- ✔ Bad eject drive gear(tape may jam)
- ✔ Bad gear on the cassette housing
- ✔ Gear-sense assembly (does not accept tape either)
- ✔ Defective cam/loading motor driver IC
- ✔ Bad supply side sensor (cassette light on)
- ✔ Open mode switch
- ✔ Dirty cassette leaf switch (clean or replace it if the contacts are bad)
- ✔ Loading assembly (also does not accept tape - loading gears are jammed)
- ✔ Loading motor driver IC (no capstan motor rotation, the machine shuts off, and the display flashes)
- ✔ Defective arm-gear assembly

No Loading or Slow Loading - Zenith TVSA1320

Sometimes the cassette in this Zenith unit would not load and at other times the loading was very slow. The loading motor voltage was monitored at the loading motor terminals and varied as the motor rotated. Critical voltage measurements were made on the loading motor driver IC204. The voltage measurements were incorrect on terminal pins 5 and 7. It was determined that IC204 was bad. Replacing IC204 solved the loading problem (*Fig. 9-2*).

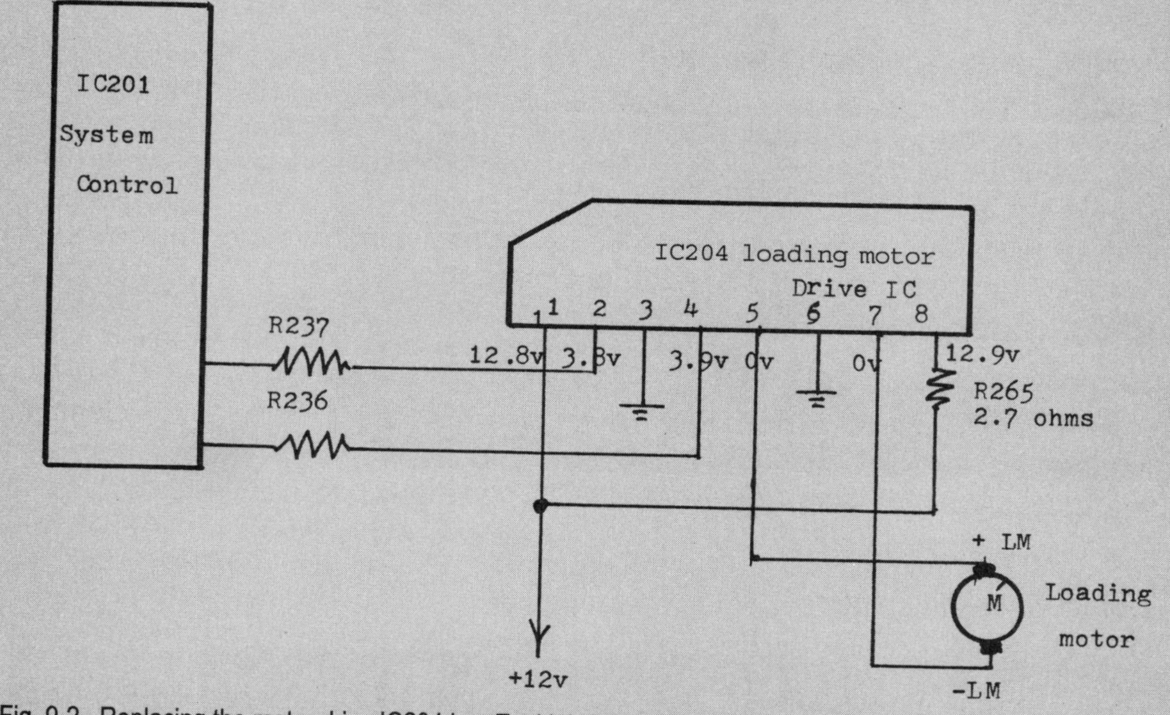

Fig. 9-2. Replacing the motor drive IC204 in a Zenith VCR solved a no loading or slow loading problem.

- ✔ F/L gear assembly rack (no eject either)
- ✔ Leaf switch on the F/L PCB (clean it)
- ✔ Open in-switch
- ✔ Carriage assembly (*Fig. 9-3*)
- ✔ Gear assembly rack and cam (incomplete load and eject)
- ✔ Bottom worm gear
- ✔ Cracked or broken loading gear
- ✔ Broken loading motor belt (*Fig. 9-4*)
- ✔ Bent pin to the right side of capstan motor shaft (tape jams)
- ✔ Tape-down switch (also tries to reload in eject mode - clean the contacts or, if points are bad, replace)
- ✔ Poorly soldered joints at start sensor
- ✔ Dirty loading gears (timing also off – clean)

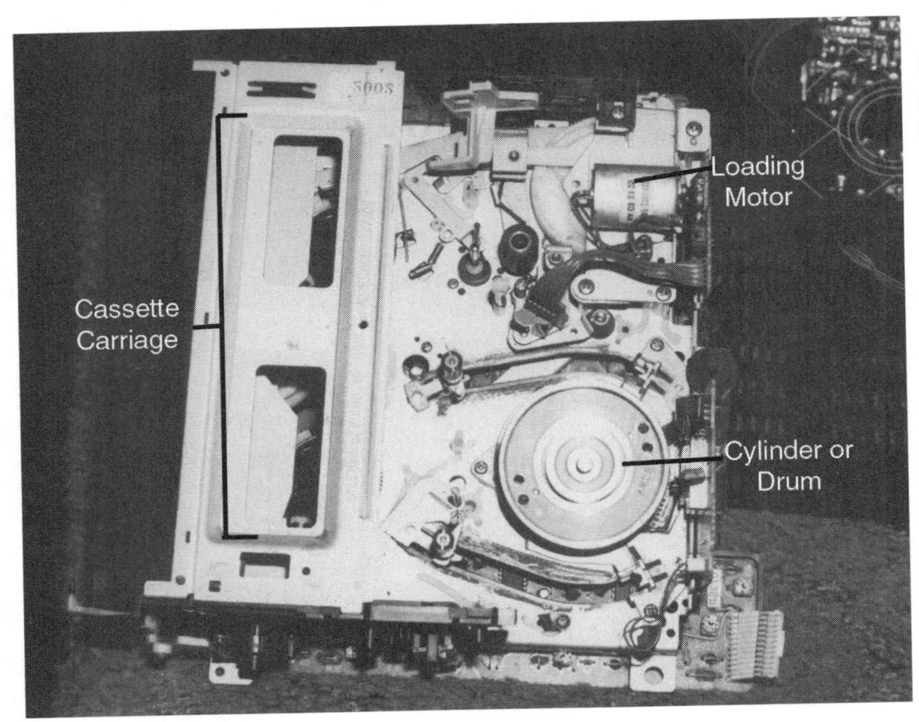

Fig. 9-3. Check the cassette carriage when it will not accept a tape or load a tape.

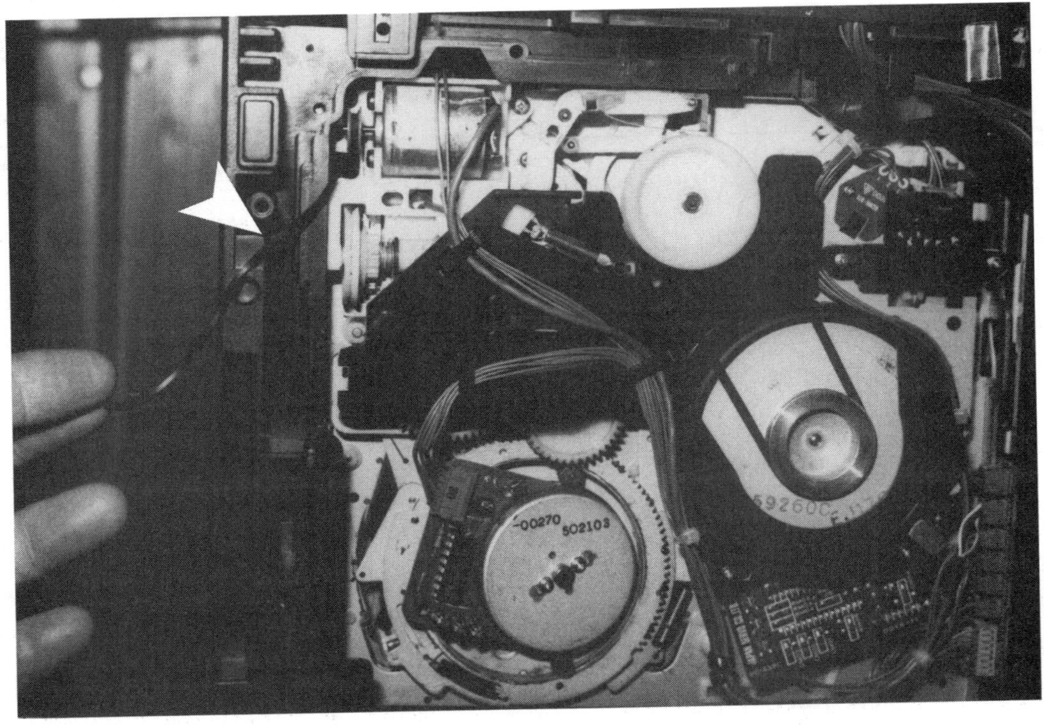

Fig. 9-4. A broken loading motor belt prohibits the cassette from loading a tape.

- ✔ Defective 13-volt zener diode in the power supply (measure the voltage at the loading motor)
- ✔ Loading motor driver IC
- ✔ Defective IC 12-volt regulator in the power section on the main PCB (all other functions normal)
- ✔ Voltage regulator IC in 13-volt line (also may be dead and drum motor may spin)
- ✔ Misaligned timing gears or a broken gear (cassette housing may be loose)
- ✔ Bad brush assembly (mode switch)

Loads Improperly

Loading improperly may include loading with an empty cassette and loading and ejecting slowly.

Possible Causes:

- ✔ Reel drive, idler plate, and cam gear
- ✔ Slipping capstan drive belt (after slight hesitation, unit ejects tape)
- ✔ Defective carriage assembly (tape may jam)
- ✔ Large electrolytic capacitors (1000 to 2200 μF) in the motor loading line (voltage drop to the loading motor – test)
- ✔ Defective 5-volt regulator IC in the power supply
- ✔ Large electrolytic capacitors in the power supply 14-volt line (VCR may also shut off before cassette has ejected)
- ✔ Defective loading motor (also may quit in middle of loading/ejecting) (*Fig. 9-5*)
- ✔ Leaky zener diode in 16-volt line
- ✔ Large filter capacitors (1000 to 3300 μF) in the power supply (check electrolytics with ESR meter)
- ✔ Tape-end/begin sensors and capacitors (ejects immediately) (*Fig. 9-6*)
- ✔ Defective leaf switch (then ejects)
- ✔ Defective diodes in the power section on the main PCB

Improper Loading – Symphonic VCR1381

Improper tape loading was a problem in a Symphonic VCR1381 TV/VCR. When the tape was inserted, the VCR shut off. After checking the loading motor operation and voltage measurements on the loading motor, the loading circuits appeared to be normal. Replacing the tape-end sensors corrected the problem.

Fig. 9-5. Mechanical top view

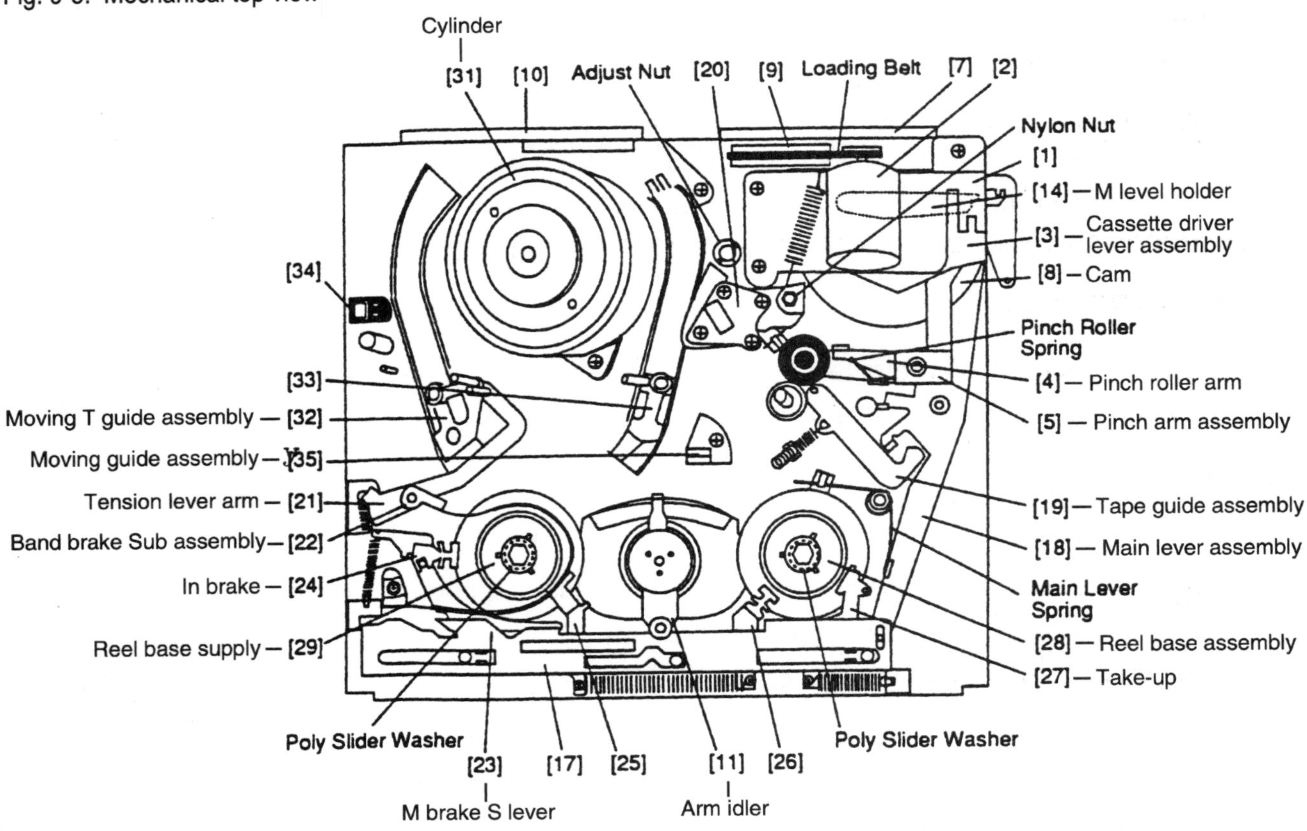

Fig. 9-6. Cassette carriage assembly in eject mode.

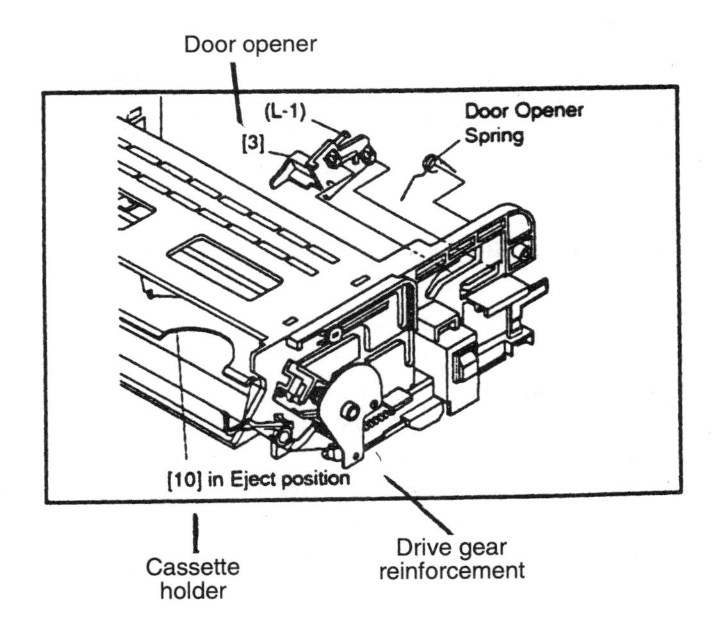

Tape Creases

Possible Cause:

- ✔ Stiff arm load rotation (clean and lubricate)

Cassette Tries to Load, VCR Powers Off

Possible Cause:

- ✔ Poorly soldered connections of transistor take-up sensor

Door Does Not Work

Possible Cause:

- ✔ Defective carriage assembly (with tape stuck inside)

Tape Arm Does Not Move

Possible Cause:

- ✔ Defective mode switch

Tape Light Does Not Shut Off

Possible Causes:

- ✔ Ring crack on the take-up of tape-end sensor (sensor tests okay)
- ✔ Bad supply side sensor (cassette does not load)

Cassette Loads, Ejects

Possible Causes:

- ✔ Defective or dirty leaf switch on top of the housing and a broken wire on the switch terminals (happens when no tape is inserted)
- ✔ Slipping capstan drive belt (tape loads improperly then ejects tapes after a slight hesitation)
- ✔ Leaf switch on the F/L PCB (clean it)

Load, Shut Down - Emerson VT0950

In an Emerson VT0950 TV/VCR the tape would load and then the VCR would shut down. Critical voltage tests were made on the servo control IC. Replacing IC2001 solved the shutdown problem.

In another TV/VCR of the same model, the VCR had mechanical problems loading. Replacing the rotary switch cured this problem.

- ✔ Bad connections on the ribbon cable (clean and reseat the cable)
- ✔ Bad tape end sensor
- ✔ Defective tape end sensor
- ✔ Transistor take-up sensor (powers up, loads empty tray, unloads, and rewinds)
- ✔ Mode select switch (readjust or replace)
- ✔ Badly soldered connections on (MPU) microprocessor pin terminals
- ✔ Defective 12-volt regulator IC (also audio/video tuner does not operate)
- ✔ Leaky capacitor on the cylinder motor drive PCB

Continually Loads or Ejects

Possible Cause:

- ✔ Bad cam gear on the right side of the chassis and cassette housing assembly
- ✔ Defective basket gear assembly

No Tape Indicator Light When Cassette Loads

Possible Cause:

- ✔ Defective cassette up/down switch (located below the carriage)

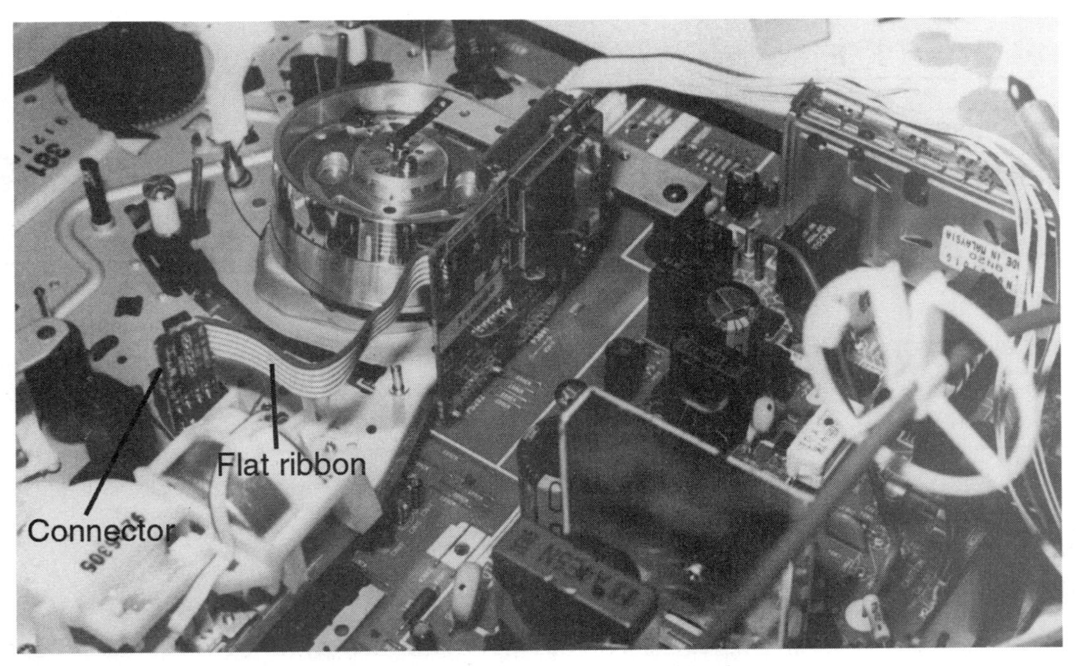

Fig. 9-7. Bad flat ribbon connections can cause cassette to sporadically eject the tape.

Intermittent Loading

Possible Cause:

✔ Mode switch (replace) and flat ribbon cable connector (clean)

Incomplete Loading, Ejects

Possible Causes:

✔ Gear assembly and F/L rack

✔ Gear assembly and cam

Randomly Ejects

Possible Cause:

✔ Damaged flat ribbon cable (*Fig. 9-7*)

Intermittent Loading – Panasonic Models

Sometimes on the Panasonic models, the cassette loads and other times it does not. By monitoring the loading motor voltage at the loading motor terminals, it was determined that there was not a constant voltage at pin 3 of IC6003 of the loading motor drive IC. The cassette would unload without any problems though.

When IC6003 was moved, the loading motor voltage was normal and then it was missing. The +14 volt DC was discharged at pin 6 to ground and a continuity measurement was made from pin 3 of the loading motor IC to R6075 with the ESR meter. It was a bad connection at pin 3 that caused the intermittent loading problem (*Fig. 9-8*). All nine pin terminals were resoldered and this solved the intermittent loading problem.

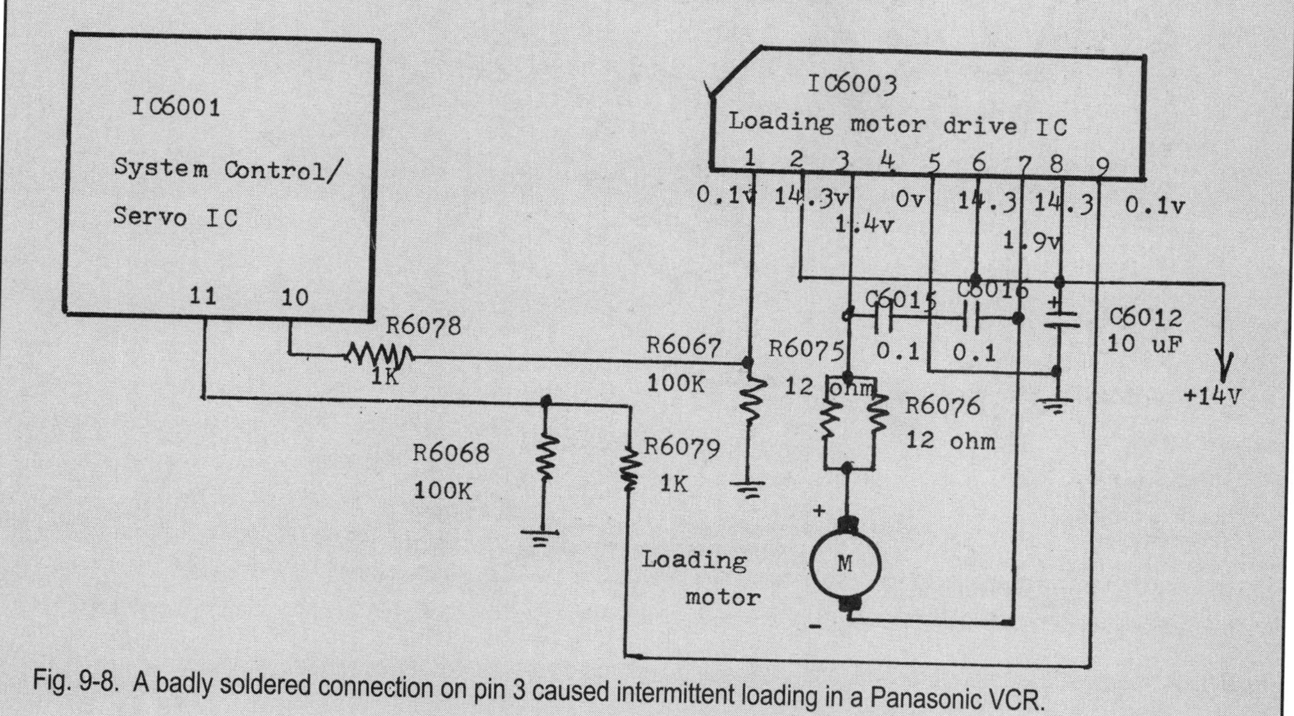

Fig. 9-8. A badly soldered connection on pin 3 caused intermittent loading in a Panasonic VCR.

Eject Function Eats Tape

Possible Cause:

- ✔ Clutch and soft brake assembly (decrease spring tension)

No Eject

Possible Causes:

- ✔ Sensor located on the bottom of assembly
- ✔ Eject gears and side plate (replace all, gears may have jumped from correct timing)
- ✔ Carriage assembly loading cam gears
- ✔ Defective or dirty loading switch
- ✔ Carriage assembly or lower side cam gear out of alignment
- ✔ Dirty or corroded eject switch (also may cause no rotation of loading motor - check the eject switch terminals for normal continuity with an ohmmeter or ESR meter)
- ✔ Bad plate assembly, broken tooth, and loading arm (*Fig. 9-9*)
- ✔ Capstan motor losing torque (dismantle and clean clutch assembly and increase the tension spring)
- ✔ Defective lift cam assembly

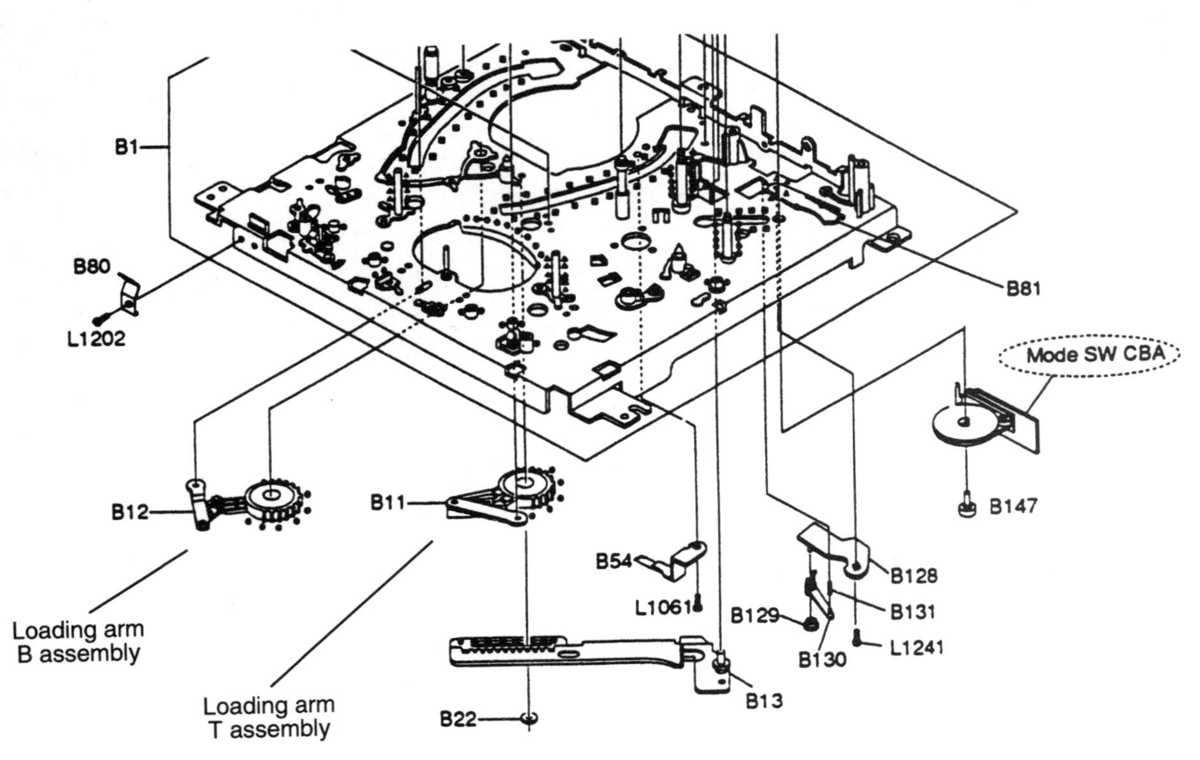

Fig. 9-9. A bad plate assembly, broken tooth and loading arm can cause the cassette to eject.

- ✔ Worn or loose loading belt
- ✔ Capstan drive belt slipping (hesitates, ejects immediately after loading)
- ✔ Defective basket gears
- ✔ Worn wheel unit
- ✔ Leaky diode
- ✔ Open or misaligned solenoid
- ✔ Defective lift cam assembly
- ✔ Loose ground connection at terminal deck board
- ✔ Defective loading motor IC
- ✔ Cracked function mode switch
- ✔ Bad loading motor and worm gear (Fig. 9-10)
- ✔ Bad mode motor (also no play, fast-forward, or rewind, but capstan motor rotates)
- ✔ Bad reel motor

Fig. 9-10. A defective loading motor and a worn drive belt can cause slow, random, and intermittent cassette loading as well as a no eject symptom.

Intermittent Eject

Possible Causes:

- ✔ Bad solenoid driver transistor on the main PCB (plays fine)
- ✔ Intermittent mode motor
- ✔ Defective pause switch
- ✔ Poor ground or loose connection at terminal deck board

- ✔ Defective clutch base assembly.
- ✔ Defective bracket right side assembly (also causes binding)
- ✔ Bad carriage assembly gear
- ✔ Defective safety switch (ejects immediately after loading)
- ✔ Pinch roller assembly (ejects during record mode) (Fig. 9-11)
- ✔ Tape-end sensors (ejects immediately after loading, fast-forward is erratic, and cassette carriage repeats loading motion)
- ✔ Inverter transistor in the power supply (ejects slow or jams and capstan motor has little torque)

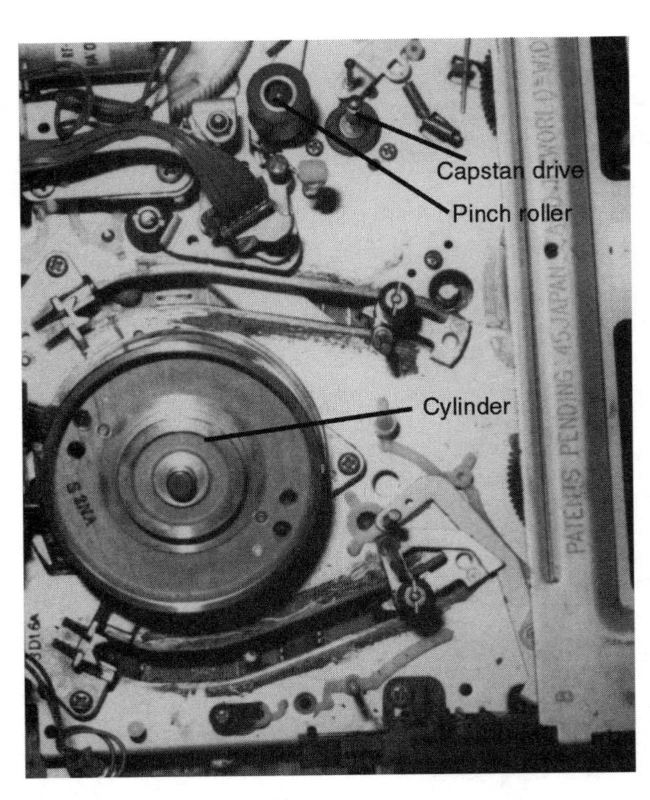

Fig. 9-11. Replace a pinch roller in the transport when the roller assembly ejects the tape during record mode (as shown in a Magnavox TV/VCR)

Ejects Tape and Shuts Down

Possible Causes:

- ✔ Dirty or bad leaf switch on the load basket assembly (clean it)
- ✔ Weak rack spring (only ejects half way - cut off and shorten old spring)
- ✔ Poorly soldered connections
- ✔ Bad head amp
- ✔ Blown fuse due to defective zener diode
- ✔ Open or burned 2.2-ohm resistors off the switching regulator IC (tape may jam and the unit powers up for two to three seconds and shuts off)
- ✔ Loading motor driver IC (also reel does not turn or eats tape)

Eating, Pulling, or Spilling Tape

When the tape is eaten or pulled, it is because the excess tape out of the cassette is wound around the various tape movement components.

Eats Tape

Possible Causes:

- ✔ Excess on the spring near the take-up reel (it may stick up – trim off excess)
- ✔ Cam stopper (shuts down)
- ✔ Stripped mode motor pulley
- ✔ Defective take-up clutch
- ✔ Binding brass motor bearing (also garbled audio and a noisy mechanism in playback mode)
- ✔ Defective transistor in the power section on the main PCB (with no SW +9 volts - power shuts off and play mode ejects)
- ✔ Defective left-end sensor (also does not stop at end of rewind mode)
- ✔ Defective supply reel brake (with a new idler, belts and pressure roller - replace felt or entire brake system)
- ✔ Clutch and high spring tension on the soft brake assembly (in eject mode – reduce tension) (Fig. 9-12)

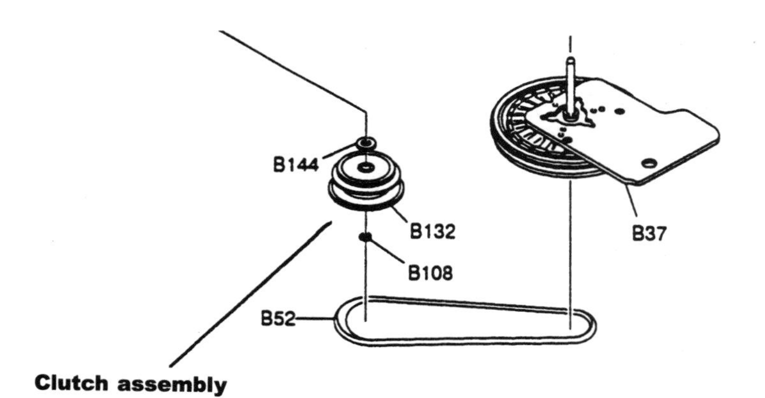

Fig. 9-12. Suspect a bad clutch assembly when the VCR eats or pulls the tape in eject mode.

- ✔ Tab on carriage assembly (rubbing sound in fast-forward and rewind modes - bend it out of the way)
- ✔ Bad base assembly
- ✔ Defective center bracket assembly
- ✔ Friction gear assembly and transmitting arm assembly
- ✔ Tension spring
- ✔ Bad function plate

- ✔ Defective pinch roller basket (also no audio or video in playback mode)
- ✔ Sticking reverse arm (no play, only fast-forward and record, eats tape in eject mode - clean and lubricate arm)
- ✔ Defective brake assembly (if problem is intermittent or if tape slacks from rewind to play mode)
- ✔ Broken idler gear spring (also makes a clicking noise)
- ✔ Take-up brake assembly (at eject or rewind)
- ✔ Bad reel motor driver IC

Pulls Tape

Possible Causes:

- ✔ Bad or sluggish reel motor (check the voltage, especially small-ohm resistors that may have increased in value)
- ✔ Take-up reel (rides up and binds, problem is intermittent)

Spills Tape

Possible Causes:

- ✔ Defective take-up reel or take-up pulley (does not rotate)
- ✔ Bad base assembly
- ✔ Defective reel motor IC and a poor ground (no reel rotation)

Tape Problems: Jams, Eats, Ejects at Shut Down

Eats Tape, Reel Does Not Turn, and VCR Shuts Down

Possible Cause:

- ✔ Defective loading motor IC

Eject Causes Shut Down

Possible Cause:

- ✔ Dirty leaf switch (clean)

No Eject, Tape Jams or VCR Eats Tape, Shuts Down

Possible Causes:

- ✔ Open or burned small resistors (2.2 ohm) in switching regulator circuits
- ✔ Stretched or excess spring near the take-up reel
- ✔ Defective cam stopper

Tape End Sensor - RCA TV/VCR

The VCR tore the tape from the cassette at the end of recording. This symptom indicated that the tape end sensor was not turning the VCR off or switching to stop mode when the leader end of the cassette was finished. A quick high-resistance measurement between the collector and emitter terminals of the tape end sensor (GT104) showed no leakage. +5 volts was measured on the collector terminal and very little voltage on pin 52 of IT001 (see *Fig. 9-13*). A resistance measurement between pin 52 and the tape sensor was open. Broken foil or bad component connections can also be located with an ESR meter. Simply resoldering the jumper connections solved this problem.

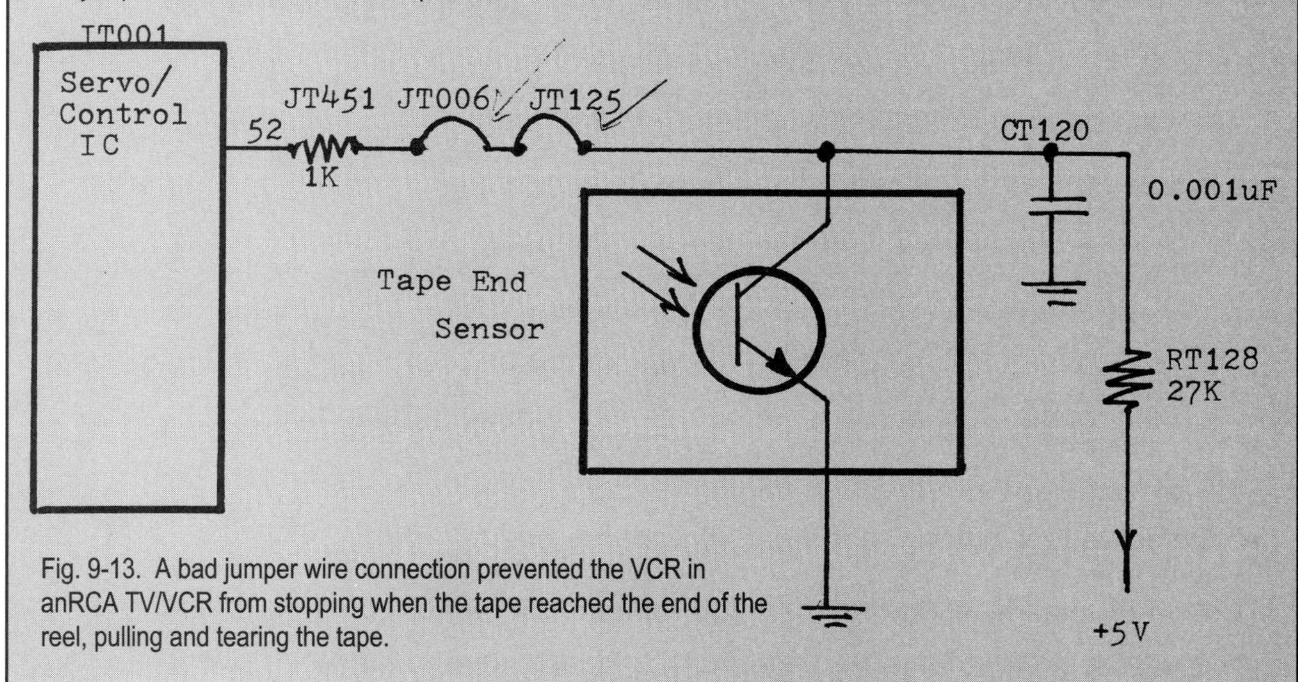

Fig. 9-13. A bad jumper wire connection prevented the VCR in anRCA TV/VCR from stopping when the tape reached the end of the reel, pulling and tearing the tape.

Eject Stops and Shuts Down

Possible Cause:

- ✔ Weak rack spring (*Fig. 9-14*)

Ejects Randomly and Shuts Down

Possible Cause:

- ✔ Bad solder connections at connectors and head amp

Loads, Tape Jams

Possible Causes:

- ✔ Defective right side plate in the carriage assembly
- ✔ Bad cassette holder

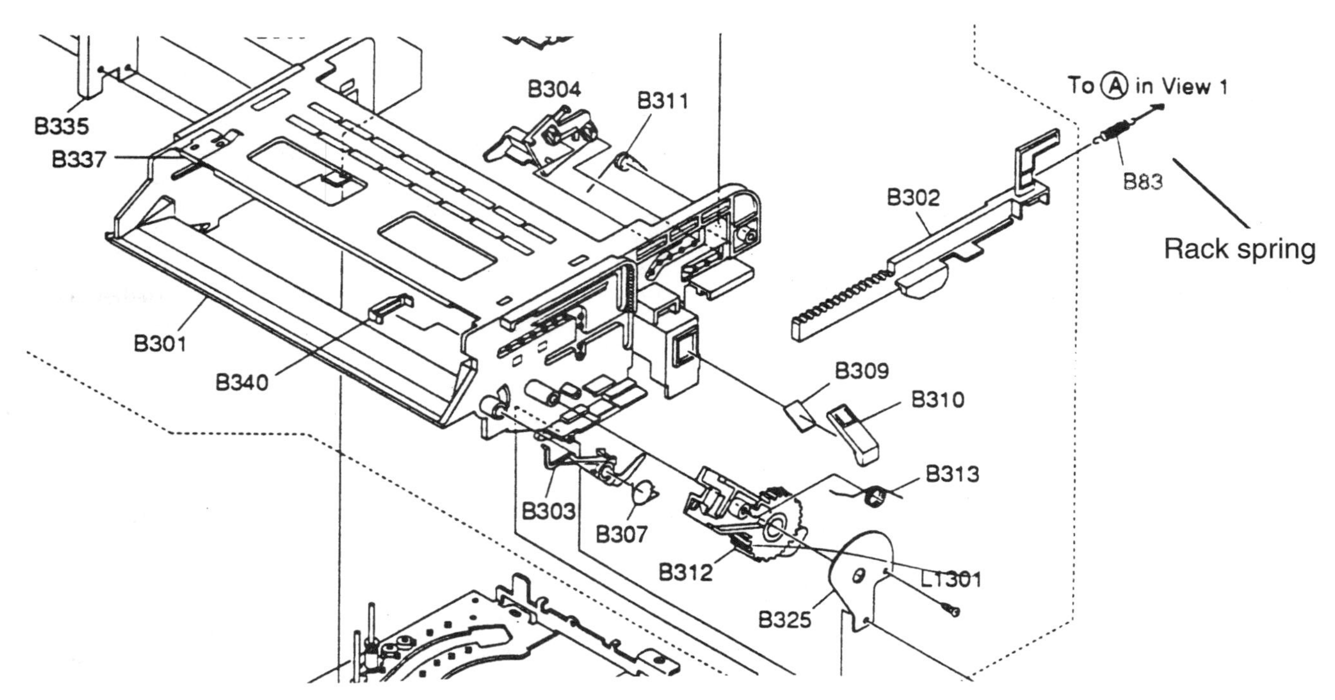

Fig. 9-14. Shorten the weak rack spring when eject stops half-way through and VCR shuts down.

No Eject, Tape Jams

Possible Causes:

- ✔ Bad tape-end sensor
- ✔ Mode motor out of position (realign, reburnish, and clean the mode switch)
- ✔ Defective motor driver IC in the Syscon circuit
- ✔ Bad plastic slide on the supply side of the housing
- ✔ Defective carriage assembly (tape may load improperly)
- ✔ Loading motor driver IC
- ✔ Bent pin to the right side of capstan motor shaft
- ✔ Bad 2.2- or 3.9-ohm resistor in 12-volt power voltage line
- ✔ Leaky or shorted capacitors in the loading motor circuits (motor cycles)
- ✔ Defective reel motor
- ✔ Eject drive gear (replace)

Blown Fuse, Tape Jams

Possible Cause:

- ✔ Rubber belts, tires, cam gear and slant guide tracks (clean and re-lubricate)

Tape Jams, Tape Will Not Load

Possible Cause:

- ✔ Bent reverse pin (near the capstan shaft)

Tape Jams, VCR Shuts Down

Possible Cause:

- ✔ Open resistor in power source caused by a leaky diode (connect the ESR meter to all large electrolytics (1000-2200 μF) on the CTL board with tape still in VCR)

Tape Jams, No Power Up

Possible Cause:

- ✔ Play and Fast-forward tactile switches (stuck in on-position – clean or replace)

Tape Jams

Possible Causes:

- ✔ Bad mode switch
- ✔ Bad reel motor
- ✔ Capstan motor (check for a frozen motor bearing)

Tape Jams at Eject

Possible Causes:

- ✔ Supply reel (capstan motor may also be slow)

Tape Jams, No Play, Fast-forward, Rewind, and Eject

Possible Causes:

- ✔ 5-volt switching transistor

Tape Jams, No Play, Fast-forward, and Eject

Possible Cause:

- ✔ Defective brake solenoid transistor

Intermittent Shutdown – Sylvania VCT190

A Sylvania VCT190 TV/VCR intermittently shutdown in play mode and record mode. Simply cleaning the surface reel reflector solved the intermittent shutdown problem.

Tape Jams, No Play and Rewind, VCR Shuts Down

Possible Cause:

✔ SMD diode off of the on/off +9 volt IC

Tape Jams, No Display

Possible Cause:

✔ Defective transistor regulator in power supply

Tape Jams, No Play, Fast-forward, and Rewind

Possible Cause:

✔ 5-volt transistor in power supply

Mode Motor and Switch Problems

A defective or bad mode motor can cause several different problems in a VCR. For example, if the VCR is eating tape, a stripped gear or pulley in the mode motor may be the problem.

The mode motor is often bad if any of the following occur.

✔ Tape jams, no eject, no play, no fast-forward, and no rewind
✔ Intermittent cassette loading
✔ Ejects in play mode
✔ Shut down after loading
✔ Loading motor keeps rotating, it squeals, and VCR shuts off
✔ All functions are erratic
✔ Tape loads, no playback, and VCR shuts down in two to five seconds
✔ VCR shuts down with heat or vibration
✔ VCR powers up, supply reel rotates counter-clockwise with guides fully loaded and VCR shuts off
✔ No fast-forward or rewind, plays okay
✔ No playback, tape loads, and shuts down in 2 or 3 seconds.
✔ Tape runs too fast in play mode (may look like fast-forward search)
✔ Jumpy picture with poor video (lines in picture) and audio
✔ Tape pulled and eject makes grinding noise

Some part of the mode motor may be dirty or misaligned if any of the following symptoms occur.

- ✔ Intermittent cassette loading, no eject (also check the flat ribbon cable connector)
- ✔ Starts fast-forward mode after loading, no eject
- ✔ Incomplete loading, switches modes
- ✔ Fast-forward play mode with high-pitched, garbled sound
- ✔ Tape loads, no video playback, shuts down
- ✔ Switches from record to rewind and fast-forward by itself

Cassette–In Switch Problems

The Cassette-In Switch can cause many different problems in the loading circuits. The cassette-in switch is activated once the cassette has been entered into the VCR. It tells the large microprocessor that the cassette is seated within the VCR. A defective cassette-in switch can refuse to allow the cassette to go into a particular mode or might eject the tape immediately.

The cassette-in switch is usually located near the cassette left mechanism. No operator controls will function if the service-in switch is not activated.

A dirty or broken leaf cassette-in switch can cause erratic or no tape loading. Clean up the leaf switch and take a continuity measurement across the switch terminals with the ohmmeter or ESR meter. Replace a cassette-in switch with bad points or no continuity reading on the meter. The cassette might not load if there is a defective capstan in-switch transistor or if the cassette-in switch is open.

Chapter 10: Dead VCR Power Circuits

Dead VCR circuits can be caused by a variety of problems, including improper voltage or lack of voltage from the AC power supply. Although some TV/VCR units employ a separate power supply for the VCR circuits, most TV/VCR combinations have one power supply with several different voltage sources. The AC switching power supply has both hot and cold power supply circuits. Usually the hot source is in the input circuits and the cold voltage source is in the output circuits.

RCA TV/VCRs have a Switched Mode Power Supply (SMPS), which provides 33-, 17-, 9.5-, 6.2- and -5-volt source to the system control/servo sections, while a Panasonic TV/VCR switching power supply furnishes 44-, 14-, 5- and -30-volt sources. Most electronic technicians check the various voltage sources first in order to locate a dead circuit in TV/VCR combinations (*Fig. 10-1*).

Other causes of dead power circuits include a defective line voltage IC regulator, blown fuses, burned or open resistors, and open or loss of capacitance in filter capacitors. A defective line voltage IC regulator can also be the cause of a dead chassis, no power up, no functions, and shutdown problems. An open fuse might indicate an overloaded circuit or a defective component in the power supply, while a blown, blackened fuse can indicate

Fig. 10-1. Find the power supply section of the VCR in order to locate the defective component in the switching power supply.

a shorted component. Burned or open fusible resistors can indicate a shorted or leaky component with no voltage source. Electrolytic capacitors can be open, leaky, lose capacity, or have an ESR problem. An ESR meter can quickly locate the defective capacitor because it finds those with equivalent series resistance (ESR), while a regular capacitor tester might test the electrolytic as a normal filter or decoupling capacitor. So remember to keep the ESR meter handy and also a DMM. A quick in-circuit test of transistors with the diode test of the DMM can locate a defective transistor.

Electrolytic Problems

Shorted or leaky electrolytics in the power sources can destroy fuses, resistors, and diodes. When an electrolytic is suspected as the problem, check all filter and decoupling electrolytics in the power supply and voltage sources with the ESR meter.

Power Surge and Lightning Damage

A strong power line surge or a lightning strike can blow or open fuses, diodes, resistors, and semiconductors in the VCR. Sensitive microprocessors and system control IC's can be damaged by lightning, which causes a no power up-shutdown symptom. Suspect burned or lifted foil traces caused by excessive lightning damage in the power supply. Some other results of lightning damage include a dead VCR due to a cracked IC voltage regulator and burned connections and feed through wires. Power on switches and transistors can be damaged by lightning or strong power surges.

Dead Chassis

When a VCR chassis is dead, it usually indicates a chassis that has no functions.

Possible Causes:

- ✔ Defective system control and servo IC
- ✔ Defective line voltage regulator or IC transistor (*Fig. 10-2*)
- ✔ Open fusible resistors or fuses
- ✔ Broken or cracked foil on PC wiring (This is generally a result of dropping the unit.)
- ✔ Poorly soldered connections (An ESR meter is ideal for locating broken traces or foil wiring and connections.)
- ✔ Combination of several different defective IC's, transistors, and diodes in the power supply (no voltage source)
- ✔ Blown parts due to lightning or a power surge
- ✔ Shorted diode (no power but LED lights work)

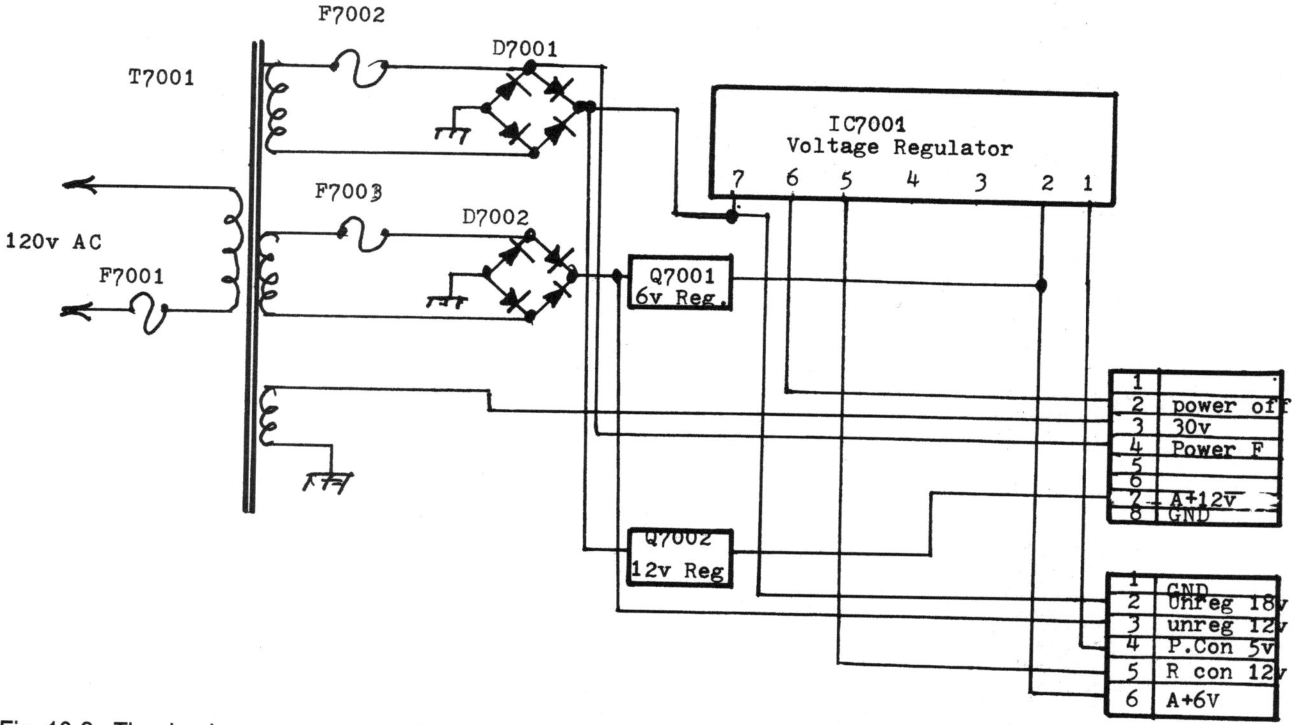

Fig. 10-2. The dead no power-up symptom can be caused by a defective line voltage regulator circuit.

Dead with No Power Up

Usually a dead chassis accompanied by another symptom that corresponds with the dead chassis, such as no power up or an open fuse and no functions, can provide a clue to the electronic technician, helping find the defective circuit or component. There are several causes of a dead-no power up symptom:

- ✔ Defective 12-volt regulator transistor switch in the power supply
- ✔ Shorted or open line voltage regulator IC
- ✔ Defective IC in the power supply (improper or no voltage source)
- ✔ Open fuse in the power supply
- ✔ 5-volt filter capacitor in the power supply
- ✔ Defective regulator transistor in the 5-volt source
- ✔ Shorted or leaky diodes in the power supply
- ✔ Low voltage supplied to the microprocessor or system control IC
- ✔ Defective buffer transistor for the SW IC (can also affect power up on the TV)
- ✔ Combination of several defective ICs and shorted transistors and diodes in the power supply
- ✔ Diodes in the power supply breaking down under load

In addition to not powering up at all, sometimes a unit will power up and then Shutdown or even come on by itself after appearing dead.

- ✔ Defective regulator IC or transistor (shuts down after three to five seconds with no functions)
- ✔ Shorted or defective diode in the power supply (comes on by itself and clock display flashes)
- ✔ 12-volt zener diode
- ✔ Defective mode switch (supply reel rotates counter-clockwise, guides are fully loaded, shuts down)
- ✔ Low value electrolytic in the power supply source (comes on by itself)

There are many other problems that can occur in conjunction with a dead chassis. These are a few others I have encountered and not yet discussed.

Dead, Intermittent Drum or Reel Spin

Possible Causes:

- ✔ Defective Syscon IC
- ✔ Shorted diode a low-voltage source with a defective electrolytic filter capacitor

Dead, Low Level Whining Noise

Possible Causes:

- ✔ Defective switching block and bad power transformer

Dead, No Power Up

Possible Causes:

- ✔ Defective tuner assembly (loads down the power supply)

Dead with Fuse Problems

Sometimes a fuse opens for no reason. Fuse holders can have poorly soldered connections. If a VCR goes dead immediately after power up, make sure to check all diodes and transistors in the power supply. This section addresses the causes of fuse problems.

Dead, Apparent Blown Fuse

Possible Causes:

- ✔ Open line fuse causing a dead transport (*Fig. 10-3*)
- ✔ Shorted diodes in the power supply (dead chassis, which blows small amp fuses)
- ✔ Leaky diode transistor and a secondarily derived defective diode in power supply

- ✔ Badly soldered connection on the 2-amp main fuse holder (the display and power shuts off)
- ✔ Loose fuse holder (intermittent TV/ VCR operation – it may appear to be dead some of the time)
- ✔ One or more defective transistors and small (10-22 μF) electrolytics (causes dead 1.6-amp blown fuse in the power supply)
- ✔ Failed switching regulator transistor and shorted over-voltage protector transistor
- ✔ A shorted transistor and open electrolytic in the power supply
- ✔ A badly soldered joint on the fullwave diodes (causes fuses to keep blowing)
- ✔ Hardened grease on the loading gear
- ✔ A jammed carriage, alignment of gears, and reset timing in the power supply
- ✔ A blown thermal fuse in the primary winding of a power transformer
- ✔ A defective electrolytic on the OSD PCB (causes repeated fuse failure with other parts testing okay)
- ✔ 47μF electrolytics in the power source (causes repeated failure of diodes in the power supply; check for ESR problems)
- ✔ Large filter capacitors in the 12- and 14-volt line (check with the ESR meter; you might find several bad electrolytics with ESR problems)

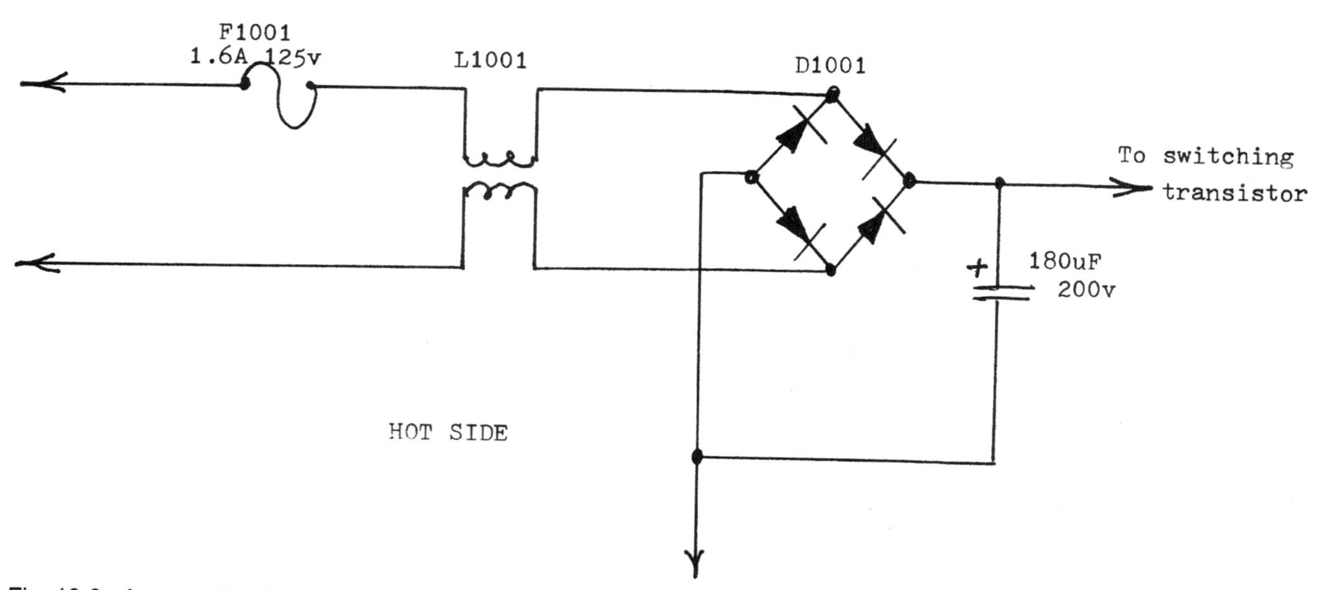

Fig. 10-3. An open line fuse (1.6A, 125V) can become open for no reason at all, which causes a dead transport.

Dead with the Fuse Okay

Always check all electrolytic capacitors in the power supply with the ESR meter when this appears to be the problem.

Possible Causes:

- ✔ Badly soldered joint on the fuse holder
- ✔ Burned, open isolation resistors in power line
- ✔ Defective voltage regulator IC
- ✔ Shorted diode in the power supply

Dead, Fuse Okay, No Eject

Possible Cause:

- ✔ Bad 6-volt regulator IC in main PCB

Dead, Fuse Okay, No Functions

Possible Cause:

- ✔ Defective voltage regulator IC in power supply

Dead, Fuse Okay, Shutdown

Possible Cause:

- ✔ Open low 2.2-6.8 ohm or fusible resistor in the power source

Dead, Fuse Okay, No Display

Possible Cause:

- ✔ Defective filter capacitor

Dead - No LEDs

Possible Causes:

- ✔ Leaky diode in the power supply (see *Fig. 10-4*)
- ✔ Defective power IC regulator (no power LED)
- ✔ Voltage to system control IC (LEDs come on then shut off – check all pins)

Dead - Orion TVCO900

The TV/VCR was dead and the fuses checked okay. After checking all the power supply components, a bad solder joint was found on one end of R501 (2.2 ohm) 7-watt resistor. The soldered joint was double-checked with the ESR meter on both leads of R501. Resoldering the bad resistor leads solved the dead TV/VCR problem.

- ✔ Defective 5-volt standby zener diode on main PCB (power LED stays on when power off is intermittent)
- ✔ Leaky 5-volt standby zener diode on main PCB (LED stays on with intermittent dead symptoms)
- ✔ Leaky voltage regulator IC (no clock display)
- ✔ Leaky diode in power supply
- ✔ Bad connections on transistors in power supply (no clock)
- ✔ Broken gears (clock only, no functions - realign with reset timing)
- ✔ Bad cam gear (no clock, partial loading of tape)
- ✔ Bad connections at the power transformer (no clock, no power)
- ✔ Open 0.22-0.33-ohm resistors in power sources (no power, no clock)

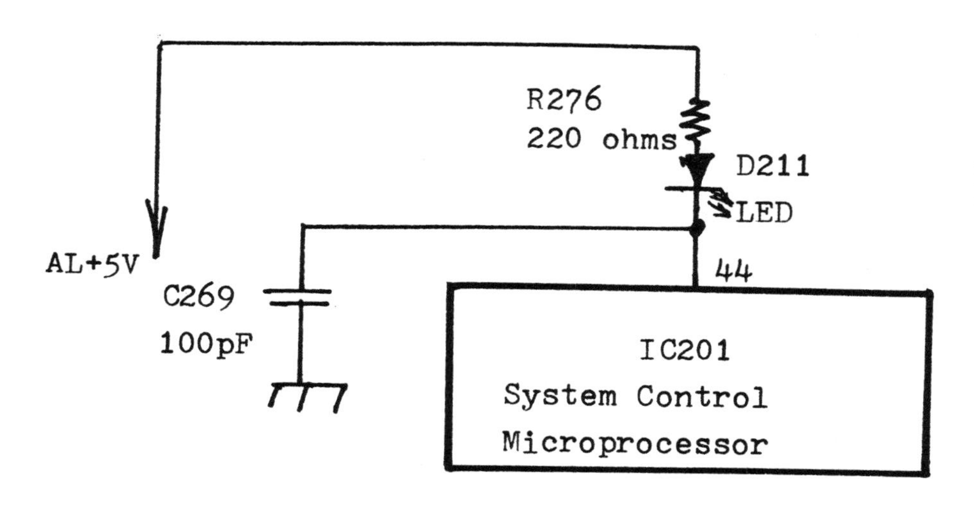

Fig. 10-4. A dead or lighted LED can indicate if the +5-volt line is dead or normal in the VCR section.

Dead with Display Problems

Sometimes the dead symptom, when accompanied by a no display or clock, indicates certain defective components in different circuits.

Dead, No Display

Possible Causes:

- ✔ Open electrolytics with excessive ripple (*Fig. 10-5*)
- ✔ Open fusible resistors in power supply
- ✔ Open or increase in resistance of a resistor on the AC side of the switching regulator
- ✔ Open or increased resistance of resistors off of the timer IC on PCB
- ✔ Open loading motor IC
- ✔ Open large electrolytic on the secondary side of power supply

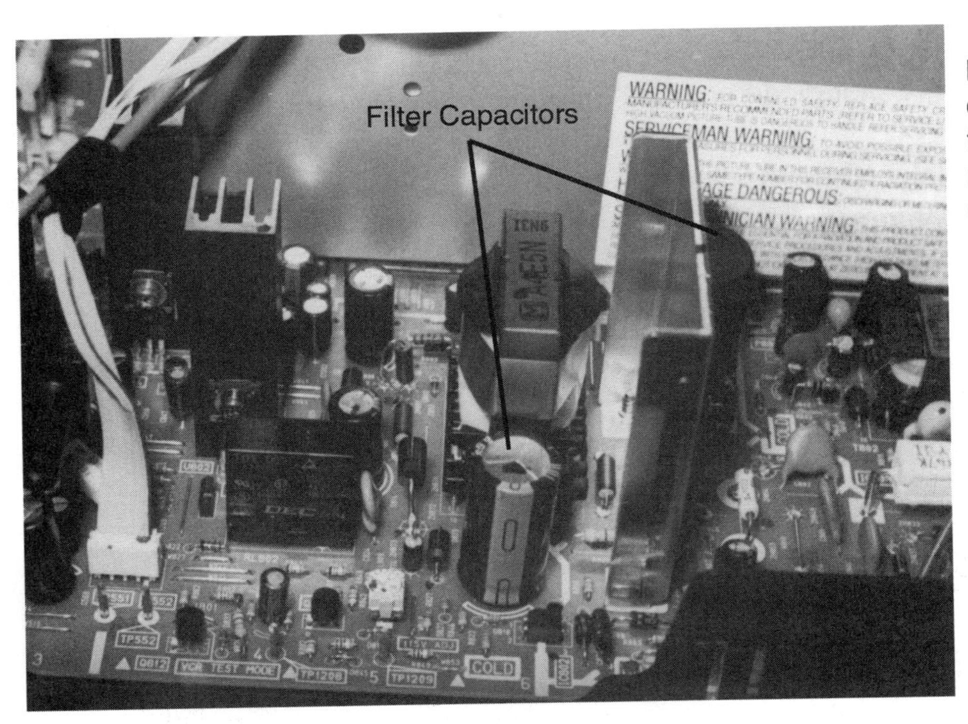

Fig. 10-5. Scope each filter capacitor in the power supply to determine if excessive ripple is present indicating a defective electrolytic.

- ✔ Open resistors and a bad 30-volt zener diode
- ✔ Small SMD 0.22-0.47 ohm 1/10 watt resistors
- ✔ Bad regulator unit
- ✔ Bad mode switch
- ✔ Leaky or open transistor on timer board (normal power supply voltages)
- ✔ Check 330-470 μF electrolytics in power supply

Dead, No Display, No Functions

Possible Causes:

- ✔ Open 6-volt regulator transistor or IC in power supply
- ✔ Defective zener diode on power supply secondary side
- ✔ Defective zener diode on main PCB next to supply sensor
- ✔ Defective 120 μF filter in power supply
- ✔ Bad 16 MHz crystal

No Power Up, LEDs Okay – Emerson VT0950

The LEDs worked fine in an Emerson VT0950 TV/VCR, but there was no power. Diodes can be checked in-circuit with the diode test of the DMM or with a diode tester. Remove one terminal for accurate diode tests. After testing, it was determined that the no power up problem was caused by a leaky D408 diode.

Dead, No Power Up

Possible Cause:

- ✔ Open resistor in the power supply or voltage source (cause by a shorted or leaky capacitor)

Dead, No Power, Dark Display

Possible Causes:

- ✔ IC1 on display board
- ✔ Large 3300-4700 μF power supply filter capacitor (*Fig. 10-6*)

Fig. 10-6. A dead VCR with no functions can be caused by open or dried-up electrolytics in the power supply.

Dead, No Display, Power Supply Voltages Okay

Possible Causes:

- ✔ Voltage on reset IC at the timer PCB
- ✔ Large electrolytic in voltage source

Dead, No Display, Squeal

Possible Cause:

- ✔ Large 1000-2000 μF electrolytic on the SW+5 volt line

Dead, No Display, No Loading

Possible Causes:

- ✔ Increase in resistance of a 1-megohm resistor on the primary side of power supply
- ✔ Shorted bypass capacitor 0.47-0.1 μF on the 5-volt line in power supply

Dead, No Display, Cylinder Motor Runs Continually

Possible Cause:

- ✔ Shorted diode and the 1000 μF electrolytic in power supply

Dead, No Display, Oscillating Power Supply

Possible Cause:

- ✔ Large 820 μF electrolytic on secondary side of power supply

Dead, Flashing Display, No Power Up

Possible Cause:

- ✔ 4.19 MHz crystal on timer PCB

Dead, No Functions

No functions with a dead VCR transport means no functions of any type are working.

Possible Causes:

- ✔ Sensor lamp (sometimes dew light stays on)
- ✔ Reset mechanical mode switch
- ✔ Bad mode switch and broken gears on shaft connection (motor runs, unit turns itself off)
- ✔ Defective side switch on the right side of the carriage assembly (starts with loss of function intermittently in play or rewind)
- ✔ Leaky diode on 5-volt line
- ✔ Leaky zener diode in the power supply

No Functions, Power On/Off Remote Okay

Possible Cause:

- ✔ Bad microprocessor IC (check for open connections with ESR meter)
- ✔ If dew light flashes, check for open connections around the system control IC using the ESR meter.

Dead, No Functions, No Power Up

Possible Causes:

- ✔ Play and fast-forward tactile switches stuck in ON position
- ✔ Bad 32.768 kHz crystal

Dead, No Functions, No Display, Tape Stuck

Possible Cause:

✔ Open 1.6-amp fuse

Intermittent or Erratic No Function Symptom

A bad sensor can cause intermittent loss of functions or no functions (*Fig. 10-7*). Clean up the cassette door switch on the main PCB of carriage when all functions are erratic. Intermittent loss of functions can also be caused by a leaky 5.6-volt zener diode.

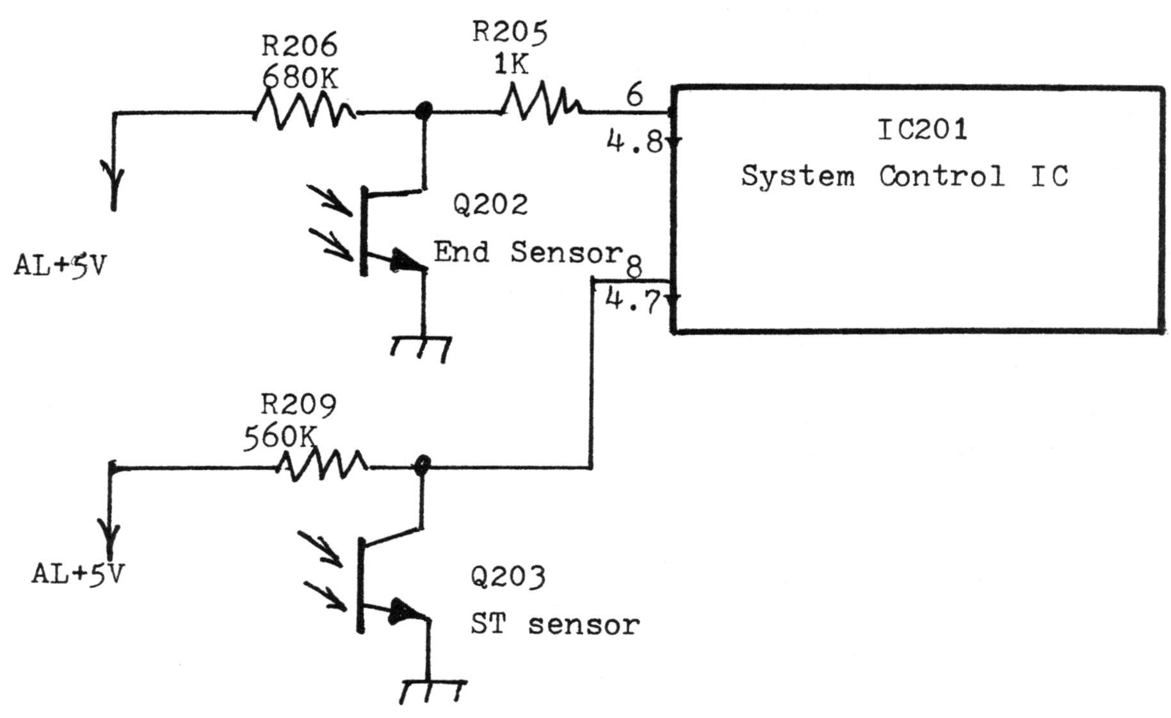

Fig. 10-7. Defective sensors can cause intermittent functions or no functions.

Bad Tape-End Sensor and Sensor Lamps

There are several ways in which a bad sensor can negatively affect VCR operation. Defective sensor lamps can cause loss of functions and no loading. A bad sensor lamp between the reels causes loss of all functions except eject. An open sensor lamp can cause a no function symptom with nothing else operating. A bad reel sensor might cause intermittent loss of function with no counter action. An open sensor lamp can cause all functions to not operate. When there is no cassette loading, the unit powers up, but does not accept tape, it can be caused by a defective sensor lamp. Infrared LED lamps can be tested with an infrared strip tester or an infrared meter. Replace failed sensors with the exact manufacturers part number match.

Power Supply Voltage

No Power Up, Power Supply Voltages Okay

Possible Cause:

✔ 4.19 or 4 MHz crystal can cause a no power up (dots flash in the display) (*Fig. 10-8*)

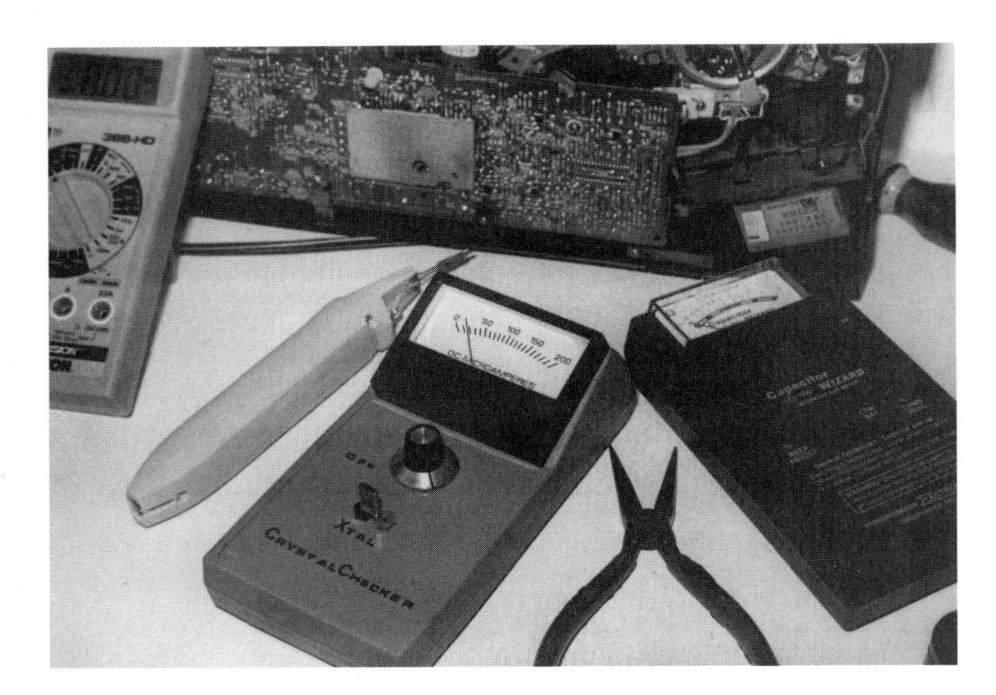

Fig. 10-8. Check the suspected crystal in the crystal checker when there is no power up and the power supply voltages are okay.

No Power Up, No Display, No Eject, Power Supply Voltages Okay

Possible Cause:

✔ Badly soldered connections on main timer board

Dead, No Display, Power Supply Voltages Okay

Possible Cause:

✔ Reset IC

✔ Loading motor does not reset mode switch (defective loading motor drive IC)

Erratic Loading, No Rewind – Orion TVCPO900

The loading was quite erratic and there was no rewind mode function in an Orion TVCPO900 TV/VCR. The VCR had mechanical problems - replacing the rotary switch solved the erratic loading and rewind symptom.

Dead, No Display, No Functions, Power Supply Voltages Okay

Possible Causes:

- ✔ Leaky diode on 5-volt line to the Syscon CTL IC
- ✔ Defective diode on the main PCB

Problems with Power Up

Random Power Up, Display Flashes

Possible Causes:

- ✔ Defective diode in power supply

Powers Up, Shuts Down

Possible Causes:

- ✔ Leaky 12-volt zener diode
- ✔ Shorted capacitor off of servo IC
- ✔ Leaky IC line regulator (shuts down after three to five seconds)
- ✔ Shorted capacitor at the ON switch of the 5-volt line

Tape Loads then VCR Shuts Down – Emerson VT-1920

The tape would load and then the unit would shutdown in an Emerson VT-1920 TV/VCR. Critical voltage measurements were made at the servo control IC2001. The supply voltage at pin 21 (VCC) was very low - 2.7 volts. Because IC2001 controls the capstan and cylinder motor circuits with a low supply voltage, IC2001 was replaced, which solved the shutdown problem (*Fig. 10-9*). In another TV/VCR of the same model and same symptoms, IC2001 was defective and replaced.

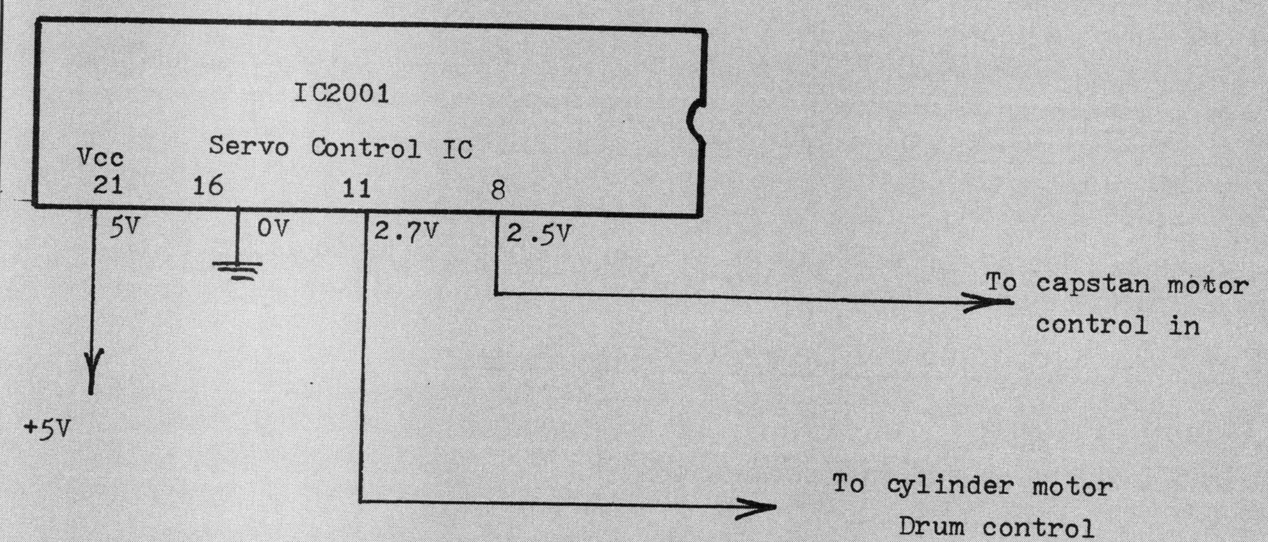

Fig. 10-9. Suspect a servo IC2001 in an Emerson VT-1920 TV/VCR when the tepe loads then the VCR shuts down.

- ✔ Power line IC regulator
- ✔ Leaky capacitor off 5-volt line in power supply
- ✔ Defective 12-volt zener diode
- ✔ Bad voltage line regular IC
- ✔ Leaky or shorted bypass capacitor in 5-volt line.
- ✔ Bad tape-end sensor (shuts down after four or five seconds)

Powers Up to Full Load then Shuts Down

Possible Causes:

- ✔ Leaky diode off of a system control IC
- ✔ Bad mode switch.

Powers Up, Loads Empty Tray

Possible Causes:

- ✔ Bad LED on the mode PCB and under the prism
- ✔ Defective tape-end sensor
- ✔ Defective error voltage detector transistor (fast-forward display is lit)
- ✔ Loading motor IC (might appear to be a bad sensor)

Powers Up, Empty Reel Rotates Counter Clockwise, Shuts Down

Possible Cause:

- ✔ Bad mode switch (*Fig. 10-10*)

Powers Up, Cylinder Spins, Squeals

Possible Cause:

- ✔ Increased resistance of a resistor off of an open base transistor in power supply

Powers Up, Tape Loads, Unloads, Rewinds

Possible Cause:

- ✔ Badly soldered connection on the take-up sensor transistor on deck of PCB (resolder all connections)

No Power Up, Display Okay, No Functions

Possible Cause:

- ✔ 220-330 µF electrolytic in power supply

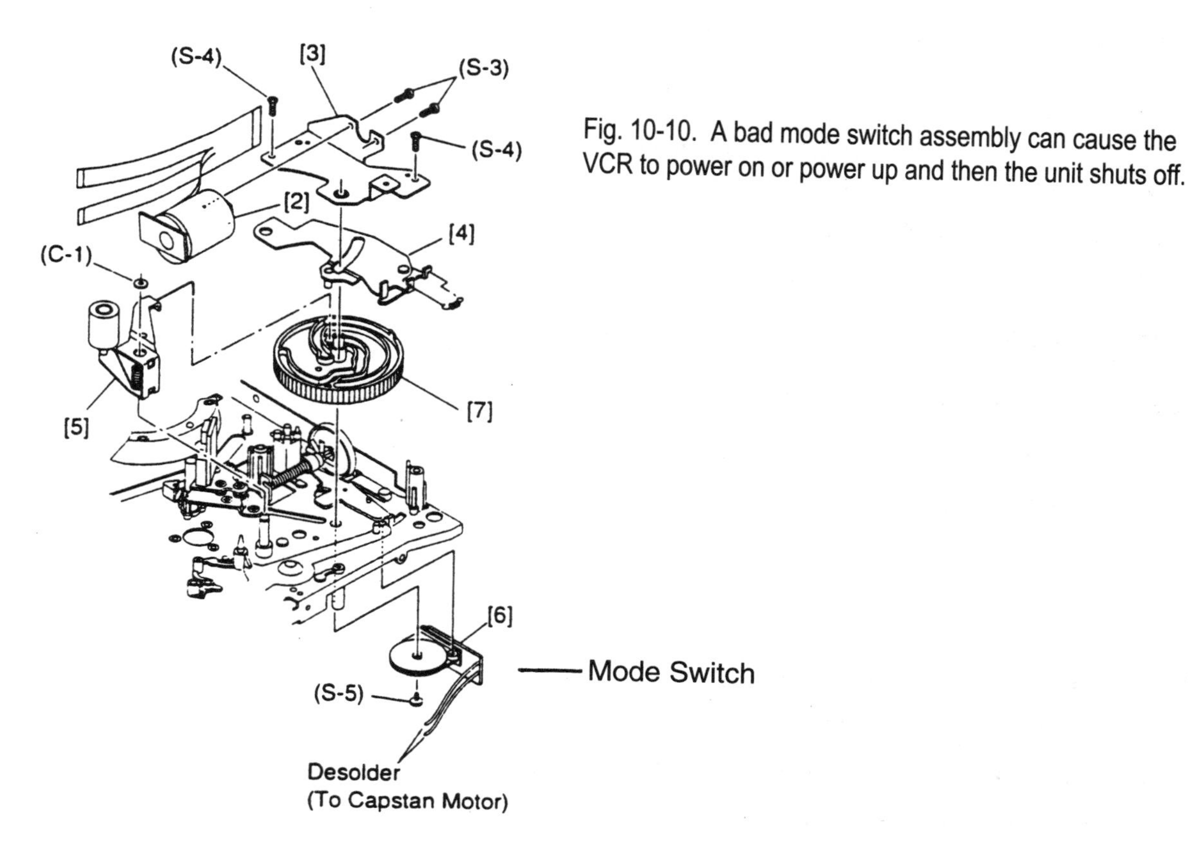

Fig. 10-10. A bad mode switch assembly can cause the VCR to power on or power up and then the unit shuts off.

No Power Up, No Clock Display

Possible Causes:

- ✔ Bad connection on the power transformer
- ✔ Shorted power supply IC

No Power Up

Possible Causes:

- ✔ Defective microprocessor or system control IC
- ✔ Bad diode in power supply (when first turned on, low voltage in the 5-volt line)

Powers Up, Loads Empty Tray, Shuts Down

Possible Causes:

- ✔ Bad cassette up and down switch on deck of PCB (replace)
- ✔ Leaky error voltage transistor detector on main PCB
- ✔ Defective cassette up and down switch on deck PCB

Powers Up, Loading Motor Runs, VCR Shuts Down

Possible Causes:

- ✔ Bad CN connectors and socket connections
- ✔ Poorly soldered connection on cable connector and socket

Powers Up, Supply Reel Rotates Momentarily, Loading Motor Runs Continuously

Possible Cause:

- ✔ Bad cam switch assembly

Intermittent Power Up

Possible Cause:

- ✔ Badly soldered connections on IC1

Clicking Sounds and Capstan Jerks – Panasonic PM-2028

The Panasonic PM-2028 TV/VCR has one power supply for the VCR section and a different power supply for the TV section. Suspect an input filter capacitor when a clicking sound is heard as the tape is inserted and the capstan begins to jerk and become erratic.

After turning the set off, discharge the electrolytic with a test lead across the capacitor terminals. Jumper clip another 120 to 200 μF at 200 volts across the old capacitor terminals to see if the symptom clears up. If it does, replace the capacitor. Replace C1004 (120 μF, 200 volt) electrolytic filter capacitor when these symptoms occur (*Fig. 10-11*).

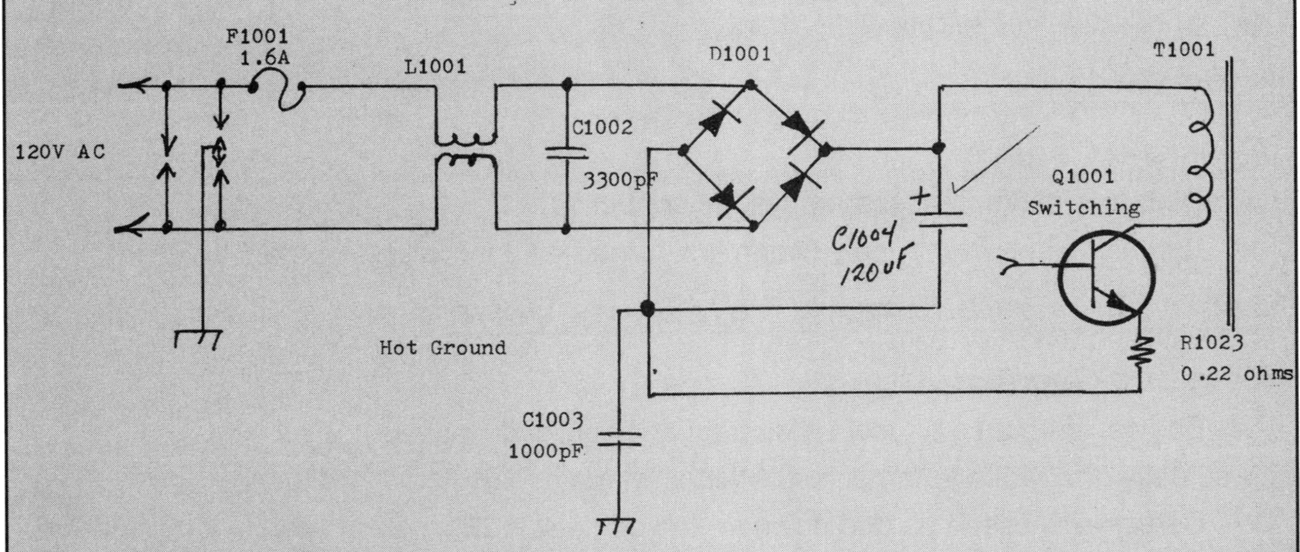

Fig. 10-11. Replace filter C1004 (120μF, 200V) in VCR power supply for a clicking noise and a jerky capstan motor.

Intermittent Power Up, No Clock

Possible Cause:

- ✔ Bad voltage regulator

Intermittent Power Up, No Functions

Possible Cause:

- ✔ Defective sensors

Intermittent Power Up, Display Flashes

Possible Cause:

- ✔ Open 10-22 μF electrolytics in power supply (check with ESR meter)

No Power Up, Clock Display On

Possible Cause:

- ✔ Defective reset transistor (check for oscillations on microprocessor)

No Power Up, No Clock Display

Possible Cause:

- ✔ Badly soldered joint at the power transformer

No Power Up, Display On, No Functions

Possible Causes:

- ✔ 4-MHz resonator
- ✔ Bad power supply filter capacitor in the 5-volt source

No Power Up, No Display, No Eject

Possible Cause:

- ✔ Poorly soldered connections on timer PCB

No Power Up, No Display, Main Power Supply Okay

Possible Cause:

- ✔ Open fuse on timer-tuner display board

No Power Up, Display Okay, Does Not Accept Tape

Possible Cause:

- ✔ Defective 100-330 μF electrolytic in power supply

No Power Up, Low Whine

Possible Cause:

- ✔ Entire switching block with bad power transformer (replace entire switching block and transformer)

Powers Up By Itself

Possible Cause:

- ✔ Defective diode in power supply

Powers Up, Fast Forwards, Rewinds, Shuts Down

Possible Cause:

- ✔ Defective Syscon IC

Shuts Down – Shuts Off

Tape Loads, Shuts Down

Possible Cause:

- ✔ Defective zener diode on servo/system control PCB

Powers Up, Shuts Off

Possible Cause:

- ✔ Defective line voltage regulator IC

Shuts Down in Stop Mode, No Eject

Possible Cause:

- ✔ Defective zener diode off of driver or system control IC

Shuts Down, No Eject

Possible Cause:

- ✔ Leaky diode in 12- or 14-volt line to loading driver IC

Shuts Off in Play Mode After Only a Few Seconds

Possible Cause:

- ✔ Photo coupler

Shuts Off After Four or Five Seconds

Possible Causes:

- ✔ Bad transistor regulator in 9-volt power supply
- ✔ Bad tape-end sensor

Shuts Down After One to Three Hours

Possible Cause:

- ✔ Defective capstan motor IC

Loads, Shuts Down

Possible Causes:

- ✔ 9-volt regulator IC
- ✔ Defective servo IC

Does Not Accept Tape, Shuts Down

Possible Cause:

- ✔ Loading motor IC and loading motor

Shuts Down Immediately

Possible Cause:

- ✔ Defective syscon and servo IC

Shuts Off After Five to Seven Seconds, No Loading

Possible Cause:

- ✔ Bad IC on the Syscon/Servo PCB

Plays, Shuts Down

Possible Causes:

- ✔ Bad bracket assembly
- ✔ Bad photo coupler

Shuts Off, No Display

Possible Cause:

- ✔ Defective power transformer

Intermittent Shutdown – Magnavox

Sometimes the VCR would play or record for a few minutes and then shut down. At other times, the unit might play for over one hour before shutting down. The tape end reel sensor was checked and appeared normal. Although the servo IC was suspected, all other parts were checked and cleared, since removing and replacing the servo IC is quite a lot of work.

The reel turntable was removed and the surface reflector was cleaned with alcohol and a cleaning stick. Excessive dirt and smoke can cloud up the reflected surface and cause the VCR to shutdown. The reflector reel pulses were not reaching Pin 40 of the servo microprocessor. Cleaning up the reflected surface solved the shutdown problem.

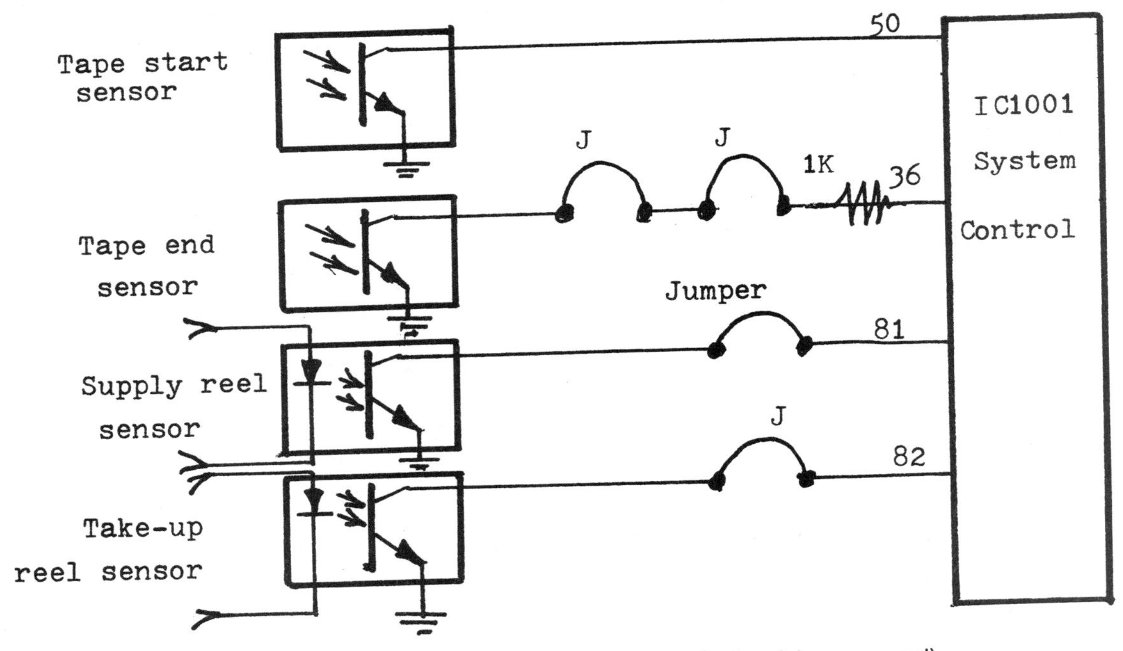

Fig. 10-12. The take-up reel with no reel pulse can shutdown during any fuction (play or record)

Shuts Down in Any Function

Possible Cause:

- ✔ Bad take up reel sensor (no reel pulse) (*Fig. 10-12*)

Powers Up, Shuts Down

Possible Cause:

- ✔ Bad cylinder motor assembly

Shuts Off When Eject Button Pressed

Possible Cause:

- ✔ Dirty or defective leaf switch on load basket

Tape Begins to Load, Shuts Down

Possible Cause:

- ✔ Defective motor block assembly (loading motor and mode switch)

Shuts Down, Heat or Vibration

Possible Cause:

- ✔ Bad mode switch assembly

Shuts Down After 30 to 60 Minutes

Possible Cause:

- ✔ Bad large (3300 to 4700 μF) electrolytics in power supply

Voltage Regulator Problems

A bad line voltage regulator IC can cause many different service symptoms including dead, intermittently functioning, and defective loading circuits.

Problems that can be caused by a defective, shorted, or leaky line voltage regulator IC include:

- ✔ Dead, no power operation and/or no functions
- ✔ Dead, no display, no loading, and no play mode

- ✔ No power or intermittent power, no clock (*Fig. 10-13*)
- ✔ Shuts down in three or four seconds, no functions
- ✔ Dead, no clock
- ✔ Dead, no functions, fuses okay
- ✔ No LEDs
- ✔ Dead
- ✔ Dead with no power lights
- ✔ Dead, no power up, but display lights up
- ✔ Shutdown, no drum rotation, no eject, intermittent rewind and fast-forward
- ✔ Shutdown 5 to 15 seconds after power up

Fig. 10-13. A defective voltage regulator can cause no power or intermittent power.

Chapter 11: Servicing VCR Shutdown and Noise Problems

Shutdown and noise problems in VCR units can have many causes. Among them are failure to power up, shutting down, shutting down after the tape has loaded halfway, and intermittent shutdown.

A defective component in the power supply is one possible cause if the VCR fails to power up or shuts down, while a defective rotating component can be the cause if intermittent shutdown occurs. Poorly soldered connections are another cause to consider. A defective hot component can cause the VCR to shut down after operating for several hours. The drum or cylinder can load and then shut down. A defective capstan motor can cause a shutdown problem within play or record mode, while a defective clutch assembly may cause the VCR to shut down in play mode (Fig. *11-1*). This chapter covers these and other symptoms and the possible causes.

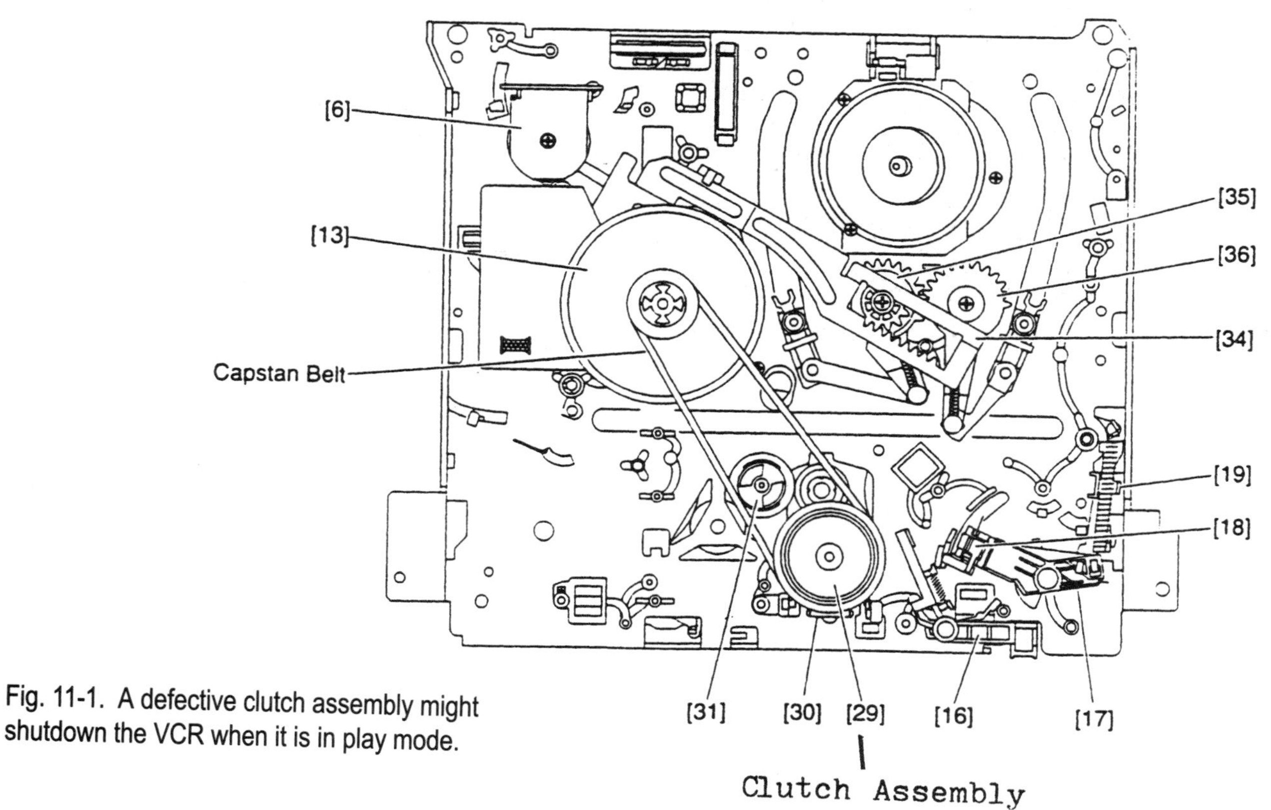

Fig. 11-1. A defective clutch assembly might shutdown the VCR when it is in play mode.

Powers Up then Shuts Down

Possible Causes:

- ✔ Defective cam switch assembly
- ✔ Bad cassette up-and-down switch on the deck of PCB
- ✔ Defective mechanism position switch
- ✔ Bad mode switch
- ✔ Bad tape end sensor. Replace the sensor when VCR powers off after three to five seconds.
- ✔ Defective bracket assembly
- ✔ Defective 1000 to 3300 μF electrolytic in the power supply
- ✔ Bad voltage regulator (unit powers up, shuts off after three to five seconds with no functions operating)
- ✔ Bad voltage regulator (may not power up, power goes dead with a no clock symptom)
- ✔ Leaky diode in the power supply (clock continues to flash)
- ✔ Defective 12-volt zener diode
- ✔ Defective LED on the mode PCB
- ✔ Defective zener diode on the servo/system control PCB
- ✔ Defective start and stop transistors
- ✔ Defective transistor in the SW5 volt line on the main PCB
- ✔ Defective loading motor IC
- ✔ Defective syscon IC when the unit powers up, goes into fast forward and rewind, shuts down

Tape Loads then VCR Shuts Down

Possible Causes:

- ✔ Badly soldered joint on the tape end sensor (tape may or may not begin to load)
- ✔ Open 10-ohm resistor caused by a shorted diode in that voltage source (tape loads and becomes stuck inside, the power shuts off in several seconds, with no UNSW+48-volt source)
- ✔ Bad leaf switch (clean the leaf switch on the load basket when eject mode causes the power to shut off)
- ✔ Bad mode switch (VCR cycles back and forth, will not take tape)
- ✔ Defective 9-volt regulator IC (tape loads, power light is lit, VCR shuts down)
- ✔ Defective loading motor IC
- ✔ Defective servo IC (may look like a bad mode switch)
- ✔ Defective system control IC

- ✔ Burned or open 2.2-ohm resistors in the switching regulator IC circuits (VCR powers up, loads, tape becomes stuck, no eject, and the VCR powers off in two or three seconds)
- ✔ Bad mode switch (the loading motor keeps rotating, squeals, then shuts off; VCR shuts down after tape loading in play mode)
- ✔ Defective loading motor (tape loads, heads spins, and after a few seconds, VCR shuts off) (*Fig. 11-2*).

Fig. 11-2. A defective loading motor can cause the tape to load and then shutdown after just a few seconds.

Intermittent Loading then Shutdown

Possible Causes:

- ✔ Bad 2.2 to 2.7-ohm resistor or a badly soldered connection of the resistor
- ✔ Missing 12 volts to the loading motor IC (tape becomes stuck, power shuts off in a few seconds with the display blinking)
- ✔ Defective loading motor driver IC
- ✔ Defective loading motor IC (VCR loads, shuts down in a few seconds, also eats tape as the reels will not turn)
- ✔ Intermittent transistors within the power supply (power shuts off intermittently under load)
- ✔ Defective loading motor
- ✔ Defective cam stopper (eats and spills out tape, no eject, and shuts down the VCR)

Starts to Play then Shuts Down

Possible Causes:

- ✔ Bad mode switch (tape starts to play, squeals, then shuts off)
- ✔ Replace the mode switch and all belts if the tape loads but there is no video playback and VCR shuts down in two to four seconds
- ✔ Defective loading motor IC (VCR shuts down in a few seconds, eats tape, reels will not rotate; or unit plays, display flickers, and shuts down).
- ✔ IC pulls extra current, runs hot when in play mode, eject leaves tape pulled out, display flashes, then shuts down (*Fig. 11-3*).

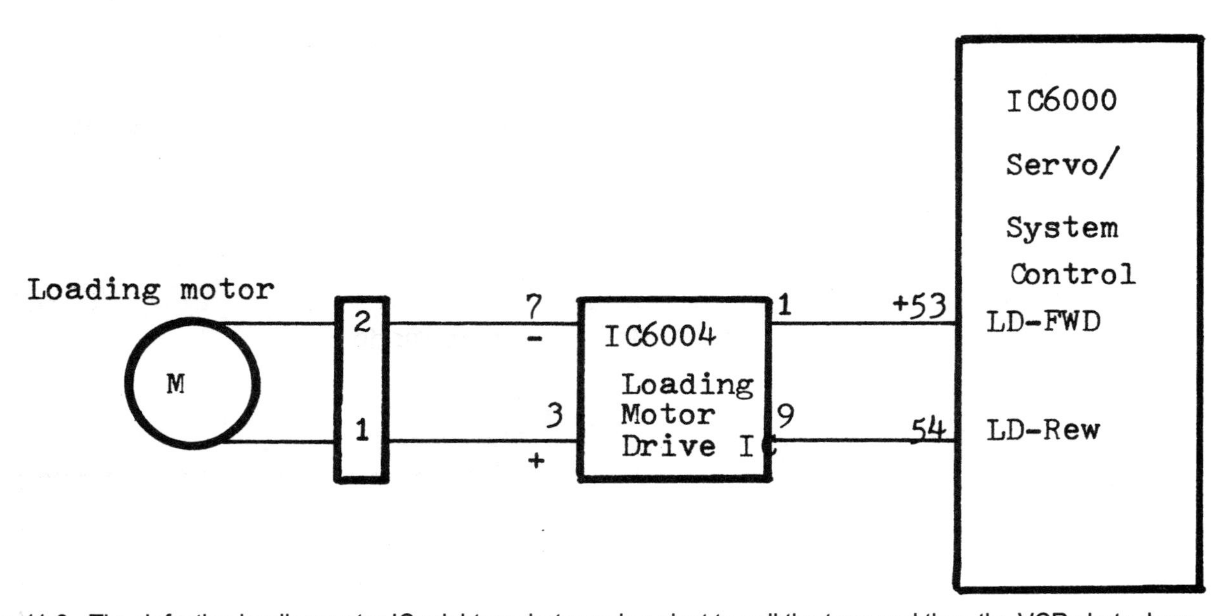

Fig. 11-3. The defective loading motor IC might run hot causing eject to pull the tape and then the VCR shuts down.

- ✔ A reset IC (play mode operates by itself, then shuts down)
- ✔ Defective diode in the system control IC circuits (VCR loads, plays video, powers off)
- ✔ Defective line voltage transistor or IC regulator (VCR shuts down in PB or REC mode)
- ✔ Defective bracket assembly
- ✔ Defective mechanism position switch
- ✔ Bad gear drive assembly
- ✔ Poor solder connections on the drum stators (play mode shuts down while recording)
- ✔ Take-up reel magnet (replace and re-glue if it falls off and causes play mode to shut down after a few seconds)

- ✔ Hardened grease on the pulley shaft or a dry shaft
- ✔ Bad capacitor in the regulator circuits (play mode shuts down while taping
- ✔ Reel sensor assembly (VCR plays for a few seconds, stops, shuts down)
- ✔ Replace the photo sensor under take up reel (VCR shuts down in play or rewind mode)
- ✔ Bad loading sensor (play mode starts, shuts down)

Play Mode Shutdown

Possible Causes:

- ✔ Hardened lubricant, dry pulley shaft (re-lubricate when play mode shuts down halfway through the tape movement)
- ✔ Bad mode switch
- ✔ Bad take-up reel sensor (replace if VCR shuts down in play, fast-forward, and rewind)
- ✔ Mode switch and all motor belts (replace when the tape loads, plays, and shuts down in a few seconds (*Fig. 11-4*)
- ✔ Defective bracket assembly
- ✔ Bad gear drive assembly
- ✔ Defective loading motor IC (play mode shuts down, eject leaves the tape pulled out, display flickers). Replace when shuts down in record mode.

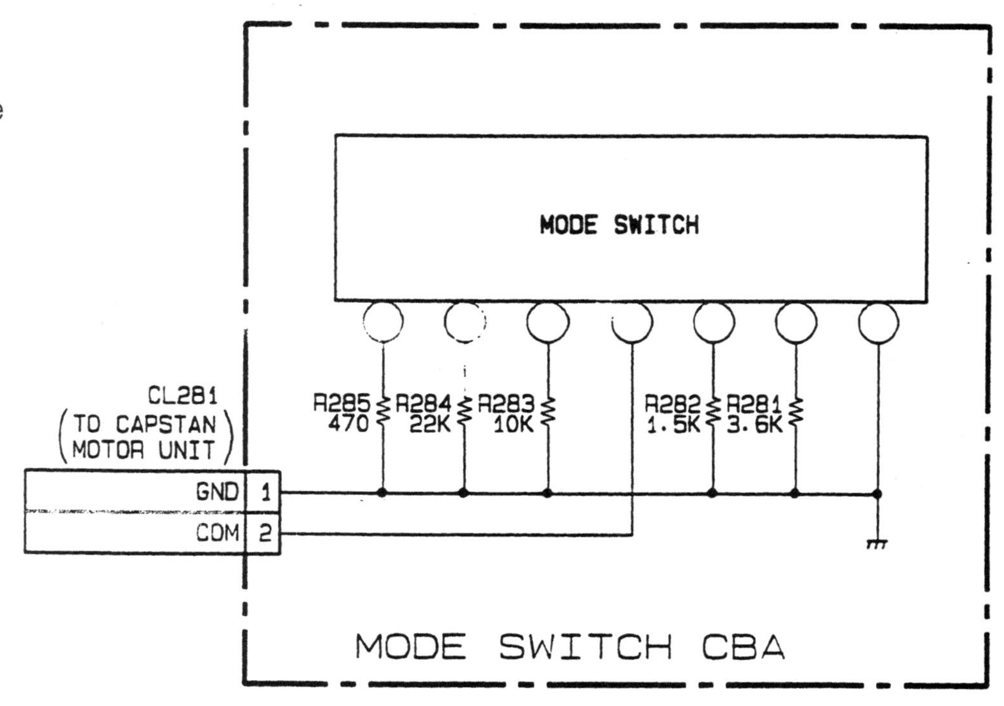

Fig. 11-4. You may have to replace a bad mode switch when the tape loads and then shuts down.

- ✔ Leaky diode in power supply with no UNSW+37 volts (no playback video, no tuner audio, no playback audio, unit shuts down)
- ✔ Open capacitors in the regulator section (play mode shuts down in record mode)
- ✔ Fusible resistors in the main power PCB (audio okay, no play mode, shuts off in three or four seconds)

Intermittent Shutdown in Play or Record

Possible Causes:

- ✔ Bad idler assembly (VCR shuts down after running in play mode for awhile, reel motor loses torque)
- ✔ Bad mode switch (intermittent tape loading, shuts down in two to four seconds, no playback video)
- ✔ Bad gear drive assembly VCR starts to play, then shuts down)
- ✔ Bad reel clutch (play mode goes intermittently into stop mode)
- ✔ Bad diodes in voltage source (capstan rotation intermittent, stops in play/record mode)
- ✔ Open zener diode in B+ voltage source (play mode intermittently goes to stop after five minutes

Plays and then Stops

Possible Causes:

- ✔ Defective mechanical position switch
- ✔ Replace the reel photo sensor (VCR stops after a few minutes up to one hour)
- ✔ Bad reel clutch assembly
- ✔ Replace the idler plate assembly (no eject, take-up reel does not turn, stops in play and record mode) (*Fig. 11-5*).
- ✔ Bad zener diode (6-volt line raises in voltage, play goes to stop in five minutes)
- ✔ Leaky zener diodes
- ✔ Bad transistor in microprocessor system (take-up reel stops after 8 or 10 seconds, guides retract, rewind okay)
- ✔ Defective voltage regulator (VCR shuts down in playback or record mode)
- ✔ Replace the photo coupler (any mode goes into stop in five seconds or less)
- ✔ SMD transistor reel pulse amp (play goes into stop in a few seconds)
- ✔ Bad wave shaper IC on the main deck PCB (stops in five or six seconds in both EP and SP modes)

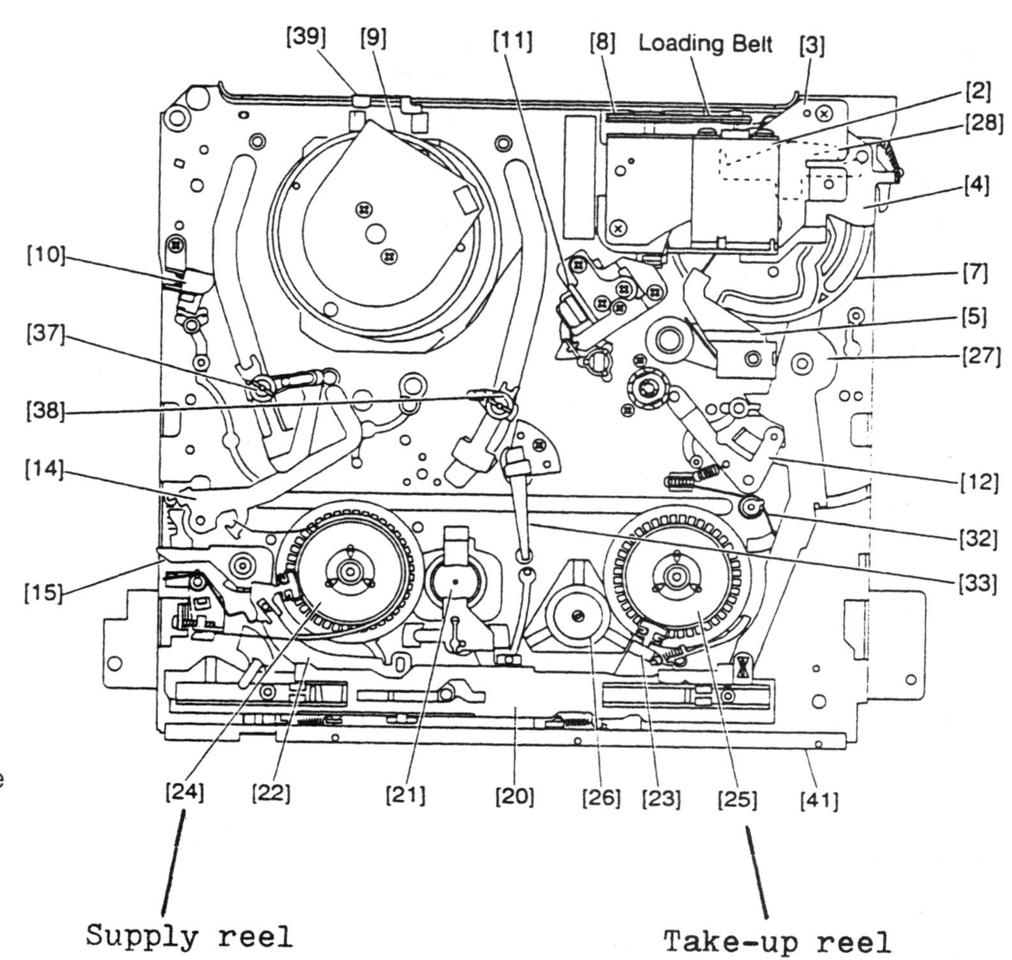

Fig. 11-5. Replace the plate assembly when the take-up reel does not turn and stops in play and record modes.

Capstan Motor Shutdown

Replace the defective capstan motor for an intermittent shutdown problem or when play mode intermittently powers up and then shuts down. Also replace the capstan motor if the motor continues to run when the power is off.

Possible Causes:

- ✔ Replace capstan motor IC and low ohm resistors in the IC circuit (capstan shuts down after 30 minutes to a few hours of playing time)
- ✔ Replace capstan motor IC (VCR shuts down after 30 minutes in playback)
- ✔ Bad voltage regulator IC (play mode stops with no capstan rotation)
- ✔ Defective IC on the servo PCB (capstan motor runs very slow or stops)
- ✔ Defective diodes in the capstan motor voltage circuits (capstan motor rotation might be intermittent, stops and shuts down in play or record)
- ✔ Bad solder joint in the power supply voltage source (capstan runs fast and shuts down)

Cylinder or Drum Shutdown

Use the ESR meter to check out all jumpers and poorly soldered connections around the regulator and drum when the VCR loses drum sync, becomes warm, and shuts down in record mode. Resolder the drum stators on the PC board when the cylinder or drum runs fast and shuts down after loading.

Possible Causes:

- ✔ Defective zener diode on the regulator PCB (cylinder motor rotates when powered up, shuts down in play mode)
- ✔ Bad drum stators (cylinder rotates fast, then shuts down in play mode)
- ✔ Resolder or replace cylinder stators on the PCB (drum runs fast, shuts down after loading tape)
- ✔ Check all 1000-3300 μF electrolytics in the **5-volt source** when drum motor voltage is around 4.1 volts (tape loads, unloads, shuts down; check for no SW30 pulse)
- ✔ Bad lower cylinder assembly (VCR shuts down in play and record modes) (*Fig. 11-6*)
- ✔ Missing 5-volt source from the power supply (cylinder does not rotate, no eject, tape loads, shuts down)
- ✔ Leaky diode on the cylinder motor PCB (intermittent drum motor PG, intermittent and shuts down)
- ✔ Loose pads on the drum motor stators (loads tape around the drum, shuts down)
- ✔ No release of the supply reel brakes, excessive back tension (drum slows down or stops)
- ✔ Defective diode in the tape loading circuit on the main PCB (drum starts to spin, tape loads sometimes, power shuts off)
- ✔ Badly soldered joints at the drum motor (cylinder runs fast, then shuts down)
- ✔ Bad cylinder motor assembly (VCR shuts down as power is turned on) (*Fig. 11-7*)

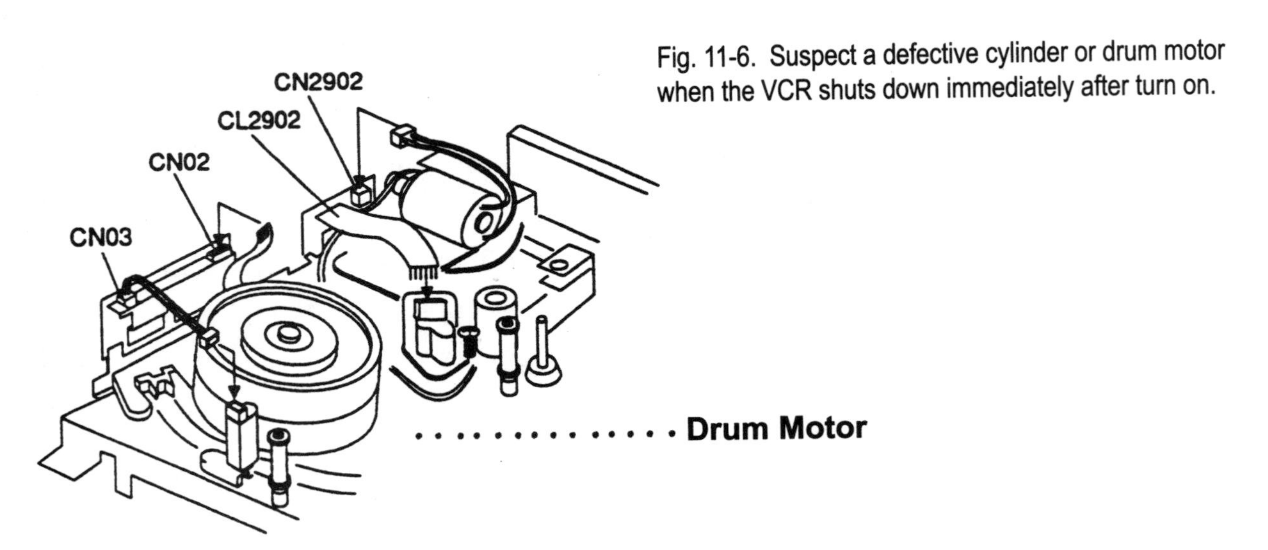

Fig. 11-6. Suspect a defective cylinder or drum motor when the VCR shuts down immediately after turn on.

Fig. 11-7. Suspect a bad cylinder motor assembly when the VCR shuts down as the power is turned on.

Bad Cylinder Motor Drive IC

Replace the cylinder motor drive IC when the cylinder motor operates erratically and sometimes shuts down in a Panasonic PV-M2021 TV/VCR combo. The IC bipolar, linear cylinder driver (IC2601) was replaced with an AN3813K (*Fig. 11-8*).

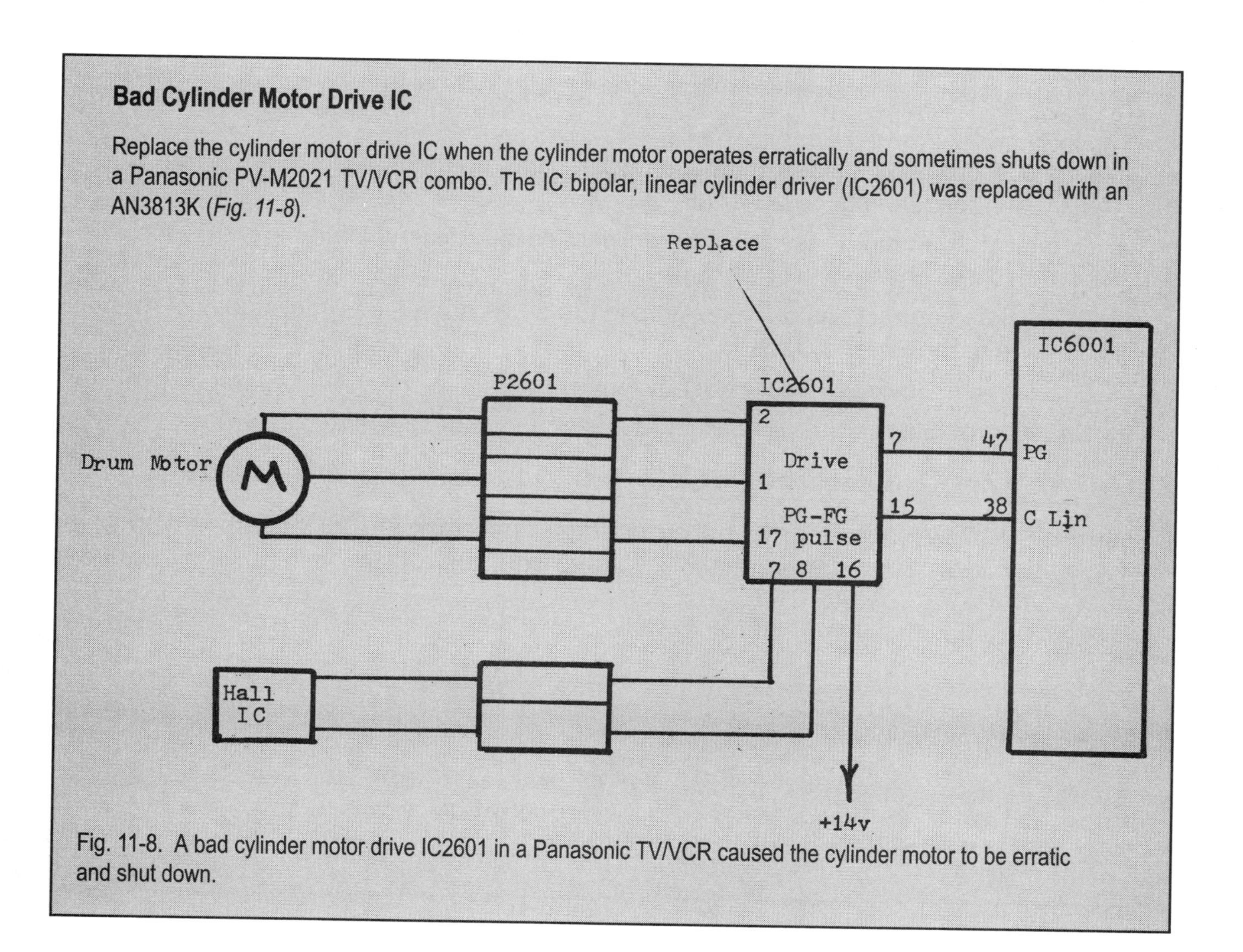

Fig. 11-8. A bad cylinder motor drive IC2601 in a Panasonic TV/VCR caused the cylinder motor to be erratic and shut down.

- ✔ Replace the motor driver IC (cylinder will not rotate, no tape loading, powers off in a few seconds)
- ✔ Voltage regulator IC (shutdowns with no eject, fast-forward okay, drum does not turn)
- ✔ Replace the drum motor IC (tape loads around the drum, shuts down)
- ✔ Replace the servo IC (drum motor does not rotate, tape unloads, VCR shuts down)
- ✔ IC protector on the servo PCB (cylinder does not spin, shuts down)
- ✔ Defective regulator IC (play mode goes into stop, no cylinder rotation)
- ✔ Leaky diode on the regulator PCB (cylinder motor runs at power up, shuts down in play mode)
- ✔ Defective drum motor (drum loads, shuts down)
- ✔ Bad solder connections at the cylinder motor (tape loads with no head switching pulse from the cylinder, shuts down)
- ✔ Both tape end sensors (cylinder does not rotate after tape loads, then shuts down)
- ✔ Bad drum motor (cylinder runs fast, shuts down)
- ✔ Lower drum assembly (VCR shuts down in play and record mode)
- ✔ A leaky diode can cause intermittent drum motor PG pulse
- ✔ Leaky diode on the cylinder motor PCB
- ✔ One or more leaky diodes in tape loading circuits on the main PCB (drum starts to spin, no tape loading, power shuts off)
- ✔ Defective drum motor assembly (video shakes intermittently; after pressing the stop button, video changes with startup)
- ✔ Bad solder connections on transistors in the 5-volt source (cylinder does not rotate, no drive to the lower cylinder)
- ✔ Bad servo IC (cylinder motor will not rotate)
- ✔ Bad IC protector on the servo PCB (cylinder does not spin, shuts down)

Drum Voltage Regulator Problems

Measure the voltage regulator IC output voltage and compare to the schematic when the drum servo and capstan control IC voltage sources are low or showing very little applied voltage (*Fig. 11-9*).

Drum Shakes - Sears 13-inch

The video shook intermittently in a Sears 13-inch TV/VCR. Replacing the drum motor assembly was the only answer. Most drum or cylinder motor assemblies can be removed by first removing three metal screws under the bottom area of the cylinder assembly.

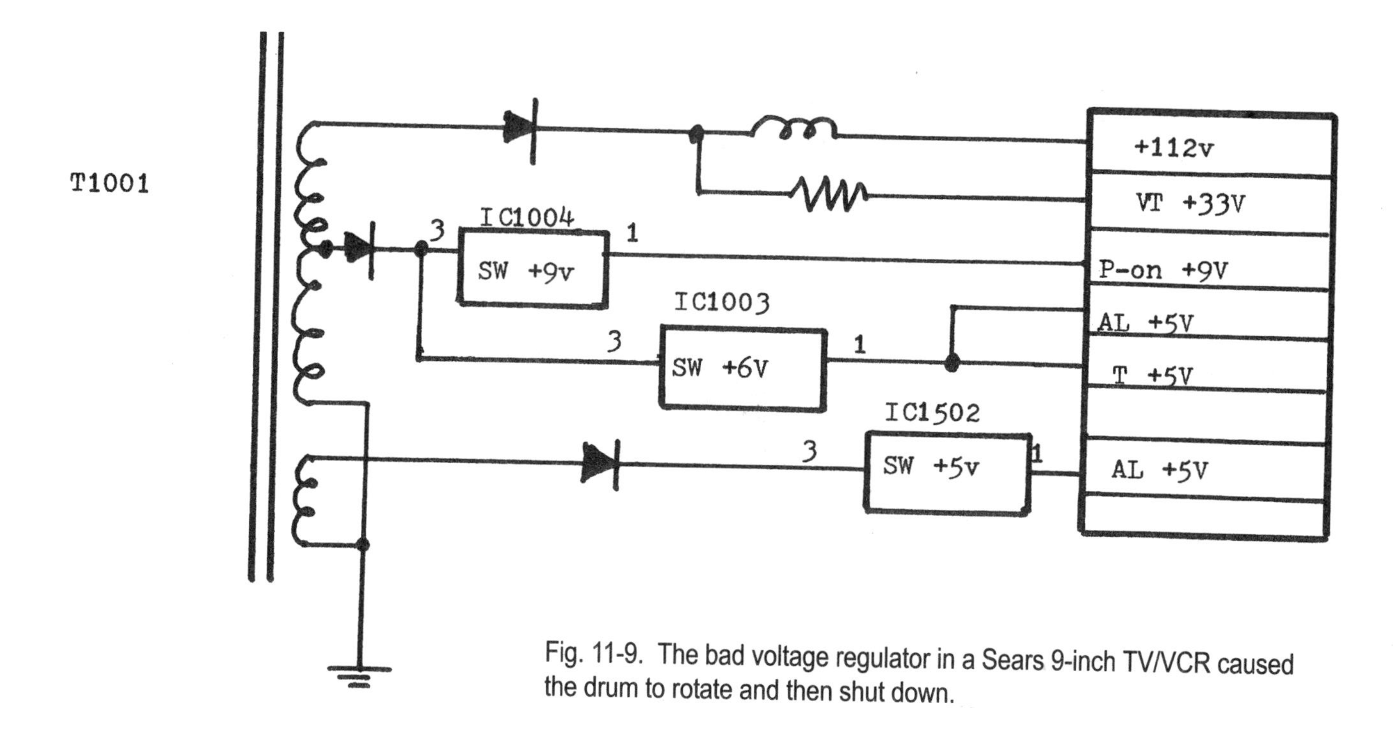

Fig. 11-9. The bad voltage regulator in a Sears 9-inch TV/VCR caused the drum to rotate and then shut down.

Symptoms

- ✔ No capstan or drum rotation in play, then shuts down
- ✔ Drum will not rotate, VCR shuts down in rewind, fast-forward is OK
- ✔ Drum spins, power light is on, no other functions
- ✔ Tape is stuck, loaded around the heads, will not unload tape
- ✔ Drum will not turn, no EE, fast-forward okay, shuts down

Cylinder Motor Runs Fast then Shuts Down

Possible Causes:

- ✔ Shorted diode in the servo section and a bad filter capacitor in the drum IC voltage source (cylinder motor runs constantly, will not load tape, no display)
- ✔ Bad electrolytics (22 to 100 μF) in the servo circuits with the ESR meter (no picture, horizontal lines in playback, drum motor runs too fast all the time)

Mechanical Problems – Sansui CM2500

Replace the rotary switch in a Sansui CM2500 TV/VCR when mechanical problems, such as erratic loading, no playback, or no rewind occur. Mark down the switch leads before removing so they can easily be replaced; order critical switching parts from the manufacturer. Also check the mode switch for mechanical problems.

- ✔ Bad bypass capacitors (0.015 to 0.047 μF) in the servo circuits (capstan runs at high speed in play back, fast-forward, rewind)
- ✔ Bad servo IC (cylinder operates slower than normal, video playback will not sync, lines in picture)
- ✔ Defective 4.19 MHz crystal (drum spins at high rate of speed, no cassette loading, no display, power lights are on)
- ✔ Bad drum motor
- ✔ Poorly soldered connections on the cylinder PCB
- ✔ Replace the drum stators if there is fast cylinder speed and rotation
- ✔ Replace or resolder the drum stators when the cylinder runs too fast, runs intermittently, and shuts down in play mode
- ✔ Replace the cylinder stators assembly when the cylinder runs fast and shuts down after loading the tape
- ✔ Bad drum motor IC (cylinder speed is erratic)
- ✔ Bad drum assembly (irregular cylinder speed)
- ✔ Bad cylinder motor IC (cylinder motor runs with the power off, no capstan rotation, no cassette loading, shuts down)
- ✔ Intermittent voltage regulator IC (drum motor rotates erratically with capstan motor in fast-forward mode; spray the suspected IC with coolant or freeze spray to make it act up)
- ✔ Bad diode on the cylinder motor PCB (intermittent drum motor PG and shutdown)
- ✔ Leaky diode in the power source of the cylinder motor (runs too fast)

Defective Servo IC

A defective servo IC can cause the following problems:

- ✔ Shutdown after loading the tape
- ✔ Tape loads three-quarters of the way, then unloads
- ✔ Tape loads, VCR shuts off (might appear to be a bad mode switch)
- ✔ No function operations except power on and off, remote functions okay
- ✔ No speed select
- ✔ VCR shuts down after one or two seconds (replace the servo IC and syscon IC)
- ✔ Intermittent capstan and drum speed (*Fig. 11-10*)
- ✔ No drum or capstan operation

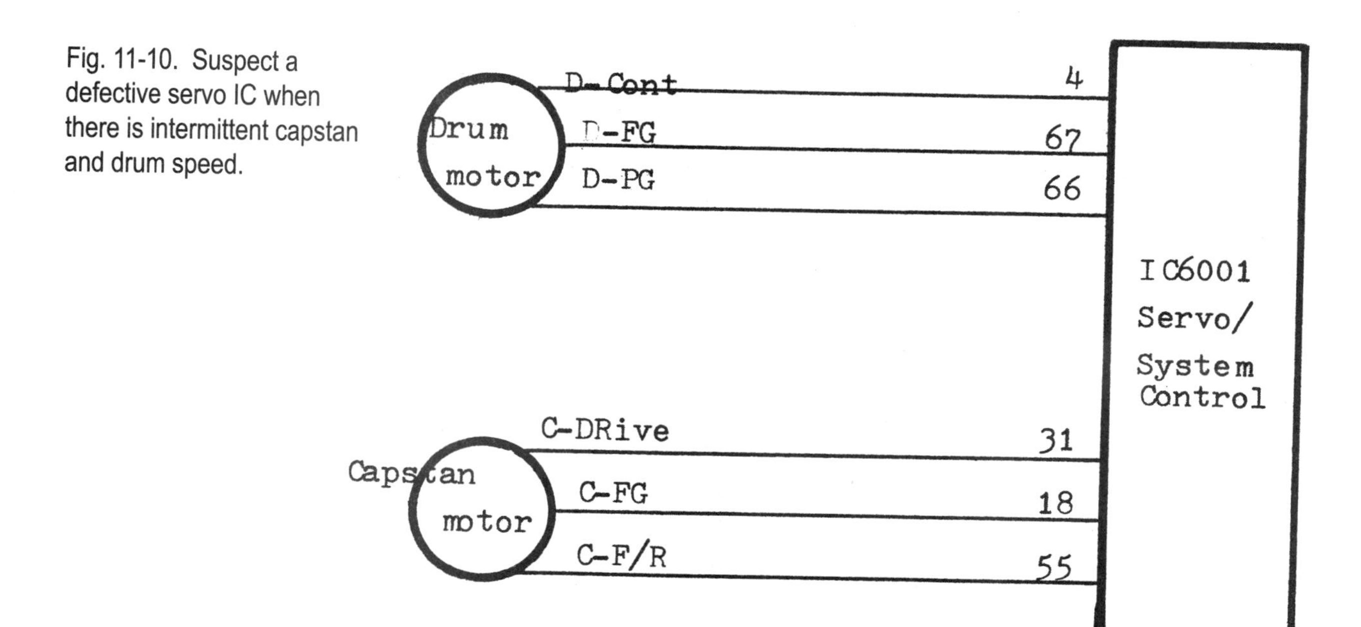

Fig. 11-10. Suspect a defective servo IC when there is intermittent capstan and drum speed.

Loads, Shuts Down – Emerson VT1920N

An Emerson VT1920N TV/VCR was loading then shutting down immediately. When a VCR does this, check for a defective servo IC2001. IC2001 controls the capstan motor, cylinder motor, control head, head amp, and tracking VR signals. Replace with original servo control part number 197D49010A (*Fig. 11-14*).

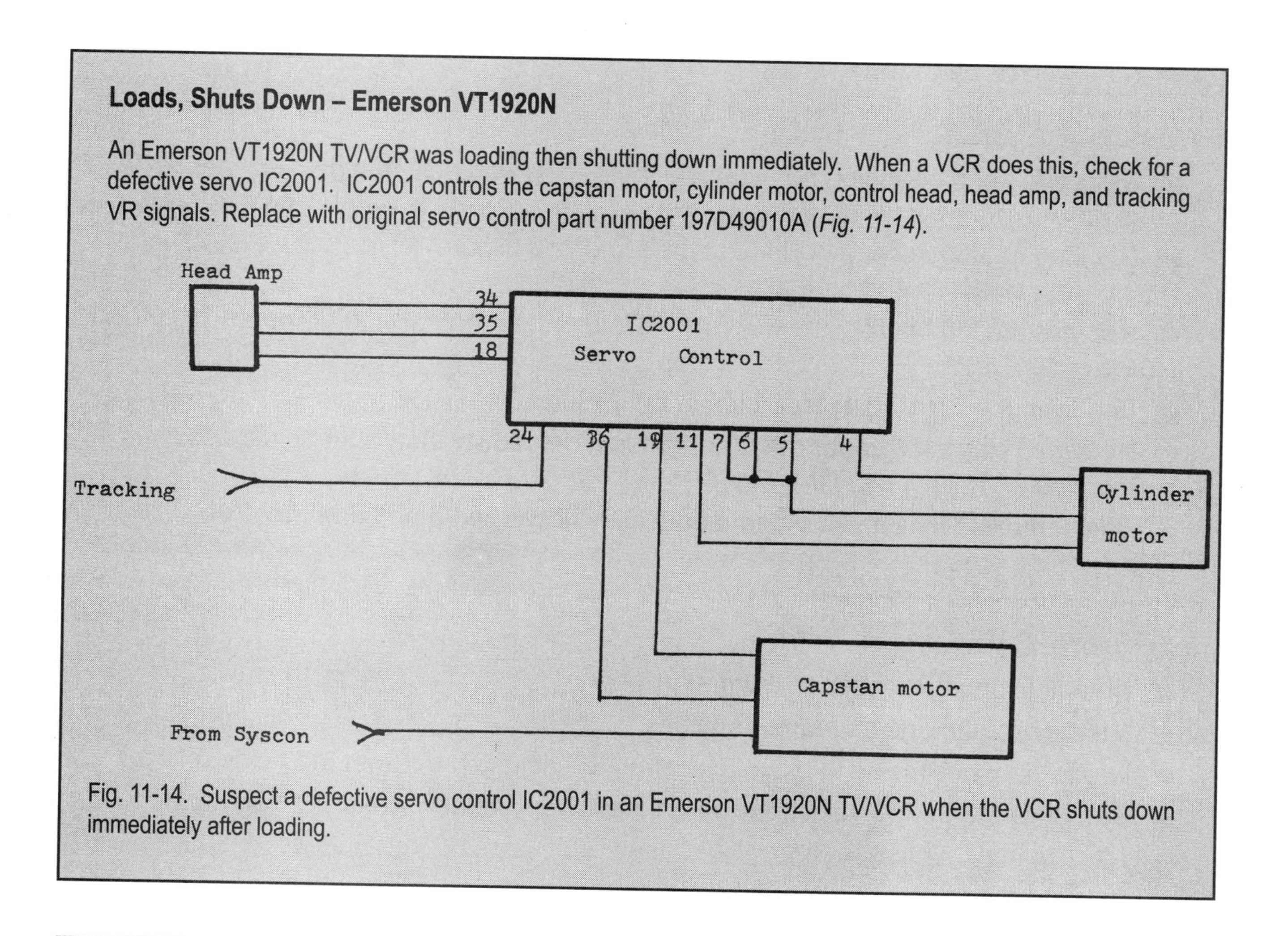

Fig. 11-14. Suspect a defective servo control IC2001 in an Emerson VT1920N TV/VCR when the VCR shuts down immediately after loading.

- ✔ Drum motor runs very slow (looks like a bad drum motor) and has video playback lines through the picture
- ✔ Capstan motor always on
- ✔ Capstan motor does not rotate
- ✔ Drum motor does not rotate, tape unloads, VCR shuts down
- ✔ Drum motor is dead, unloads, shuts down VCR

Noise Lines

Observe the TV screen and listen to the small speaker to see and hear the various lines and noises from the TV/VCR. Lines and bars can appear in the picture. Noise bars also can appear in the picture in review mode; noise might be heard in the speaker. This may be caused by a defective servo IC (snow bands through the picture, with wow and flutter sounds).

Hum bars in the picture might also be heard in the small speaker. A loud grinding noise might be heard from the rear of the TV/VCR. Sometimes the sounds you hear and the noise that is seen on the screen can be caused by some unusual, broken mechanical components.

Lines in Picture

Possible Causes:

- ✔ Defective head switching relay (stuck or has dirty contacts that causes lines in half of the picture)
- ✔ Dry loading ring assembly (lines at the top of the picture, left guide will not load all the way; clean up and lubricate
- ✔ Replace the left loading link lever for an intermittent noisy band at the top of the picture
- ✔ Bad cylinder motor (diagonal lines in the picture)
- ✔ Defective servo IC (upper cylinder rotates more slowly than normal, causes lines in the picture, will not sync in) (*Fig. 11-11*)
- ✔ Bad servo IC (lines in the video, audio okay; looks like a bad drum motor)

Lines in Video Playback

Possible Causes:

- ✔ Defective upper cylinder or drum assembly
- ✔ Misaligned and unsecured tape guides
- ✔ Loose slant bar (noise in playback cannot be adjusted out with the guide roller)
- ✔ Bad SMD 3.3-μF, 50-volt capacitor on the lower cylinder; looks like bad heads

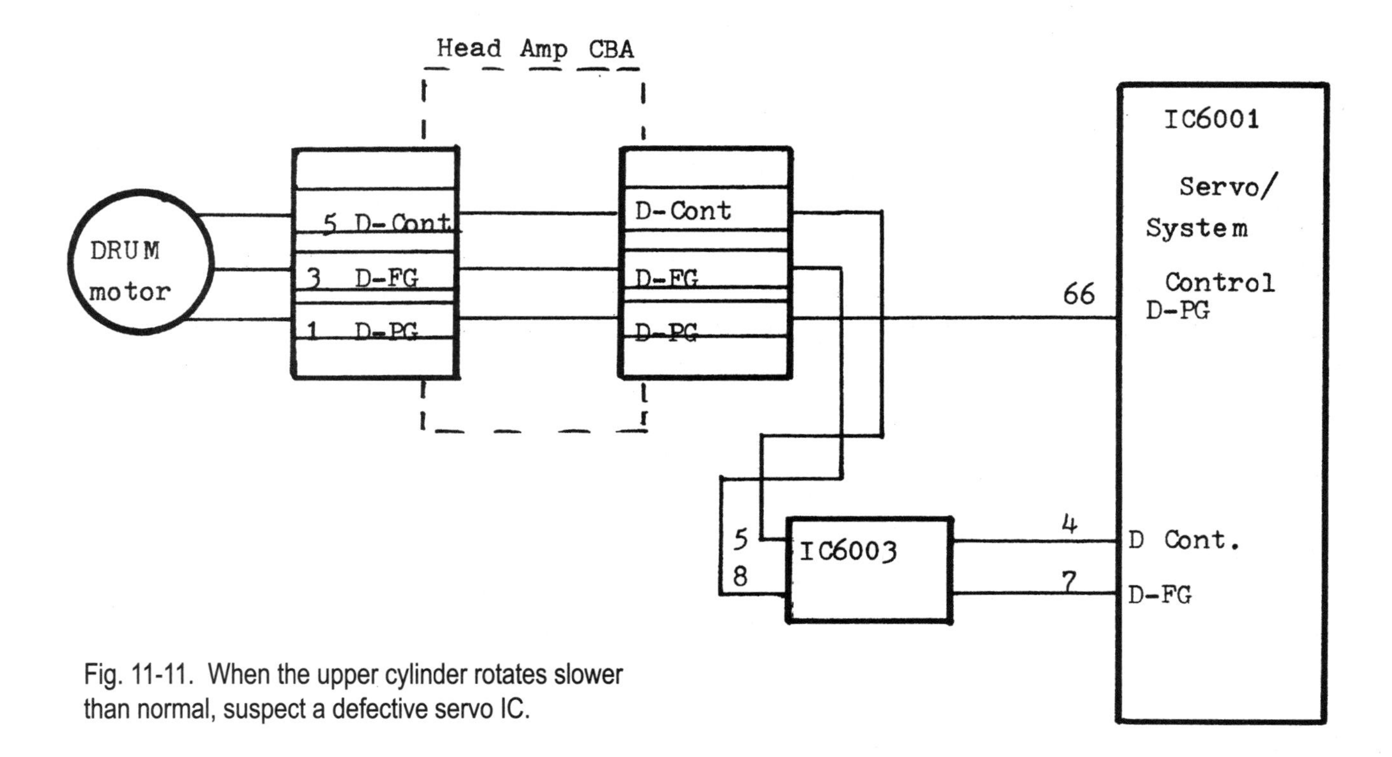

Fig. 11-11. When the upper cylinder rotates slower than normal, suspect a defective servo IC.

- ✔ Bad 100-220 µF electrolytics in the power supply; check with ESR meter (wavy lines and noise in video playback mode)
- ✔ Bad take-up arm (video jumps and rolls in playback, causing pole not to seat properly)
- ✔ Bad upper cylinder (noisy lines in the picture in playback)
- ✔ Upper cylinder not properly grounded (intermittent and noisy video playback; may look like a tracking adjustment)
- ✔ Defective servo IC (video playback has lines running through it; looks like a bad drum motor
- ✔ Cylinder control IC (lines at all speeds; all other functions okay with normal power supply voltages)

Noise in Picture

Possible Causes:

- ✔ 1000-2200 µF electrolytics (lines in video, audio noise in record mode, VCR becomes intermittent, eats tape when ejected)
- ✔ Bad capacitor in the voltage regulator circuit (noise in the picture stops, capstan motor stops rotating)

- ✔ Defective drum stators (noise in the picture, intermittent, no cylinder motor rotation)
- ✔ Defective capstan motor (noise in the picture during frame advance and normal playback)
- ✔ Bad ground brush contact with weak pressure (noisy picture in special effects mode) (*Fig. 11-12*)

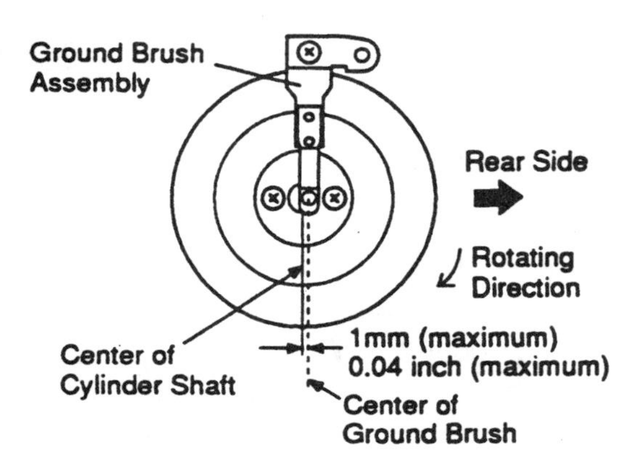

Fig. 11-12. A noisy picture might result from a poor ground brush assembly on top of the cylinder.

Noise in Playback

Possible Causes:

- ✔ Loose head relay switching assembly
- ✔ Poor grounding behind cylinder
- ✔ Bad upper cylinder assembly

Use an ESR meter to do in-circuit tests on electrolytic capacitors in the cylinder and capstan circuits that can cause noise lines in the picture. Horizontal lines in the picture during any playback speed can be caused by open electrolytics on the cylinder drive PCB, causing RF interference. Check electrolytics in the servo section when the cylinder motor runs too fast and video has (horizontal lines) in playback mode.

White Lines and Snow

Possible Causes:

- ✔ Bad head amp IC
- ✔ Defective servo IC (snowy band moves through picture, some audio wow and flutter)
- ✔ Bad electrolytics off of the servo IC circuits (check with the ESR meter when the video alternates between a normal picture and then snow; VCR might change speeds)
- ✔ Defective trigger gear assembly IC (lines and snow in the picture during playback in EP, SP modes)

- ✔ Bad upper cylinder (white lines on the tape after recording)
- ✔ Improper alignment of the ACE head might cause snow or horizontal lines in the picture
- ✔ Dirty video heads (snowy pictures, horizontal lines in playback mode)
- ✔ Defective video head (snowy picture in playback; replace the defective drum or cylinder)
- ✔ Defective upper drum assembly (white lines on the tape after recording)
- ✔ Bad joint at the head amp connections (snowy, poor video in EP and SP speeds; looks like bad heads)
- ✔ Broken capstan brake spring (snow noise in SP, slow-motion mode)
- ✔ Stretched tension band (picture is good and then bad)
- ✔ Bent static guard (intermittent snow shooting across the screen)
- ✔ Very loose VCR belt (intermittent snow in playback)
- ✔ Bad trigger gear assembly (video alternates between normal and snow, with a grinding noise)

Hum Noise

Possible Causes:

- ✔ Lever assembly on the tape deck (play looks like it's in pause mode, motor-like noise in audio)
- ✔ Dirty relay contacts (audio hum is heard in speakers, intermittent audio squeal in playback mode)
- ✔ Bad large filter capacitors in the power supply
- ✔ Bad large 2000-3300 µF electrolytics in the power supply (buzz in the audio and hum bars in the picture)
- ✔ Large 2200 µF electrolytic (noise in the 5- and 9-volt sources, black bar in the picture, intermittent audio buzz, no horizontal or vertical sync)

Scraping Noise – Panasonic PVM2021

In a Panasonic PVM2021 TV/VCR a scraping noise might be heard when the cassette is loaded or attempts to load and the unit will not accept the tape if the tape is inserted in a crooked fashion. The clearance between the worm wheel and worm unit might be small, the open angle might be bent and interfere with the pressure arm unit, or the sub-plate assembly may become loose and have had excessive force used on it, causing the cassette-up assembly to jam. Replace the complete cassette-up assembly to cure the scraping noise problem.

Grinding and Clicking Noises

Possible Causes:

- ✔ Bad trigger gear assembly (mechanism grinding noise, video alternates between normal and snow)
- ✔ Defective carriage arm with no loading of cassette
- ✔ Bad mode switch (grinding noise, no picture or reverse, pulls tape, might have intermittent operation)
- ✔ Bad drive clutch wheel loud grinding in rewind or fast-forward)
- ✔ Bad take-up clutch with binding bearing (chews up tape, audio garbled, noisy mechanism in playback)
- ✔ Bad capstan bearings (vibrating noise in fast-forward, rewind, or in mode changes; replace worn bushings or the entire motor)
- ✔ Bad holder assembly (vibrating noise in fast-forward, rewind, or mode changes)
- ✔ Bad worm gear (clicking sound, no tape loading)
- ✔ Bad wheel in the worm gear on loading assembly (clicking sound, no tape loading)
- ✔ Bad electrolytics in capstan circuits (capstan motor rattles)
- ✔ Large filter capacitors in the power supply (capstan motor erratic and chattering)

Squealing Noises

Possible Causes:

- ✔ 10-47 μF electrolytics on the power supply PCB (dead squealing noise from the power supply)
- ✔ Bad switching power transformer (no power up, low-level whine or squeal)
- ✔ Bad mode switch (loading motor continues rotating, squeals, shuts off)
- ✔ Bad filter capacitors (picture is noisy, squealing noise is heard as capstan motor runs
- ✔ Rack and bracket assembly
- ✔ Dry bottom capstan bearing (clean and lubricate)
- ✔ Bad static discharge arm on upper cylinder assembly (high-pitched squeal in play or record mode)
- ✔ Bad soldered terminals on the transistor take-up sensor on deck PCB (VCR powers up, goes to full loading; push eject and unit unloads, squeals)
- ✔ Defective capstan motor IC (capstan motor is running full speed, playback at super speed with high-pitched squeal)

Squeal in Playback – Panasonic PVM2027

The picture was very noisy and a high-pitched squeal was heard in play or record modes in a Panasonic PV-M2027 TV/VCR. Replacing the static discharge arm on the upper cylinder assembly solved the noisy picture and squealing sound in the audio (*Fig. 11-13*).

Fig. 11-13. Replace static discharge arm on the upper cylinder assembly when a noisy picture and squealing noise occur.

Picture Problems

Possible Causes:

- ✔ Bad mode switch (jumping picture, playback video has lines, very poor audio and video)
- ✔ Low power filter (LPF) circuit (prerecorded tapes play okay, self-recorded very noisy)
- ✔ Defective transistor on the preamp video PCB (only hash marks, no video recording with audio record okay)
- ✔ Servo or system control IC (no video in playback, no sync, picture jumps and jitters around)
- ✔ System control IC (noise bar in review)
- ✔ Bad 1-µF, 50-volt electrolytic in the head amp PCB (video playback jitters, will not sync in SP and EP modes, records video okay)

Intermittent Shutdown – Magnavox CRL191VCR

Sometimes a Magnavox CRL191VCR TV/VCR would play or record and the next minute the units function became intermittent. To fix this problem, remove the reel table. Clean up the reflector surface for reel pulses at a pin terminal on the microprocessor.

- ✔ Large filter capacitor 1000-3300 µF in the power supply (bad cylinder bearing, jumps and flashes are seen on the screen)
- ✔ Bad drum motor assembly (video shakes intermittently; press stop and play it again, video might change to normal)
- ✔ Bad capstan stators (play mode jerky, looks like a bad bearing)
- ✔ Bad capstan motor (power supply begins to smoke after main fuse replaced)
- ✔ Broken or worn reel band (clean or replace to correct jittery playback and garbled sound)
- ✔ Loose back tension band (flagging at top of picture)
- ✔ Left loading link lever (intermittent noise band at the top of the picture)
- ✔ Bad main filter in power supply source (black bar in picture, no horizontal or vertical sync, hum in audio)
- ✔ 1-µF, 50-volt electrolytic on the drum PG line of the servo section (use the ESR meter to check if there is a bar through the lower quarter of the picture in playback with unstable tracking)
- ✔ Cylinder assist control (noise bars drifting through the picture - adjust)

Tracking Problems

When tracking problems occur, there are several methods you can try to remedy the situation. Start by realigning and securing tape guides when adjustment of tracking does not remove lines in the video. Check for a brass pin loose from the guidepost when tracking will not adjust up. A loose guidepost may need to be realigned and secured when tracking noise cannot be adjusted out of the picture. A defective back tension band may cause poor tracking in playback with the picture pulling at the top. A bad 1-µF, 50-volt capacitor on the drum PG line in servo section may cause bars through lower part of picture in playback and unstable tracking.

Chapter 12: Troubleshooting VCR Motor Circuits

There are three basic motors found in the TV/VCR combo: the drive or loading motor, the capstan motor, and the cylinder or drum motor. Some large VCRs might also have separate reel and mode motors.

As its name implies, the loading motor loads and unloads the tape. The capstan motor rotates and pulls the tape through the various heads and guides, while the cylinder or drum motor rotates the video heads at the proper speed and phase with respect to the tape. Within the TV/VCR combo, the servo IC controls the operation of the capstan and cylinder motors (*Fig. 12-1*).

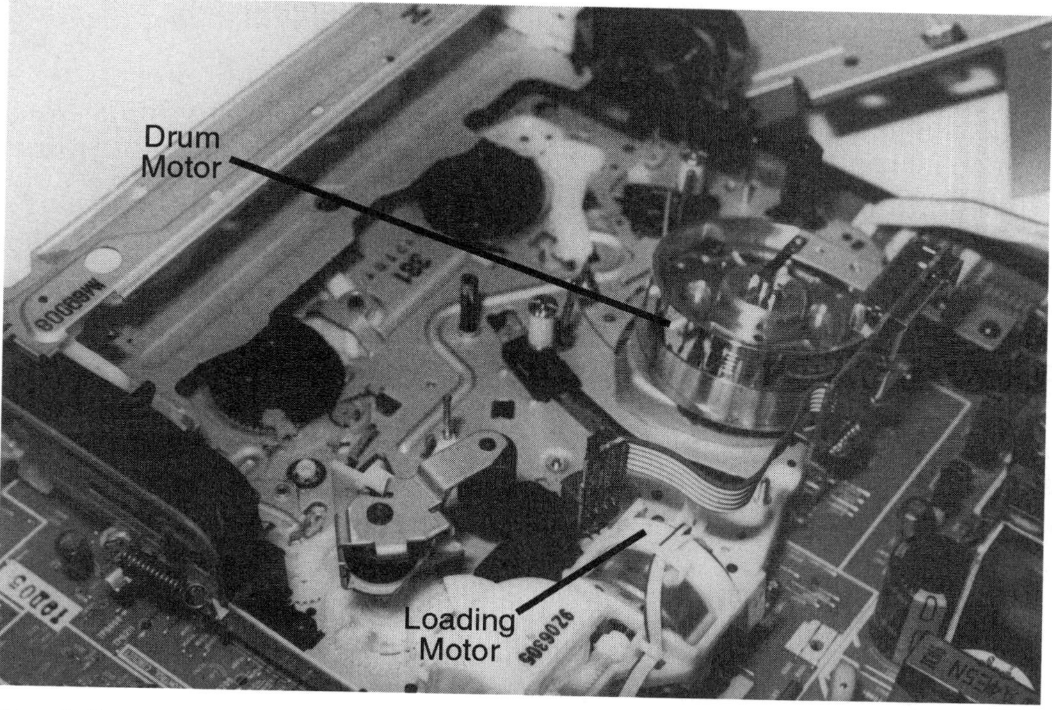

Fig. 12-1. Usually the capstan motor is mounted underneath the VCR chassis with the loading motor and cylinder on top of the transport.

The Loading Motor

The cassette-loading switch turns on the loading motor through a loading motor drive IC and supplies a low dc voltage to the loading motor terminals. The motor loads the tape or cassette by a loading belt, worm gear, or direct-drive operation. When the cassette is ejected, the polarity of voltage to the loading motor is changed, reversing direction or ejecting the cassette. The loading motor drive belt might turn a connecting rod or worm cam gear, rotating the loading assembly.

A defective loading motor can be located by measuring the voltage across the motor terminals, running a continuity test of motor windings with the ohmmeter or with an external battery test.

If loading motor gets stuck or won't move, it can be made to eject a loaded tape or rotate by using an external 9-volt battery. Simply touch the positive (red) lead to the positive terminal of the battery and the black lead to the minus battery terminal. Only touch the motor lead for a second if the motor does not start to rotate. The loading motor can also be tested to see if it rotates by connecting a 1.5-volt flashlight battery across the motor terminals.

A defective loading motor might run hot because of shorted internal windings, stuck or frozen bearings, a jammed loading assembly, and the normal aging process.

Damage also can be caused by power line surges or lightning.

A loading motor might continue to rotate even if the drive belt is broken or slipping. If that happens, replace the belt. Although, a roller/belt rejuvenator kit is available, it's best to replace all motor belts to prevent further problems. Belts are fairly inexpensive as compared to the cost of a loading motor.

Loading Motor Circuits

The loading motor signal is sent from the servo/system control microprocessor to a motor drive IC from terminals 10 and 11 (*Fig. 12-2*). A loading motor forward signal is sent from pin 11 of IC6001 to pin 1 of loading motor driver (IC6003). The loading motor reverse signal is applied at pin 10 of IC6001 to pin 9 of IC6003. The loading motor driver (IC6003) provides a positive 1.4 volts to the loading motor terminal and 1.9 volts to the minus voltage terminal of the loading motor.

A defective loading motor IC can cause some of the following problems:

- ✔ No cassette loading. A defective loading motor IC or a shorted 12- to 14-volt zener diode off of the loading motor IC can cause cassette loading failure.
- ✔ Intermittent loading.

- ✔ Cassette becomes jammed or stuck.
- ✔ Tape fails to eject, no play or record.
- ✔ No rotation of the loading motor.
- ✔ No loading motor action.
- ✔ Locks up while loading, power goes off.
- ✔ Tape will not load or play.
- ✔ Ejection failure. A bad motor driver IC or capstan servo circuits can lead to eject failure.
- ✔ Loading motor stops and is red hot, no eject or tape is stuck.
- ✔ Powers up, locks up tape, shuts off.
- ✔ Eject operation begins, then unloads.
- ✔ Loading motor does not run. Check for open resistor in series with loading motor voltage source, voltage at the motor terminals, and driver IC (*Fig. 12-3*), or defective transistors in the power supply.
- ✔ Tape loads very slowly, may stop in the middle of loading.

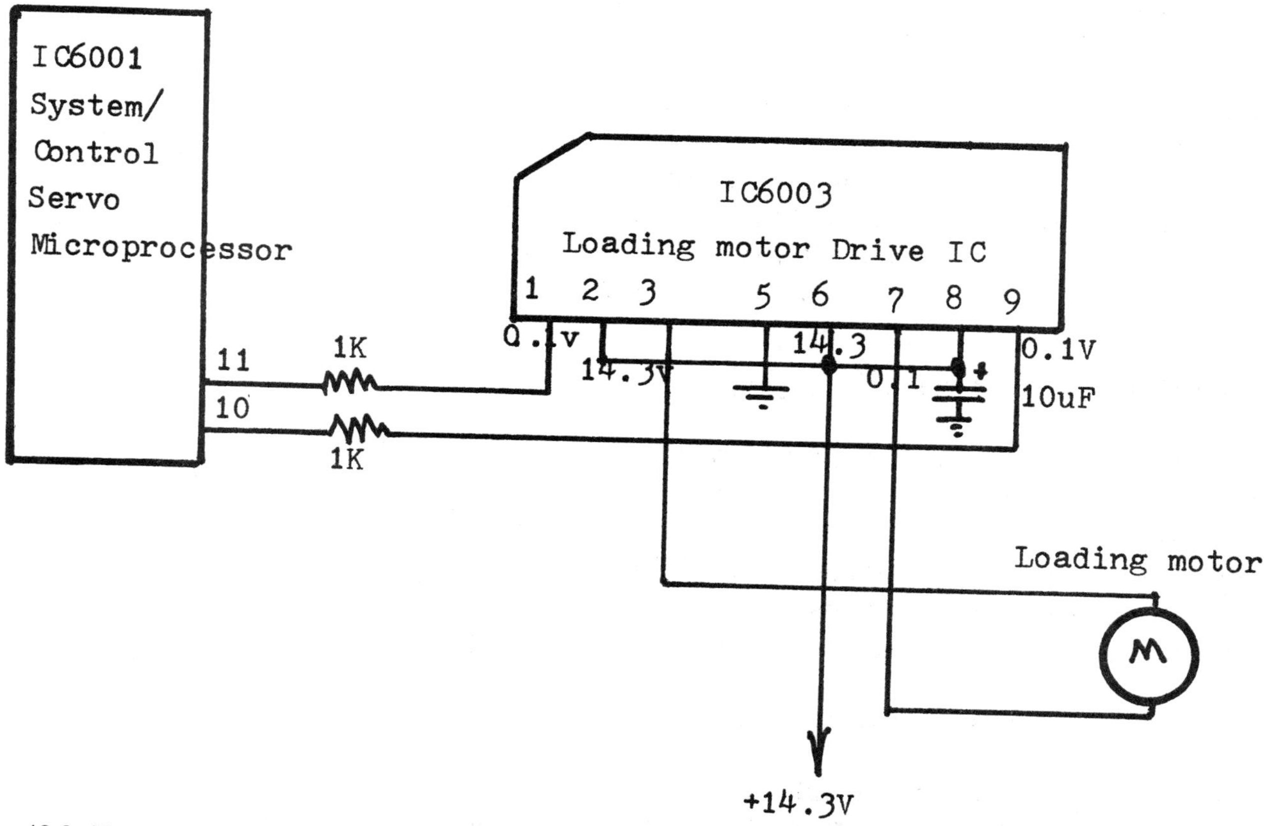

Fig. 12-2. The system control and servo IC control the loading motor through loading motor IC6003.

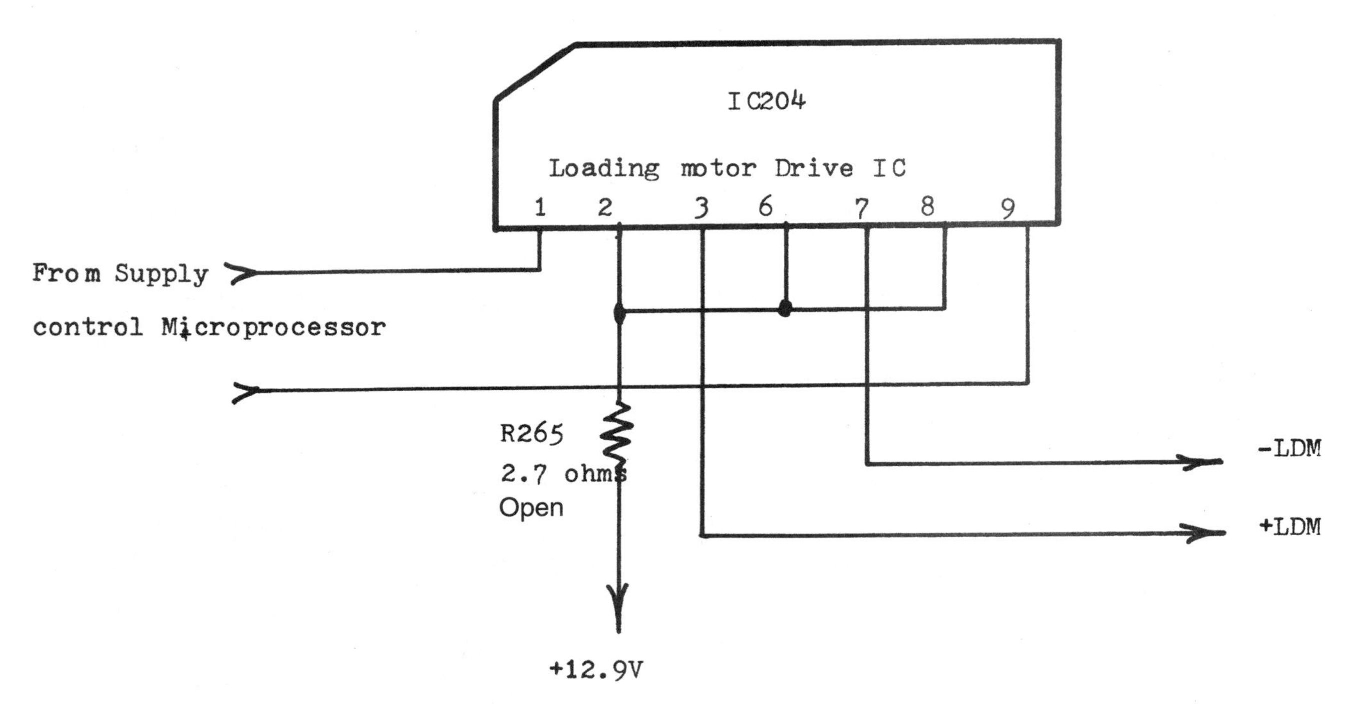

Fig. 12-3. An open resistor in the supply voltage loading motor IC can cause the loading motor to not rotate.

There are several other problems associated with the loading motor. A defective loading motor drive belt can be broken into and wound around the loading mechanism gear assembly. Check to see if the motor pulley is out of line or cracked when the belt will not stay in line and comes off of the motor pulley. A cracked pulley and bad loading motor assembly can cause no eject or cassette to not fully load, as well as the fast-forward assembly out of line.

A bad loading belt and worm clutch assembly may cause the pulley to slip and the tape to be stuck. The loading motor may also run and then shut off. A loading motor that keeps rotating, but does not complete a full cycle can indicate bad belts and cam gear assembly. If the loading motor belt is bad, eject may be intermittent or not at all, and the tape may get stuck.

When working on the loading motor, keep in mind that it might have more than one belt in the loading mechanism (*Fig. 12-4*).

Check the voltage on all servo IC terminals and compare them to the schematic because a low supply voltage (Vcc) might indicate a leaky servo IC. If the servo IC is defective, it can cause loading problems, such as loading motor and capstan motor not rotating; tape loading then power shuts off (may appear to be defective mode switch); tape only loading three quarters of the way then unloading; and poor eject operation (also check for defective capstan IC servo circuit).

Fig. 12-4. The loading motor in some VCRs may have more than one belt to operate the loading motor mechanism.

Loading Motor and Capstan Motor Run in Reverse

Begin by checking the motor terminals for an exchange of motor leads; someone may have installed a new loading or capstan motor. Always mark down the color of the motor leads before unsoldering and removing the motor. Often, the positive lead to the meter is red and the negative terminal is black.

Other possible causes include a defective servo or system control IC, defective tape end sensor, reversed motor leads, defective loading motor IC, or a defective capstan motor.

Troubleshooting Loading Motor Circuits

Use the DMM to monitor the voltage across the loading motor terminals. Suspect an open loading motor when the DC voltage is measured across the motor terminal with no operation. The loading motor voltage should be found at the loading motor terminal when the load or eject buttons are pressed. Check the continuity of the motor with the low-ohm scale of the DMM. If no continuity or measurement shows on the DMM, the motor is open and should be replaced.

Shuts Down After Loading – Emerson VT1920N

Check for a defective servo control (IC2001) in an Emerson VT1920N when the VCR shuts down after loading. Measure the supply voltage on pin 21 and the capstan signal out of pin 20 of IC2001. If the voltage is extremely low, check for a leaky servo control IC. The capstan APC signal is sent out of pin 20 to pin 10 and again out of pin 8 to the control "in" pin of the capstan motor assembly (*Fig. 12-5*).

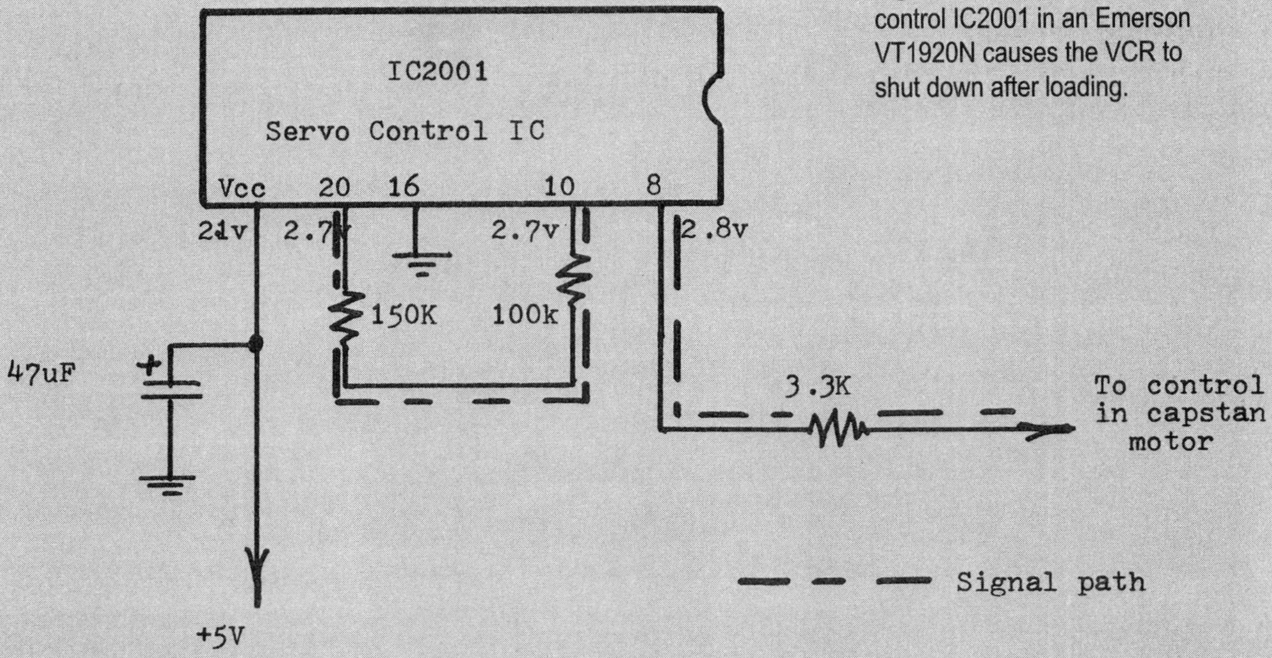

Fig. 12-5. The defective service control IC2001 in an Emerson VT1920N causes the VCR to shut down after loading.

Remove one lead of the motor terminal and clip a 1.5- to 9-volt battery across the motor terminals; see if the motor starts rotating. If the cam and drive assembly are jammed and will not let the motor rotate, remove the motor drive belt. A jammed assembly might keep the motor from rotating and can damage the loading motor.

Suspect a defective motor driver IC when no voltage is found at the loading motor terminals. Check the supply voltage fed to the loading motor driver IC, and measure all voltages on the motor driver IC and compare them to those found upon the schematic. If the loading motor fails to rotate, check for a defective servo control IC. However, do not overlook an open fuse or small resistor in the supply voltage source of the loading motor. If both the loading motor and capstan motor fail to operate, replace the servo control microprocessor.

The Capstan Motor

The capstan is driven by a belt from an accurate DC capstan motor (*Fig. 12-6*). In later VCRs the capstan has its own drive motor at the base of the capstan assembly. A capstan drive assembly pulls the tape against the pinch roller, through and around several different heads. The speed of the tape determines the rate of speed the capstan motor is rotating. Very slow speeds might cause wow and flutter in the audio. The heavy or large capstan flywheel helps to keep the tape speed fairly constant. In addition to feeding the tape, the capstan motor might also operate the take-up and supply reels that are driven by an idler tire or roller.

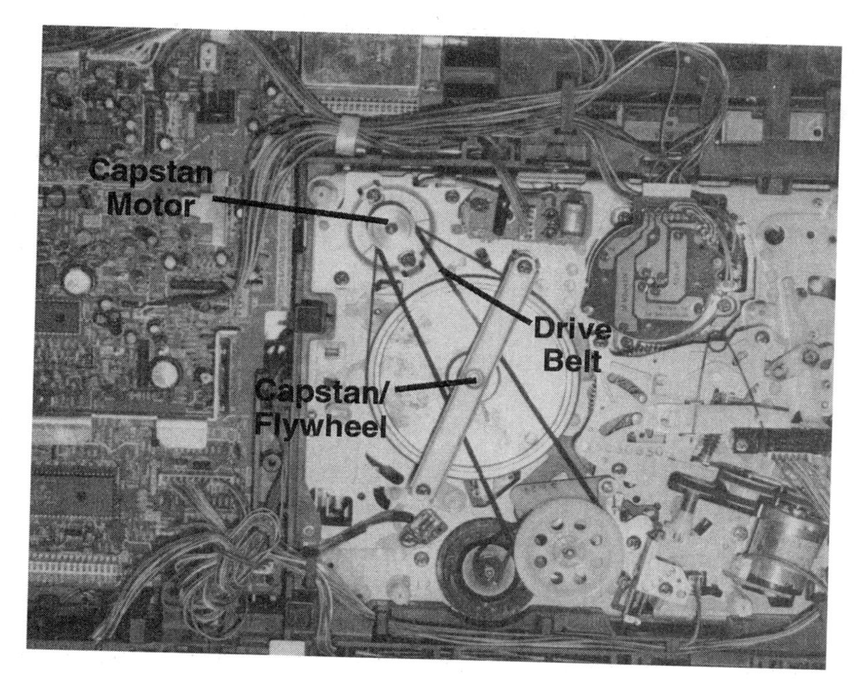

Fig. 12-6. The capstan/flywheel is driven by a small belt from the capstan motor pulley.

The DC voltage applied to the capstan motor might vary as the motor is rotated. A voltage check between 1.5 and 12 volts can indicate the servo microprocessor and capstan motor drive IC are functioning. When no voltage is found at the capstan motor, it may be an indication that either the motor driver IC or servo microprocessor is defective. A continuity measurement on the capstan motor with the DMM might indicate if the motor winding is open or normal.

Excess tape oxide seems to collect at the drive area and down around the capstan top bearing. The buildup of oxide might cause the tape to stop, shut down, and cause a dry squealing noise. Remove the capstan drive shaft and clean out the bearing and shaft with alcohol and a cloth. Lubricate the top bearing. Make sure that the excess oil or grease is wiped away from the tape drive area.

Chatter On Rewind – Emerson VT1321

To correct capstan chatter and garbled audio in EP mode in an Emerson VT1321 TV/VCR combination, replace the bronze upper sleeve bearing on the capstan motor. If the speed is slow, clean up the capstan motor bearings and remove old, hardened grease. Wash out the bronze sleeve bearing. Lubricate and then check to see if the sound is normal and the capstan runs smoothly.

Capstan Drive Circuits

In a Sears 13-inch TV/VCR, the capstan motor operates directly from the Servo System IC201, while in other VCRs the capstan motor is driven by a capstan motor drive IC (*Fig. 12-7*). The motor driver IC might power the main capstan coil drive IC and in turn controls the speed of the capstan motor. Check both the supply voltage (Vcc) on the capstan motor drive IC and the main capstan coil drive IC when the capstan motor does not rotate.

The supply voltage pins 3 and 6 on IC2501, and pin 14 of IC2500 contain the supply voltage sources. Check for a leaky IC or improper voltage source when the supply voltage is extremely low (*Fig. 12-8*). Check the voltage supply source of the power supply when no voltage is found at either voltage supply pin.

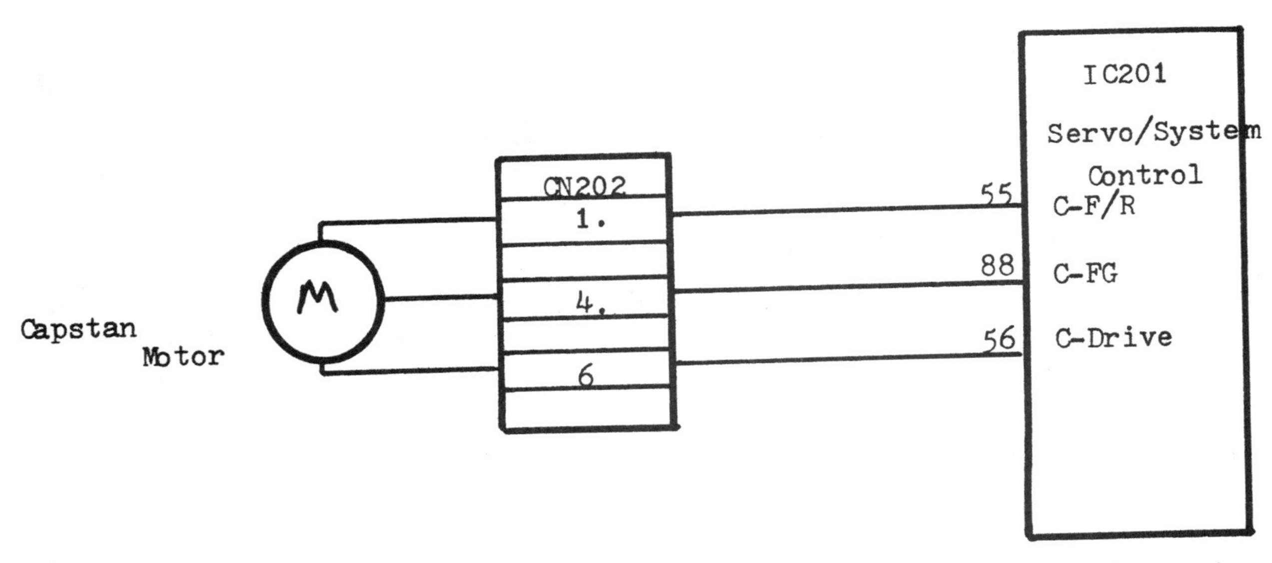

Fig. 12-7. In a Sears 13-inch TV/VCR combination, the capstan motor in the VCR is driven directly from the servo / control system IC201.

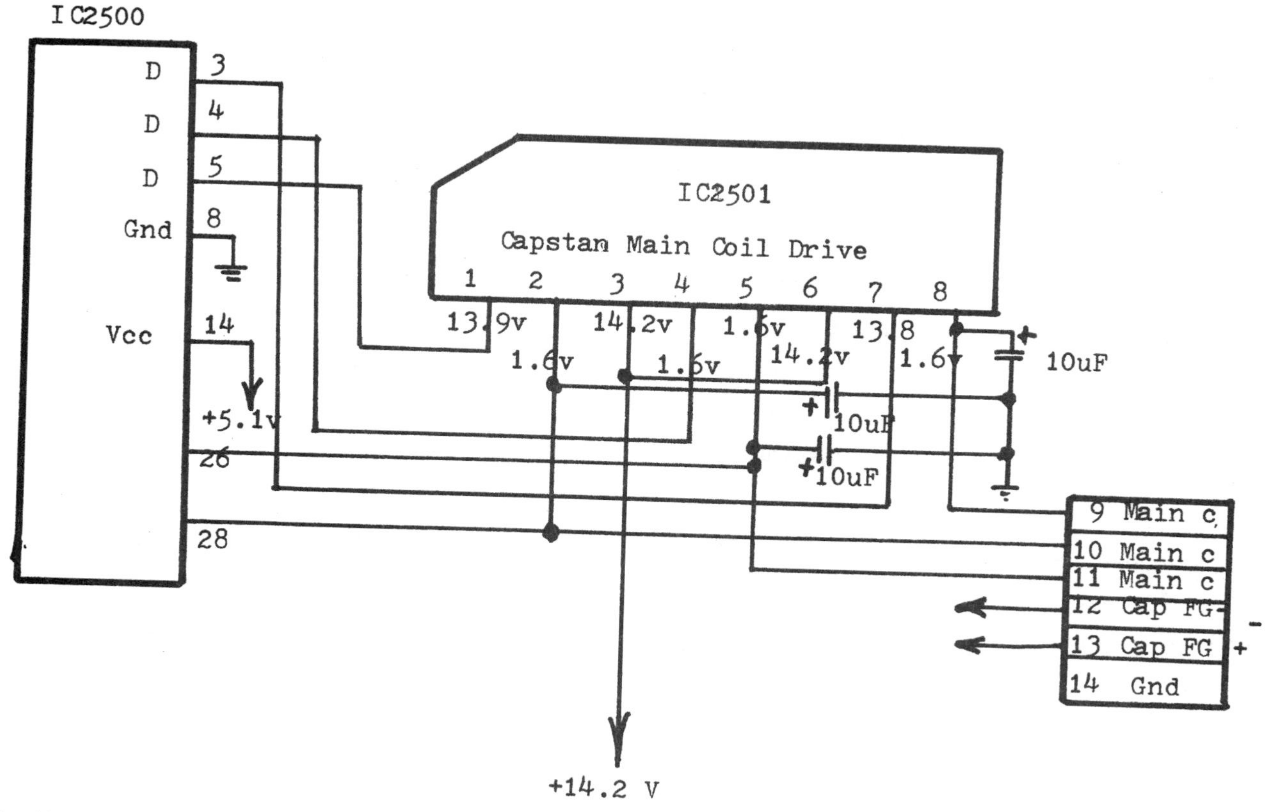

Fig. 12-8. Suspect leaky capstan drive IC2501 when the supply voltages at Pin 3 and 6 are quite low.

Defective Capstan Motor

Symptoms

- ✔ Tape is stuck and capstan motor is frozen (replace motor or check for bad bearing)
- ✔ Motor continues rotating when power is off
- ✔ Play mode appears to be in pause mode, tape unloads (also check for open small resistor (0.47 ohms) on the capstan motor PCB)
- ✔ Motor fails to rotate (replace motor)
- ✔ Squeals in play mode (clean and lubricate both top and bottom capstan bearings)
- ✔ Vibrates in load and rewind modes (bad capstan motor bearing). Rotate the capstan flywheel by hand to see if it rotates freely or is partially frozen.
- ✔ No eject, eats tape (also check for a cracked motor pulley)
- ✔ VCR shuts down immediately
- ✔ Tape loads halfway, ejects, reels will not turn
- ✔ Power supply begins to smoke (check for short in motor)
- ✔ In play mode, VCR powers up intermittently, shuts down (replace capstan motor)
- ✔ Tape is stuck inside, will not accept tape with no play or fast forward (check rotation)

- ✔ Motor will not rotate (check for defective Hall sensors in capstan motor assembly)
- ✔ Tape speed too fast in record or playback; FG pulses are missing
- ✔ 2.5-amp fuse blows with no playback, rewind, fast-forward, and record modes
- ✔ Motor runs backward, eats tape in play mode
- ✔ Excessive fast speed
- ✔ Cassette stuck inside, will not play or eject, might be intermittent trying to load
- ✔ Jerky motion in play mode (bad capstan bearing)
- ✔ Groaning or grinding noise (dry or worn capstan)
- ✔ Speed is slow, drag or wow symptoms heard in slow speeds (clean hardened, old grease from the capstan bearings, relube, and replace) (*Fig. 12-9*).
- ✔ Audio garbled in EP mode
- ✔ VCR eats tape intermittently (replace idler, motor drive belts, bad capstan motor)
- ✔ Capstan motor stops after fast search mode; replace the capstan motor if a cleanup job does not solve the speed problem.

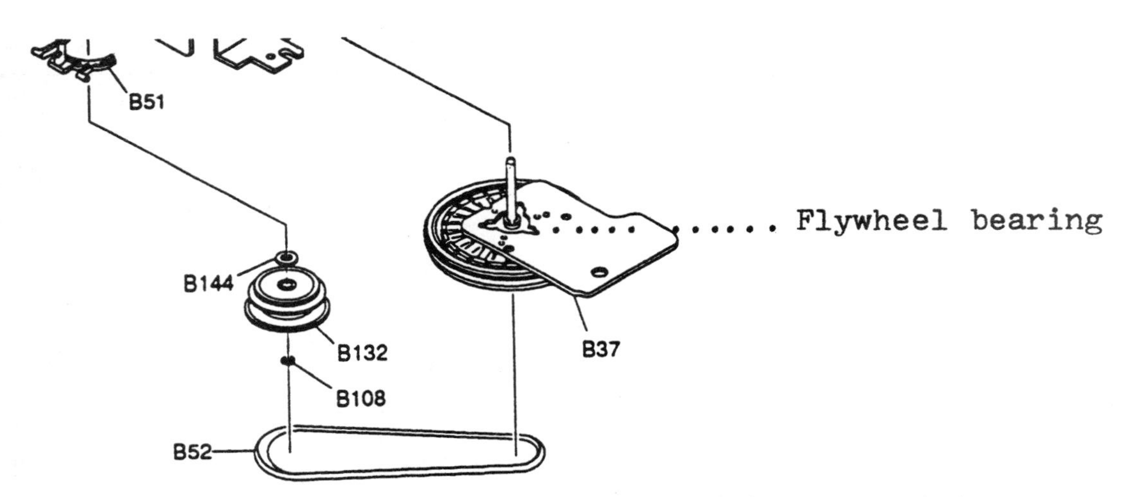

Fig. 12-9. Remove the flywheel and clean out old grease when the capstan drags or moves slowly.

Defective Supply Voltage

Possible Causes:

- ✔ Bad fuses (capstan motor fails to rotate, no voltage from power supply)
- ✔ Low or no voltage in the power supply (check all large filter capacitors (1000 to 4700 µF) electrolytics using ESR meter)
- ✔ Open or burned low ohm resistors (0.47 ohm) in the power sources to servo or motor driver IC circuits
- ✔ Bad filter capacitors (noisy picture, squeals when capstan motor rotates)

- ✔ Leaky or shorted bypass capacitors in capstan CBA (capstan motor keeps rotating, speed is erratic)
- ✔ Open electrolytics (5-volt source is low causing capstan motor to run fast in play, all other functions normal)
- ✔ Bad electrolytic in the capstan power source (capstan has rattling sound)
- ✔ Badly soldered connections in power supply (capstan runs fast, might shut down when play is pressed with a damaged tape). Check for intermittent and poorly soldered connections with the ESR meter.
- ✔ Defective diodes in power supply (capstan motor remains dead)
- ✔ 220 to 330 µF electrolytics in the power supply (slow capstan rotation in SP, no operation in LP and EP modes)
- ✔ Open low-ohm resistor in the power supply with no motor B+ voltage (no tape loading, capstan will not rotate)
- ✔ Open 2-amp fuse (unit loads, unloads, drum or capstan motor does not rotate)
- ✔ Blown fuse during play, rewind, fast-forward
- ✔ Bad diode in voltage power source (intermittent capstan start and run)

Capstan Jerks – Panasonic PVM2028

Suspect an open 120-µF, 200-volt electrolytic (C1004) in the power supply of a Panasonic PVM2028 TV/VCR when the capstan jerks while rotating (*Fig. 12-10*). Check the suspected electrolytics with the ESR meter. It's best to check all electrolytics in the low-voltage power supply with the ESR meter.

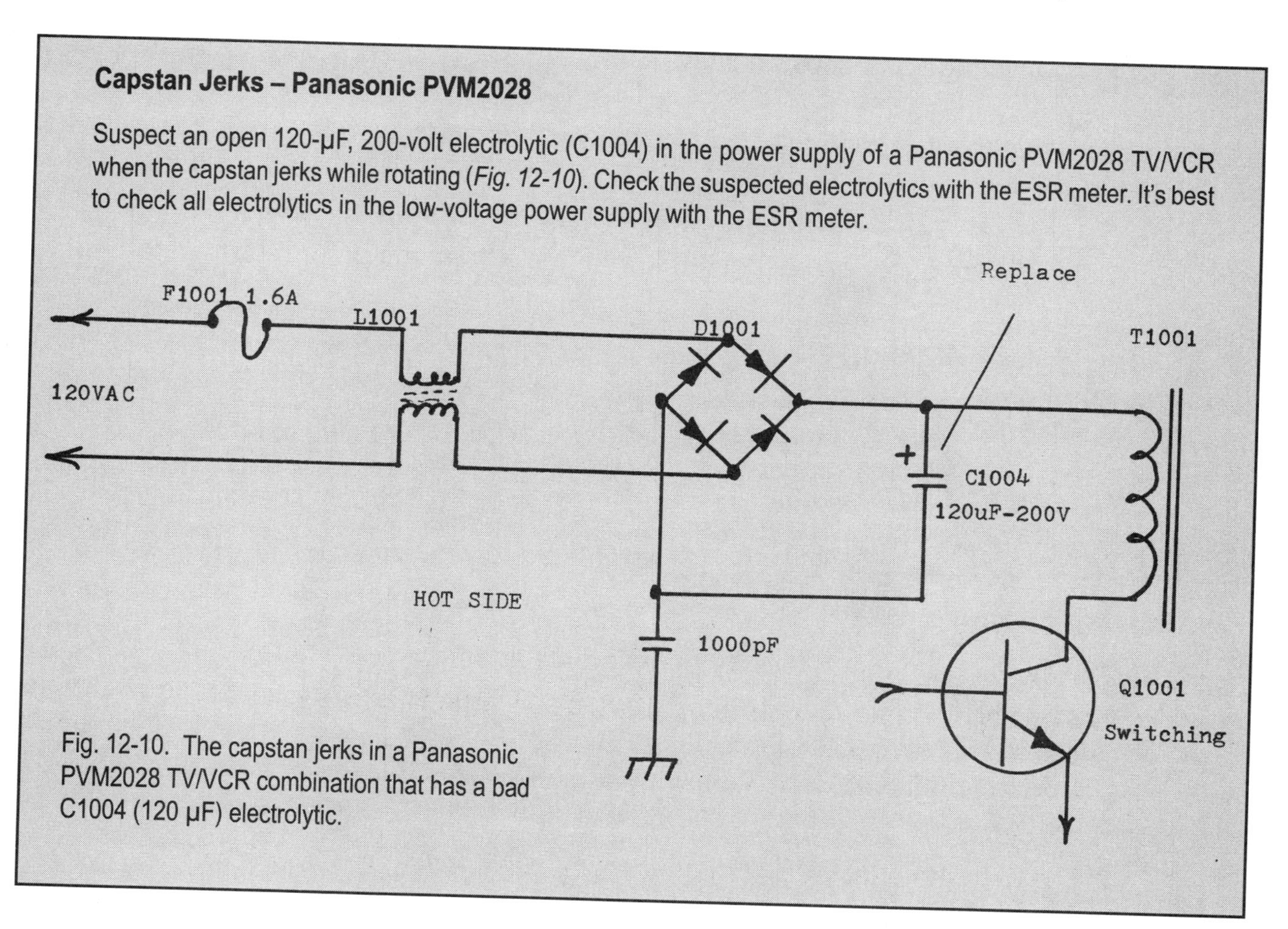

Fig. 12-10. The capstan jerks in a Panasonic PVM2028 TV/VCR combination that has a bad C1004 (120 µF) electrolytic.

- ✔ Burned resistor in the power source (capstan motor runs slow)
- ✔ Bad transistor inverter of the power supply (capstan motor has little torque, ejects slow, tape gets stuck)
- ✔ Badly soldered joint of the transistor on the mechanism PCB (capstan motor will not rotate)
- ✔ Defective filter electrolytics in the power supply (erratic and chattering capstan motor)
- ✔ Badly soldered connection on plugs of the low side of the FG head (no capstan rotation in play mode, shuts off)
- ✔ Bad tape end lamp (cylinder or capstan will not rotate)
- ✔ Dirty capstan brake arm felt pad (tape gets caught or stuck, supply reels and capstan motor run slow; clean or replace)

Bad Capstan Regulator IC

Symptoms

- ✔ No capstan rotation, no play, fast-forward, rewind, eats tape
- ✔ Erratic capstan motor in fast-forward, no drum motor operation
- ✔ Low output voltage and capstan jitters as speed lowers after playing long cassette recording (*Fig. 12-11*) (bad regulator in 13-volt supply)
- ✔ Capstan motor loses speed control, motor become overly warm
- ✔ No capstan rotation in play, shuts down
- ✔ No capstan operation, ejects tape when loaded (leaky voltage reference IC; apply coolant or freeze spray on suspected ICs to make them act up or return to normal)

Fig. 12-11. A bad capstan regulator IC can be the cause of no capstan rotation, capstan jitters, and erratic capstan operation.

Bad Servo IC

Symptoms

- ✔ Capstan motor runs slow or stops
- ✔ VCR shuts down after loading
- ✔ Drum and capstan speeds are inconsistent
- ✔ Tape loads three-quarters of the way, then unloads
- ✔ Capstan runs at high speed in playback, fast-forward, and rewind; motor does not lock in play mode (leaky capacitors off of the terminals of the servo IC)
- ✔ Capstan runs slow, no pause mode (bad transistor voltage regulator on 5-volt line)
- ✔ Capstan servo locked, no capstan rotation (bad capacitors in the servo circuits)
- ✔ Capstan motor turns slowly or stops (defective ICs on the servo PCB)
- ✔ Capstan motor does not rotate in play mode

Defective Capstan Motor Drive IC

Symptoms

- ✔ No capstan motor rotation, fast-forward, rewind, play mode seems like pause mode
- ✔ VCR powers up, capstan runs constantly, no on/off switching
- ✔ No capstan motor rotation (also check for open fuse)
- ✔ No playback, no take-up reel, capstan driver gets hot
- ✔ Inconsistent capstan speed
- ✔ No capstan rotation, shuts down after 5 seconds
- ✔ Play mode shuts down after 30 minutes or up to four hours; use DMM to monitor voltage applied to the capstan motor
- ✔ VCR shuts down after 15 to 20 minutes of operation
- ✔ VCR locks up
- ✔ No capstan motion in play or record modes
- ✔ Shorted capacitors in the drive circuits
- ✔ No cassette loading, no capstan motion, no voltage at the capstan motor terminal (leaky bypass capacitors off of the capstan motor drive IC terminals)
- ✔ Intermittent capstan speed, speed will not increase in selecting search, fast-forward, or rewind modes
- ✔ Play mode goes to stop after a few seconds, no fast-forward, supply post does not fully load (poorly soldered connections on capstan motor drive IC terminals)
- ✔ Playback button causes fast-forward search (badly soldered connections on the capstan motor driver IC)
- ✔ VCR powers up, capstan runs constantly, no on/off switching (no +5 volts, no +12 volts, the regulator low 6 volts)

Fast Capstan Motor

Possible Causes:

- ✔ Defective capstan motor
- ✔ Bad capacitors on the servo PCB (runs fast in play or cannot select record mode speeds)
- ✔ Leaky diodes in the power source of the capstan drive IC (capstan rotating very fast)
- ✔ Badly solder joints around resistors in the power supply (runs fast, shuts down when play is selected)
- ✔ Leaky 0.1 uF SMD capacitor in the FG line off of the servo IC (capstan motor goes fast in a runaway, has blue screen, pause picture okay)
- ✔ Leaky bypass capacitor (0.047 uF) off of the servo IC (runs fast in playback, fast-forward, rewind)
- ✔ Defective capstan motor with the FG pulse missing (very fast tape speed in playback and record modes)
- ✔ Glazed pinch roller, tape riding high on the AC head (fast capstan speed with a chattering and erratic capstan motor (*Fig. 12-12*)

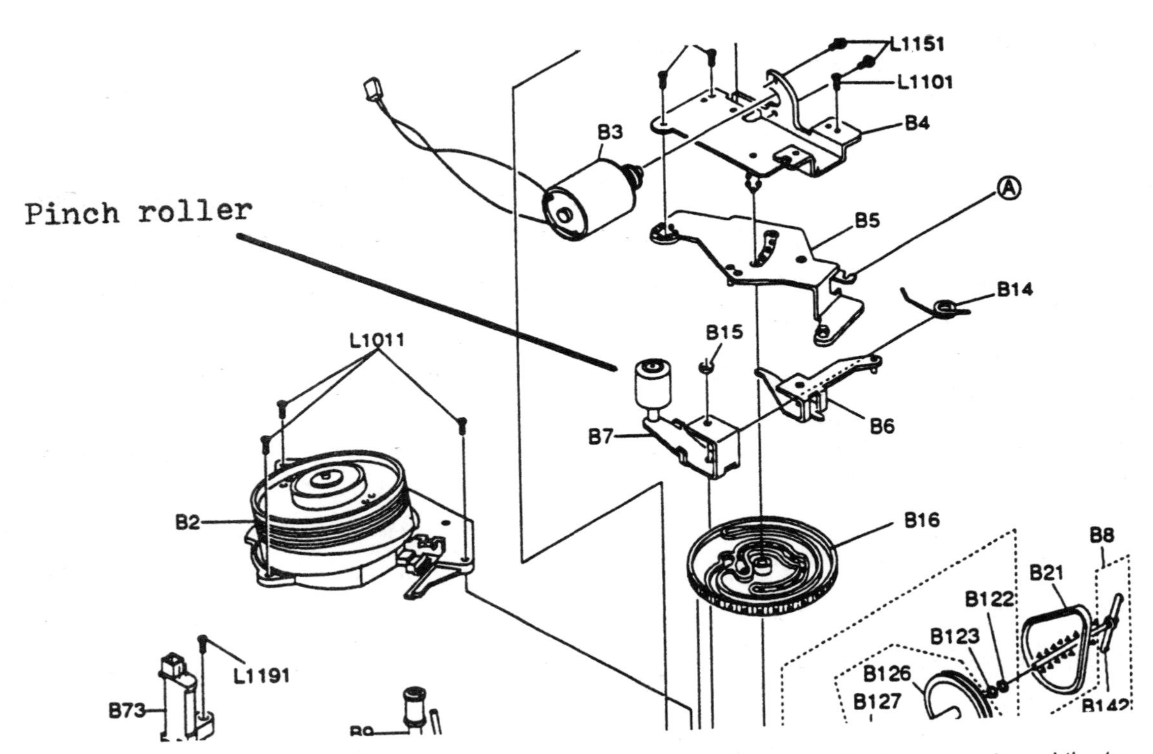

Fig. 12-12. Check for a glazed or worn rubber pinch roller, which can cause erratic tape speeds and the tape to be eaten.

- ✔ Plastic E-ring that holds capstan spindle in place (replace when capstan speed varies)
- ✔ Bad servo or capstan drive IC (playback rotates at super high speed, capstan rotating at full speed)
- ✔ Open or dried-up 1-33 µF electrolytics in servo and capstan IC circuits (capstan runs at full speed; test each electrolytic with the ESR meter)
- ✔ Capstan control transistor (capstan motor runs too fast, rewind and fast-forward shut off after a few minutes)
- ✔ Bad PG head (capstan motor runs too fast; try to resolder all leads or replace the head)
- ✔ Loose capstan magnet
- ✔ Electrolytic (1000 µF) capacitor (5-volt line is low to the capstan drive IC and the capstan motor runs fast in all functions)

Slow Capstan Motor

Check all electrolytic capacitors in the power supply for slow capstan rotation. Test each electrolytic, 180 to 330 µF, in the power supply for slow rotation in SP and no running in LP or EP modes.

Possible Causes:

- ✔ Bad capstan motor (tape speed too slow)
- ✔ Low-ohm resistors (0.47 to 2.2 ohm)
- ✔ IC on servo PCB (capstan runs very slow, sometimes stops)
- ✔ Bad 13-volt IC regulator (capstan speed goes low after a long-playing cassette)
- ✔ Defective capstan motor driver IC (capstan motor turns slow, stops)
- ✔ Bad inverter transistor in the power supply (may cause capstan to have little torque, slow eject, tape gets stuck)
- ✔ Bad 5-volt regulator transistor on the main board (capstan runs slow, no pause mode)
- ✔ Bad or broken capstan brake (capstan motor not rotating fast enough; clean the bearings or replace it)

Capstan Action Eats Tape

Possible Causes:

- ✔ Defective +5-volt regulator (no capstan action, no play, fast-forward, rewind modes, eats tape)
- ✔ Bad capstan stators assembly and Hall devices (rotate the tape by hand to remove)
- ✔ Defective capstan motor, cracked pulley (no eject, eats tape)
- ✔ Check capstan motor to see if it is running backward if it eats tape in play mode

- ✔ Dirty brake capstan pad (slows capstan, eats tape in eject mode)
- ✔ Capstan motor with a dead spot (intermittently eats tape with no fast-forward or rewind modes)
- ✔ Defective capstan motor after replacing idler and drive belts
- ✔ Stripped motor pulley (repair cracked motor pulley with epoxy or replace)
- ✔ Check for poorly soldered connections at capstan motor terminals when tape is damaged in LP mode
- ✔ Unhooked tension spring (*Fig. 12-13*)
- ✔ Poorly soldered joints around resistors in power supply board

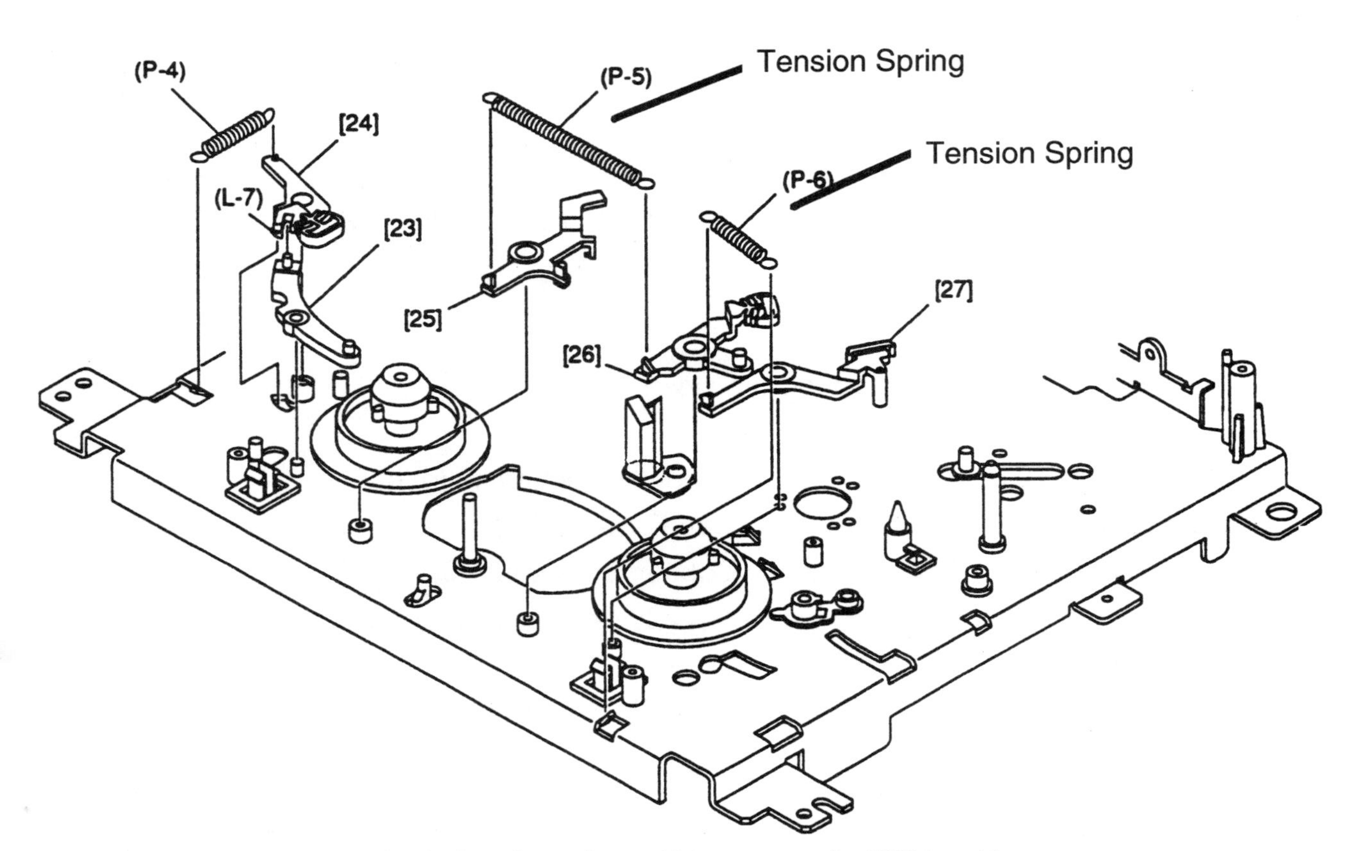

Fig. 12-13. Check for loose or unhooked tension springs, which can cause the VCR to eat tapes.

Cylinder or Drum Motor

The video head drum or cylinder rotates from a direct-drive motor or by a servo-controlled motor. The motor might consist of three separate windings, magnets, and no brushes. A servo circuit switches the three coils on and off in the drum motor. Correct timing of the switching is accomplished by a feedback mechanism supplied by the Hall-effect

ICs. The bottom half of the drum or cylinder contains the motor coils, while the top half contains the video heads. The video head drum or cylinder records and plays back the video on the tape (*Fig. 12-14*).

The Hall-effect ICs generate two types of pulses:

PG (pulse generator) pulse
FG (frequency generator) pulse

The 30-Hz (PG) signal identifies the rotation of the drum as it spins and the FG signal (180 Hz) indicates the speed of the video drum or cylinder. The Hall-effect ICs are mounted close to the spinning drum or cylinder.

Fig. 12-14. The video head drum or cylinder plays back and records video and sound on the tape.

Drum and Cylinder Circuits

In most TV/VCR transports the servo/system control IC signal is fed directly to the drum or cylinder motor terminals (*Fig. 12-15*). The PG and FG pulses are fed directly from the motor to IC201, while the D. Cont. signal is fed through OP amp (IC203) before reaching IC201. Also, IC201 controls the capstan motor circuits.

The RCA TV/VCR combo system control IC provides signals to the three drum motor coils through several inductance coils. D data out of A is found at pin 28 of IT002, D out of B at pin 29, and D out of C at pin 30 (*Fig. 12-16*). All three drum motor windings can be checked on the low ohmmeter scale for open windings. Likewise, use the DMM ohmmeter range or ESR meter to check for continuity from system control pins to corresponding motor drum coils.

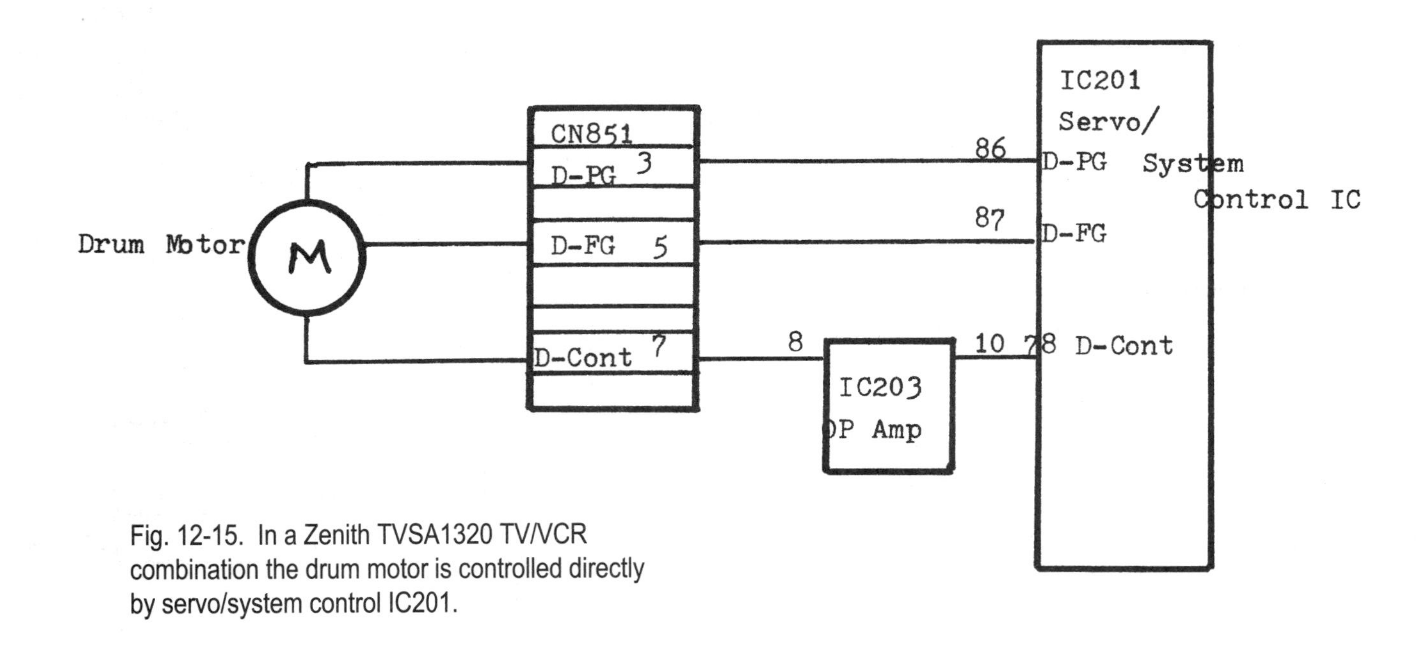

Fig. 12-15. In a Zenith TVSA1320 TV/VCR combination the drum motor is controlled directly by servo/system control IC201.

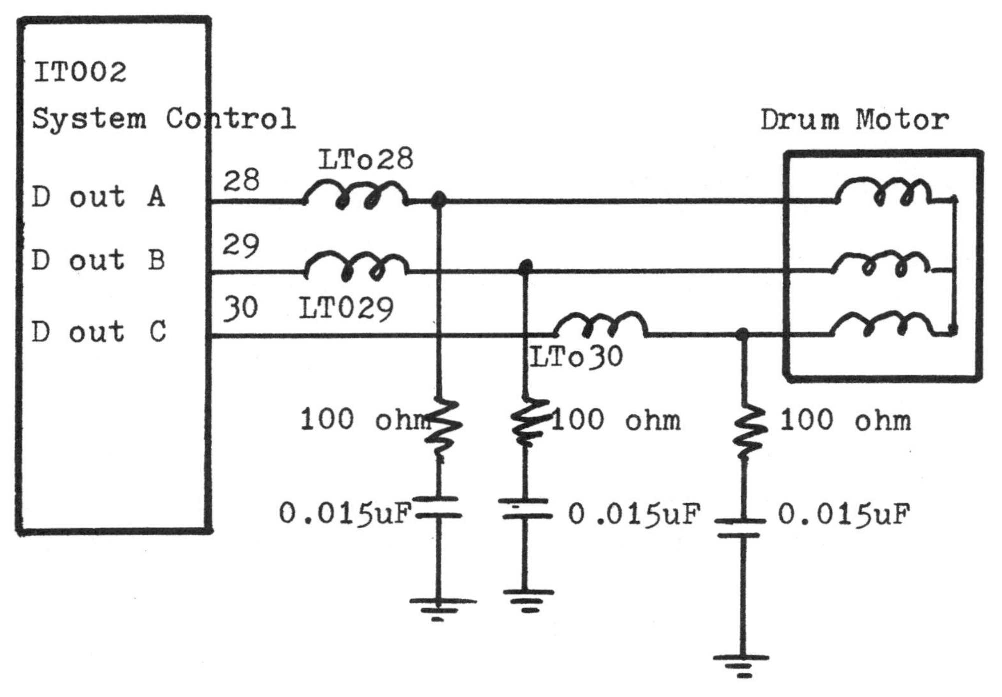

Fig. 12-16. In the drum motor in the RCA 13TVR60 TV/VCR combination, the system control IC IT002 controls the rotation of the drum motor.

The Emerson VT1920 TV/VCR transport system control IC1001 sends a signal to the Servo IC (IC2001), which in turn controls the three coils in the drum or cylinder motor assembly (*Fig. 12-17*). Critical voltage measurements on the system control and servo IC must be made to determine if the servo IC is either leaky or defective when the drum or cylinder does not rotate. An extra hot servo IC might indicate the IC is defective and should be replaced. The defective servo IC will prevent cylinder rotation, causing it to run too fast or too slow, or resulting in intermittent drum movement.

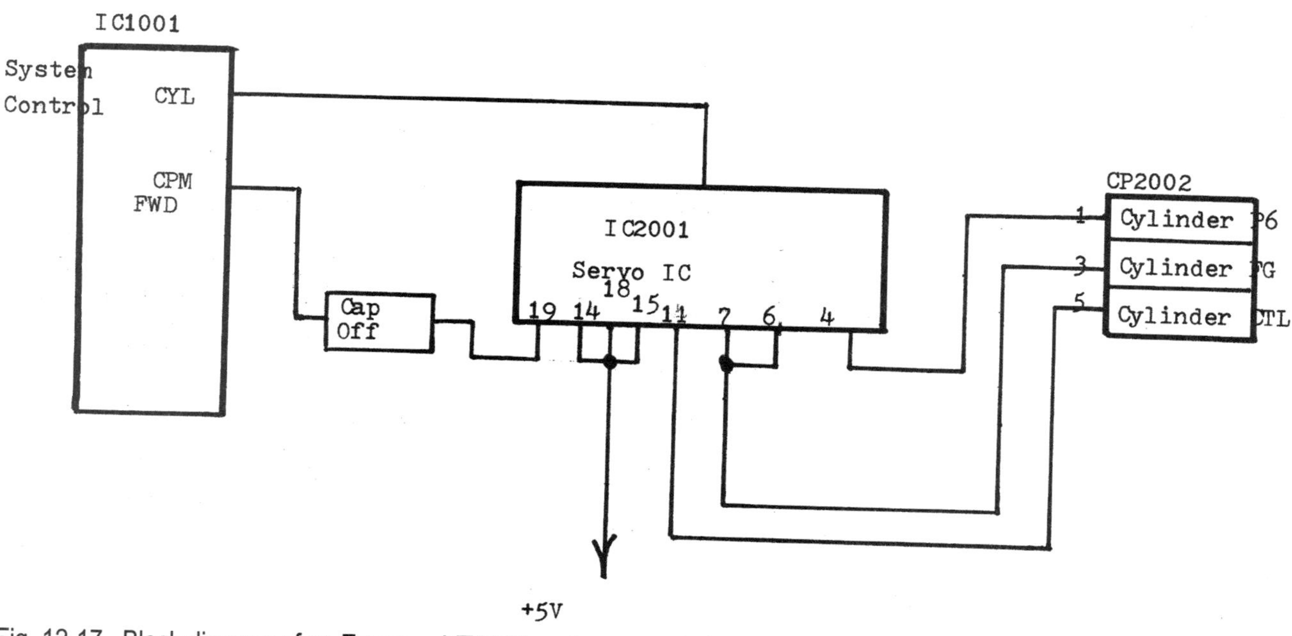

Fig. 12-17. Block diagram of an Emerson VT1920 system control IC1001 cylinder drive IC.

Drum Motor Problems

A defective drum motor can take many forms and actually cause a variety of symptoms depending on what part of the motor the problems stems from.

- ✔ If the drum motor assembly is bad, the video may intermittently shake and change after pressing stop.
- ✔ A defective drum motor may cause the VCR to load then shutdown.
- ✔ When the lower drum assembly is defective, VCR will not record video.
- ✔ If the lower drum assembly is defective, the heads may not spin and the picture playback may be poor.
- ✔ If white lines appear on the tape after recording, the upper cylinder may need to be replaced.
- ✔ If the upper cylinder is bad, there may be excessive snow in the picture.

- ✔ A bad drum CBA assembly may cause erratic speed. Solder all connections on drum stators. Replace drum stators assembly if soldering does not solve the erratic speed problem.
- ✔ Resoldering the cylinder PCB and replacing the drum motor stators may be necessary if there is no cylinder rotation, intermittent movement, and noise in the picture.
- ✔ You may need to replace drum stators if the cylinder runs fast intermittently in play mode and then shuts down.
- ✔ A bad cylinder motor can cause diagonal lines in picture when first turned on.
- ✔ A bad lower cylinder may cause the tape to load halfway and retract and no cylinder motor movement.
- ✔ A bad lower cylinder assembly could cause lines in playback and appear to be a problem with a defective head. It can also cause the VCR to shutdown in play and record mode.
- ✔ Resoldering connections at the PC board and where socket connects to cylinder servo may cure intermittent cylinder movement.
- ✔ A bad cylinder motor assembly may result in the transport shutting down when the power is turned on.
- ✔ If there are bad lower drum PCB connections, the cylinder may not rotate.
- ✔ Dirty video heads can cause noisy or bad video, or a lack of video and audio.
- ✔ Loose pads on drum motor stators assembly may cause no drum FG pulse, the VCR to load tape around the drum and then shutdown.
- ✔ An ungrounded upper cylinder may cause noisy video playback and may look like a poor tracking adjustment problem.
- ✔ A bad static discharge arm on upper cylinder can create noisy, high-pitched squeals in play and record modes.

Chapter 13: Repairing VCR Playback, Recording, Color, and Video Circuits

The video tape heads are the main components within the playback and recording circuits. The magnetic video heads provide video recording and playback as well as audio and control information. In record mode the video heads apply the signal to the magnetic head and the revolving tape. While in playback the video heads pick up the recorded signal, amplify it, and apply it to the playback circuits (*Fig. 13-1*).

Fig. 13-1. The video heads pickup the signal off of the tape, amplify it, and apply it to the video and color sections of the TV screen in when in playback mode.

The Various Tape Heads

The video heads and all components touching the moving tape should be clean at all times. Problems involving dirty video heads include:

- ✔ Low and garbled sound due to excessive buildup of tape oxide.
- ✔ Snow and lines in the recorded picture.
- ✔ A washed out and snowy picture.
- ✔ Poor color or no color in the picture due to a dirty or worn video head.
- ✔ No audio or video.
- ✔ Vertical and horizontal jitters in the screen.

Often, cleaning up the tape heads can solve many video and sound problems found in TV/VCR combos. Make sure to always clean the video heads to remove packed oxide dust that has accumulated on the video heads and remove dirt from the revolving tape. After cleaning up and repairing the heads, check the tape speed with a test speed cassette, checking the video and sound at all playback speeds (SP, LP and SLP). Demagnetize the tape heads after all repairs are made.

When the revolving tape moves too fast past the upper head drum assembly, static electricity can build on the video heads. A static guard or earth brush is located at the top of the video heads to prevent static electricity buildup and leaks of the static electricity to ground. A bent static guard can cause snow and lines across the TV screen in playback mode and a loud scraping noise might be heard (*Fig. 13-2*).

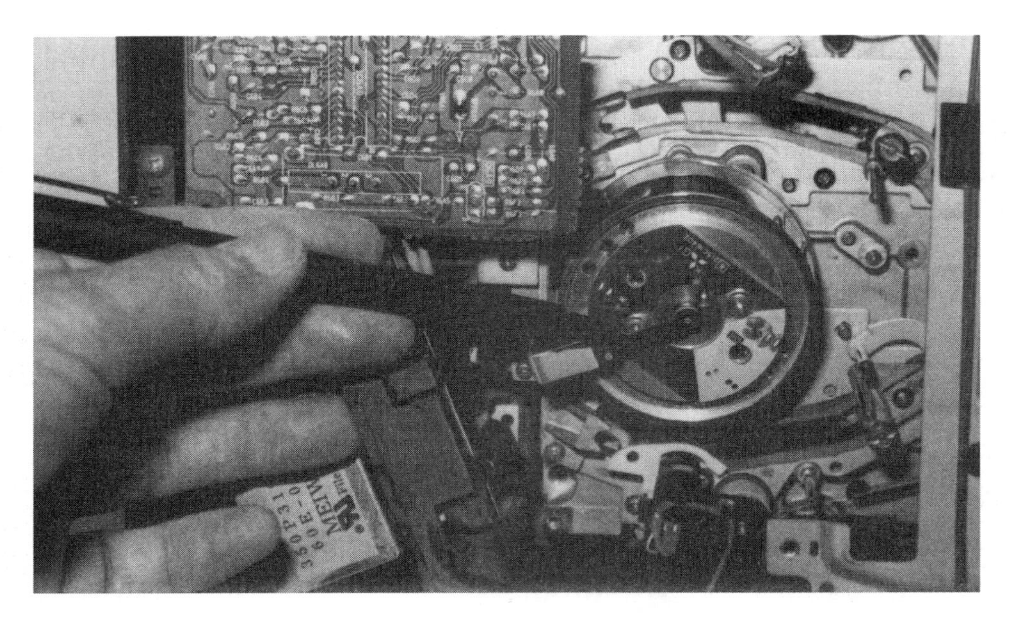

Fig. 13-2. The static guard may be bent out of alignment or have a poor grounding contact if there is a snowy picture or lines in the picture.

In addition to the video heads, the VCR might contain a stationary full erase head, fixed audio head, stationary track control head, and rotary video heads. The video heads are found in the revolving upper cylinder or drum and the other heads are stationary within the tape path. The full erase head is mounted on the VCR deck next to the video heads. A full erase head erases the entire tape during the recording mode. If a flying erase head is mounted in with the video heads in the upper cylinder, a stationary full erase head will not be found.

A stationary audio head (AC) (*Fig. 13-3*)applies sound to the tape in record mode and in playback; the audio head picks up the sound from the audio track. In a VHS VCR, the audio track is recorded to the outside of the tape in mono recording and has two different tape tracks in stereo operation. The audio control erase (ACE) control head contains both the erase and control head within the same component. A dirty control head can cause the tape to slow down or speed up. A defective AC head can cause incorrect and intermittent speed.

Fig. 13-3. The AC control head contains the erase and control signals to the servo control microprocessor.

A dirty erase or control head can result in low audio and slow rewind mode. During recording the control track head controls a series of 30-hertz pulses to sync the audio heads during playback. The ACE head must have a special tool to make adjustments of the control head assembly. Readjustment of the tracking control can help to eliminate or reduce lines in the picture.

The flying erase head is mounted before the video heads, within the cylinder or drum assembly. The flying erase head erases all video tracks and cleans them up for a new recording. The flying erase head provides a clean, glitch free picture. The flying erase head can erase two different video tracks at the same time; no rainbow effects are found in the picture with the flying erase head. The flying erase head was first used in the VCR camcorder (*Fig. 13-4*).

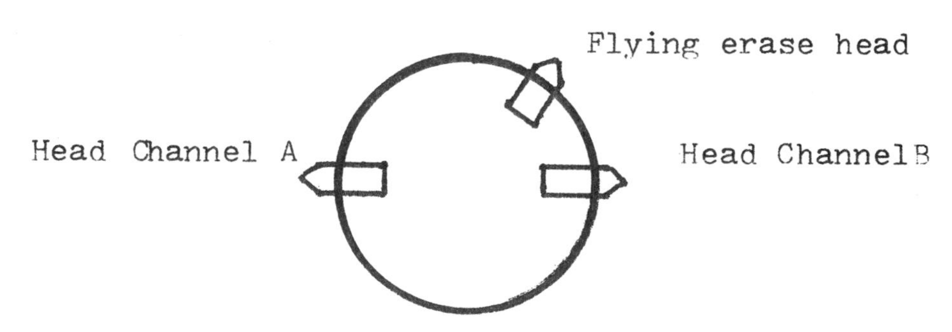

Fig. 13-4. The flying erase head erases all video tracks on the tape and is located at the cylinder drum assembly.

Video Record and Playback

In a Zenith 13-inch TV/VCR the video recording chroma (C) and video recording (Y) signal are fed into pin 44 of the Y/C process IC 401. Likewise, the record-Y (Rec-Y), record chroma (Rec-C), playback Y (PB-Y), and playback chroma (PB-C) signal are fed out of pin 41 of IC401 to the chroma/tuner circuits. After passing through the brightener and chroma (Y/C) comb filter circuit of IC401, the recording chroma signal (Rec-C) is found at pin 11 of IC401 and fed to pin 10 of the AGC circuit within the video head amp IC470. The recording Y signal (Rec-Y) from pin 14 of IC401 is found at pin 9 of video head amp (IC471).

In record mode, the video signal from each video head is switched inside the switching head amp (IC470). The video head signal is applied as the tape passes the video head and then switches off, and then applied to the next video head opposite the previous head. In record mode, both the left and right video heads apply the Rec-Y signal and Rec-C signal to the video amp (IC470). In playback mode, both heads pick up the playback (PB-Y) and chroma signal (PB-C) and connect the head signal to pins 16 and 19 of the video head amp. The timing of the video head switching and the flying erase head are controlled by the system control IC.

Audio Record and Playback

The audio playback signal is picked up by the ACE head in a Zenith TVSA1320 VCR. The playback signal from the audio head is fed to pin 6 of the audio process IC401. Likewise the record-audio signal is fed from pin 64 of IC401 to the audio head in record mode. A bias oscillator (Q853) signal to the audio erase head is found within the ACE audio control head. A full erase head (FE) erases the entire recording track with a bias oscillator signal out of Q853. The bias oscillator waveform can be scoped at the tape head to determine if the bias oscillator circuit is functioning. The audio playback (PB-audio) and audio record (Rec-audio) signal from the audio process IC is fed to input pin 50 of the input select (IC301). The audio signal is amplified, alternated, and fed out of pin 51 of IC301. From here the audio signal is sent to the audio output IC801. The audio signal is coupled to the headphone jack and pm speaker.

The AC control head circuit, found in the same component as the ACE, is controlled directly by the system control IC. The positive and negative control head signal is found at pins 92 and 91 of servo/system control (IC201) (*Fig. 13-5*). Suspect a dirty or defective AC head, when playback is fast in EP mode. A bad ACE head can cause slow speeds in record mode.

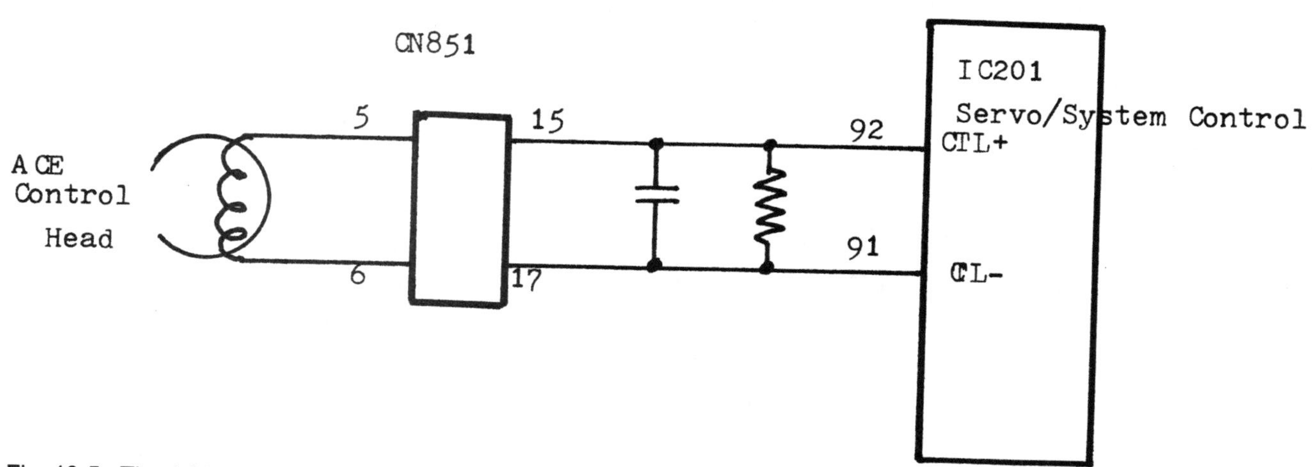

Fig. 13-5. The ACE control head feeds directly into the servo/system control IC201 at Pins 91 and 92.

Video Playback Symptoms

No Video in Playback

- ✔ Check for an open fuseable resistor in the main power supply PCB when there is no video in play mode and the audio is okay, but then the VCR shuts off in two to four seconds.
- ✔ Check for a bad fuse connection with a loss of video or audio in the VCR.
- ✔ A leaky luma/chroma transistor can cause no video with a blue screen and normal audio in playback mode.
- ✔ Check for a bad diode on the AGC line when there is no video or audio, and no cassette loading.
- ✔ No video or audio playback with an audio hissing sound can be caused by shorted diode on the 30-volt line.
- ✔ No video, normal audio, and a black screen in playback can be caused by a defective video/audio process IC.
- ✔ A bad switching transistor in the video PCB can cause no video, normal audio, and record is okay in playback mode.
- ✔ A leaky record/play switch transistor can result in no video in playback with normal audio and might appear to be a bad video head.
- ✔ Suspect a bad preamp transistor on the preamp PCB when there is no video in playback at any speed, but audio and recording is normal.
- ✔ No video in playback when the tuner video is okay can be caused by a bad tuner IC.
- ✔ A shorted video buffer transistor can cause a no video problem in playback. Check small electrolytics (4.7-10 μF) on the video PCB for no video in playback with normal audio.

- ✔ Check the small electrolytics (1 µF, 50 V) for no video in playback and at a very high rate of speed.
- ✔ Check the 4.7 µF electrolytic off of the servo IC with an ESR meter if there is no video, but the tuner video is okay.

Bad Video in Playback

- ✔ Check for several badly soldered joints in the video circuits for intermittent video.
- ✔ A bad head switching relay or a poor ground connection can cause a snowy picture in EP or SP mode.
- ✔ Poor picture quality can result from a bad head switching relay.
- ✔ Check for a head switching point poor adjustment with half of the screen black in playback.
- ✔ When the playback picture looks like bad heads, replace the lower drum assembly.
- ✔ Suspect the upper drum assembly when there is a not-so-clear picture in playback. Take a quick ohmmeter measurement of the relay and drum assembly to see if the winding is open.

Intermittent Erase - RCA 13TVR60

Double-talk and distortion was discovered on recorded tapes after another recording was made over it. The previous recording was not fully erased in a RCA 13-inch TV/VCR. All connections were checked at the full erase head and appeared normal. When moving a coil (LS030) with an insulated probe, the erase recording would cut in and out. Several attempts in recording were made before the troublesome component was found. All connections were resoldered on the bias oscillator coil (LS030) and this solved the double-talk symptom (*Fig. 13-6*).

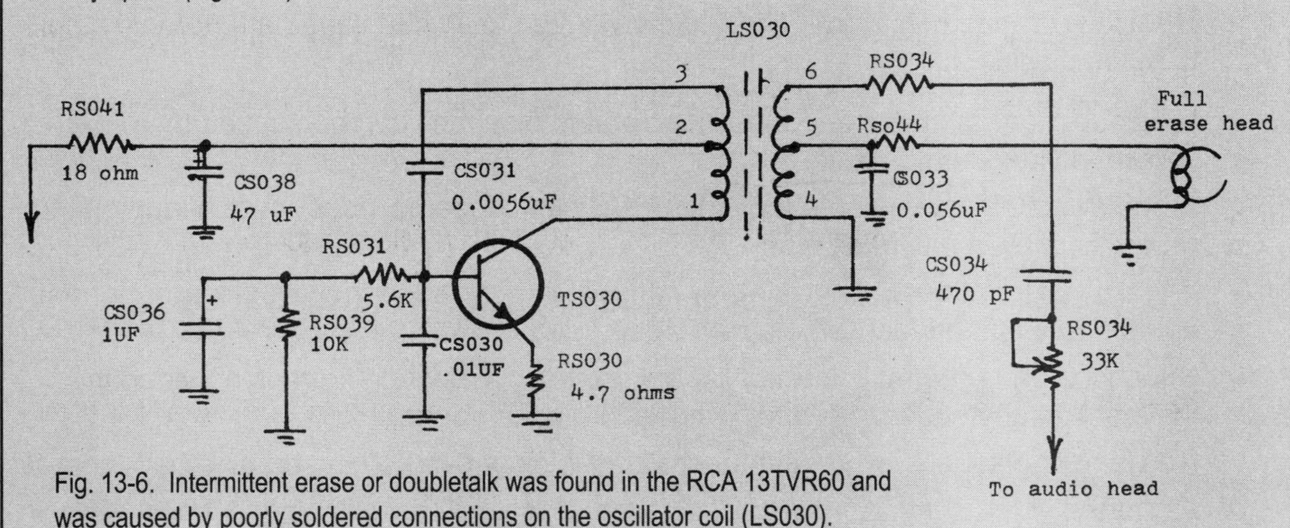

Fig. 13-6. Intermittent erase or doubletalk was found in the RCA 13TVR60 and was caused by poorly soldered connections on the oscillator coil (LS030).

- ✔ Suspect a defective IC input switch on the main PCB when there is no video and the audio is okay in playback mode.
- ✔ A bad IC on the head amp PCB can cause washed out video in playback.
- ✔ Suspect a defective CTL pulse amp IC when there is snowy video in SP and playback mode.
- ✔ A defective 3-MHz low-pass filter network can cause a washed out video tuner and no playback symptom.
- ✔ Check for a defective IC voltage regulator when the video gets weak and appears to be an AGC problem.
- ✔ Intermittent video and audio can be caused by poor solder connections on the input video IC.
- ✔ A shorted 0.1µF capacitor off of the system control IC can cause a no capstan FG pulse with high speed in playback and no color picture in playback.
- ✔ Check for a defective 10 uF electrolytic with no C+L pulses if there is poor video and audio playback at all speeds.
- ✔ Video smear in the right side of the screen can be caused by a defective 2.2- to 3.3-µF electrolytic.
- ✔ Suspect a large filter capacitor in the power supply (3300 µF), when the video is distorted in the tuner and there is no audio in playback.
- ✔ Replace the 1-µF, 50-volt electrolytic for distorted video in playback.

Audio Playback Problems

Distorted or Bad Audio

- ✔ Clean up the audio switching relay contacts for bad hum in the audio.
- ✔ Defective or dirty contacts in the audio switching relay can cause an motor-like noise in audio playback.
- ✔ A bad relay can cause cracking and static in the audio playback mode.
- ✔ The automatic head cleaner found in some VCR's can leave residue on the ACE head causing garbled sound and audio distortion in the speaker.
- ✔ A glazed pressure roller and bad belts can cause no audio in record and muffled audio in playback.
- ✔ Look for a badly soldered joint at the audio-in jack/switch when the audio is low and muffled.
- ✔ Test all small 4.7-10 µF electrolytics on the main PCB when the audio plays the wrong speed of SP at EP and EP garbled audio.
- ✔ A bad servo IC can cause garbled sound in the EP mode. Check those electrolytics (33 µF) in the servo circuit and on the capstan phase error line when audio is garbled in playback or the sound is erratic and the video playback is jumpy. It may appear to be a bad capstan motor.

No Audio

- ✔ Check for an open 4.7 ohm resistor on the switch +12-volt line when there is no audio in playback, tuner, and normal video.
- ✔ No audio playback and normal video can be caused by a 22 µF electrolytic off of the servo IC.
- ✔ No audio playback can be caused by a defective 1-µF, 50-volt capacitor in the audio circuits.
- ✔ A defective diode on the +6-volt line can cause no audio or video.

Horizontal Pulling – Panasonic PVM2078

Horizontal pulling developed at the top of the picture in a Panasonic PV-M2078 TV/VCR. The C533 (.0022 µF) capacitor was shorted. It was replaced. No sound or high voltage was found in another Panasonic PVM2078 TV/VCR. In addition to replacing C533 at 2KV, a shorted Q551 output transistor and a 10K (R004) resistor were found to be open causing the loss of sound or high voltage.

Noise Lines in the Picture

- ✔ Clean up the lower drum assembly with the Hall IC magnet when a poor picture is present in EP and SP speeds with lines in half of the picture.
- ✔ A bad 10-µF capacitor with ESR problems and no CTL pulses can cause poor audio and video with lines in the picture at all speeds of playback.
- ✔ When the video has color bars rolling through the picture in playback check all 100-200 µF electrolytics.
- ✔ A binding loading assembly can cause speed problems and lines in the picture.
- ✔ Noisy lines in record or play can be caused by a loose head switching relay and poor guides on the cylinder assembly.
- ✔ A stuck or dirty head switching relay can cause noisy lines in half of the picture.
- ✔ No video playback with flashing white lines can be caused by a 47 µF electrolytic on the main PCB; test each electrolytic on the main PCB with the ESR meter.
- ✔ Although audio playback with lines running through the picture looks like a bad drum motor, replace the defective servo IC. A bad IC can have lines through the video in playback with normal audio. Replace the servo IC causing upper cylinder to run slow and cause lines in the video playback mode.
- ✔ A bad inductance coil in the audio can cause noise and static in the audio.

Poor Tracking - Emerson VT1920

Poor tracking was noticed in an Emerson VT1920 TV/VCR. Adjustment of tracking did not remove lines in the video. After readjustments of the tape guides and securing them into position, the bad tracking problem was solved.

Color Problems

The chroma or color circuits are developed in the video and Y/C process circuits of a 13-inch TV/VCR combination. Both the recording Y and color signals use the Y/C process IC. A defective Y/C process IC can kill both color recording and playback circuits. Just about any component in the color circuits can cause a loss of color. Most of the color circuits' record and playback problems are related to the defective color ICs, transistors, capacitors, and color crystals. No color can result from an open coil (FL501) on the Y/C PCB. A bad tuner with a washed out black and white picture can cause a loss of color as well. Check the preamp IC when color fades in and out with flashing colored lines. Check all filter capacitors in the power supply with the ESR meter when there are lines in the picture and the color is bad.

No Color or Intermittent Color

- ✔ Sometimes contact cement and dirt residue upon the shield of a digital board can cause no color or intermittent color.
- ✔ A shorted or leaky capacitor in the color circuit of the Y/C process IC can also cause no color or intermittent color.
- ✔ A bad capacitor on the Y/C PCB can cause no color or intermittent color.
- ✔ A bad color module can cause a no color symptom.
- ✔ No color symptom might be caused by poor soldered connections, feed-throughs, and connectors.
- ✔ A bad IC in the head circuit can cause loss of color.
- ✔ Check all 0.01 µF capacitors in the PC signal board when a loss of color occurs.
- ✔ Check for a leaky capacitor on the Y/C PCB with the ESR meter when there is intermittent or no color.
- ✔ Badly soldered connections on the preamp/video PCB can cause no color or intermittent color.
- ✔ A shorted capacitor on the servo IC can cause playback at high speed and no color with a good picture.
- ✔ A bad 3.58 MHz crystal can cause a no color symptom; the capstan does not lock up and there is no 60-Hz square wave pulse.

No Color in Playback

- ✔ When the color in playback is saturated on SP and there is no color on EP speeds, suspect a defective delay line (DL001) on the Y/C PCB.
- ✔ Check for a defective IC on the Y/C process board, when the color quits in playback after warm-up.
- ✔ Suspect a leaky transistor on the video main PCB for no color in playback.
- ✔ A bad IC preamp can cause the color to fade out in playback with flashing red lines.
- ✔ A bad IC in the video head circuit can cause no color or flashing colors.
- ✔ Poorly soldered terminals on the Y/C IC can cause a no color symptom in playback.
- ✔ A bad bandpass filter network can cause no color in playback.
- ✔ Poorly soldered joints within the chroma section can cause a no color symptom in playback.
- ✔ Suspect a defective voltage regulator IC that loses the 5-volt source when the color fades out in playback.
- ✔ A defective chroma amp IC can cause no color in playback and the color goes to a black and white picture.
- ✔ A defective 100 µF electrolytic off of the IC line voltage regulator can cause video play back color bars to roll through the picture and the playback audio is normal.
- ✔ Check for defective transistors in the video section on the main PCB when there is color loss in playback and the tuner color is normal.
- ✔ Check for an increase in resistance of resistors in the chroma playback circuit for no color in playback mode.
- ✔ Check for a bad 82-pF, 50-volt capacitor or other small pF capacitors off of the color IC in the chroma line when there is no color in playback.
- ✔ Replace the color crystal when there is no color in playback. Check the color waveform at the crystal terminals with the oscilloscope to determine if the crystal is functioning.
- ✔ A no color symptom can be caused by a defective 0.01-µF, 16-volt capacitor on the signal PCB.
- ✔ A bad 1-µF, 50-volt electrolytic off of the luma/chroma PCB can cause no color in playback with video flagging at the top of the screen.
- ✔ Check the 100-µF, 16-volt electrolytic on the SW +9-volt line in the power supply when there is loss of color in playback and normal color in record mode.

No Color in Record

- ✔ Check all electrolytic capacitors in the power supply for no color in record modes.
- ✔ Test all electrolytics in the power supply section and the color IC for no color in record and normal playback.
- ✔ Check for a change in resistance on the head amp PCB when there is a loss of color in record and normal playback.
- ✔ Check for badly soldered connections at the color module to the main PCB when there is no color in record mode.
- ✔ Test the luminance/chroma transistor for loss of color in record and normal playback.
- ✔ Check for a bad crystal in color record circuit when the color quits after half an hour of operation.
- ✔ Replace the Y/C processing IC when the VCR does not record in color.

Blue Screen - Panasonic PVM2028

The screen went blue in a Panasonic PVM2028 TV/VCR. When this occurred, I checked for a broken (red & white) wire to the AC head (*Fig. 13-7*). This was the problem and when fixed, the video returned to normal.

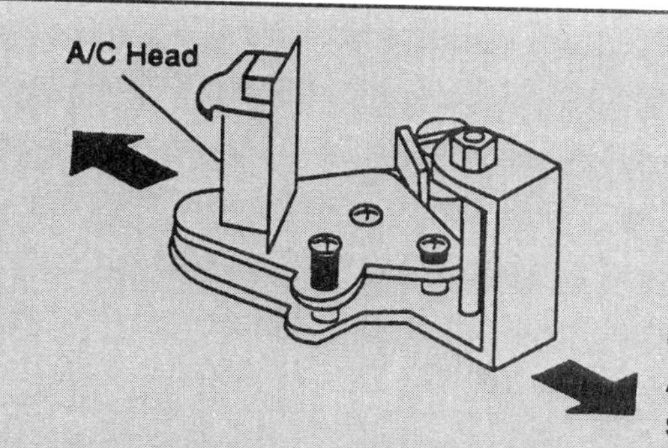

Fig. 13-7. Check for a broken wire on the AC head when the screen goes blue.

No Color in Playback and Record

- ✔ Check for an open coil in the Y/C process circuit for no color in playback or record.
- ✔ A no color symptom in playback and record can be caused by a SMD IC on the bottom side of the PCB and might appear to be a bad Y/C process IC.
- ✔ Check for a defective 51-pF, 50-volt capacitor in the luma/chroma section for no color in playback or record.
- ✔ An open 100-µH coil on the UNSW +5-volt line to the delay line can cause no color in record or playback.
- ✔ Check for a bad 3.3-µF, 10-volt capacitor in the chroma section for no color in playback or record mode.
- ✔ Test and replace the defective flatpack microprocessor when there is no color in record or playback.
- ✔ A defective chroma/luma process IC can cause no color in playback or record mode.

- ✔ A blue screen during playback can be caused by a defective 1-µF, 50-volt electrolytic on the servo/syscon PCB.
- ✔ Badly soldered grounds on the tuner can cause a blue screen in playback with a normal tuner.
- ✔ A defective transistor in the luma/color circuit can cause only a blue screen in video and playback mode, with normal sound.
- ✔ The defective head preamp can cause a blue background flashing in playback when no signal is found on the tape.
- ✔ Check for ESR problems of a 1-µF, 50-volt electrolytic on the main PCB that causes a blue screen flashing off and on during playback.

Take–up Reel Problems

The take-up reel sensor consists of a diode and light operated transistor, inside of the take-up reel sensor component. When no light is found upon the transistor section, no take-up reel pulse is found at pin 67 of IC6001. The VCR will stop at once with no take-up reel pulse applied at pin 67. Always replace the defective take-up reel sensor with the exact part number.

In a Sears 9-inch TV/VCR, the take-up reel sensor is connected directly to pin 80 of the servo/system control (IC201). When the LED light strikes the light-emitting transistor

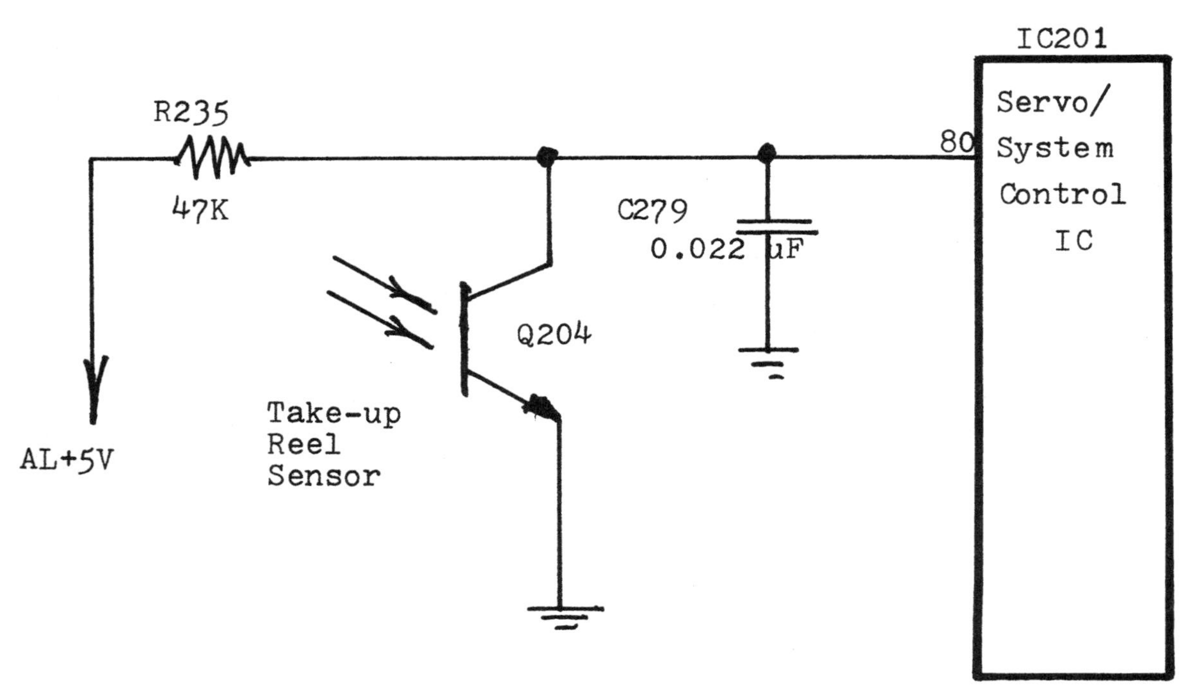

Fig. 13-8. In a Sears 9-inch TV/VCR the take-up sensor provides a pulse signal to shutdown the VCR with the servo/ system control IC.

Q204, a take-up pulse is applied to pin 80 of IC201 (*Fig. 13-8*). When the pulsating light does not strike the light-operated transistor, the take-up reel no longer sends out a pulsating signal to the servo/control IC201; the VCR shuts down, preventing tape from spilling out into the VCR mechanism.

The take-up reel sensor and the supply reel sensor supply voltage are wired in series with LEDs and the power voltage source in a RCA 13TVR60 TV/VCR. The 9-volt supply source has two 180-ohm resistors in series with both lighted LEDs through a switching transistor (TT100). The take-up reel lighted transistor section is wired directly (pin 82) to the servo control microprocessor (IT001). Both reel sensors may have to be replaced, when no voltage is found at the switching transistor collector terminal (*Fig. 13-9*).

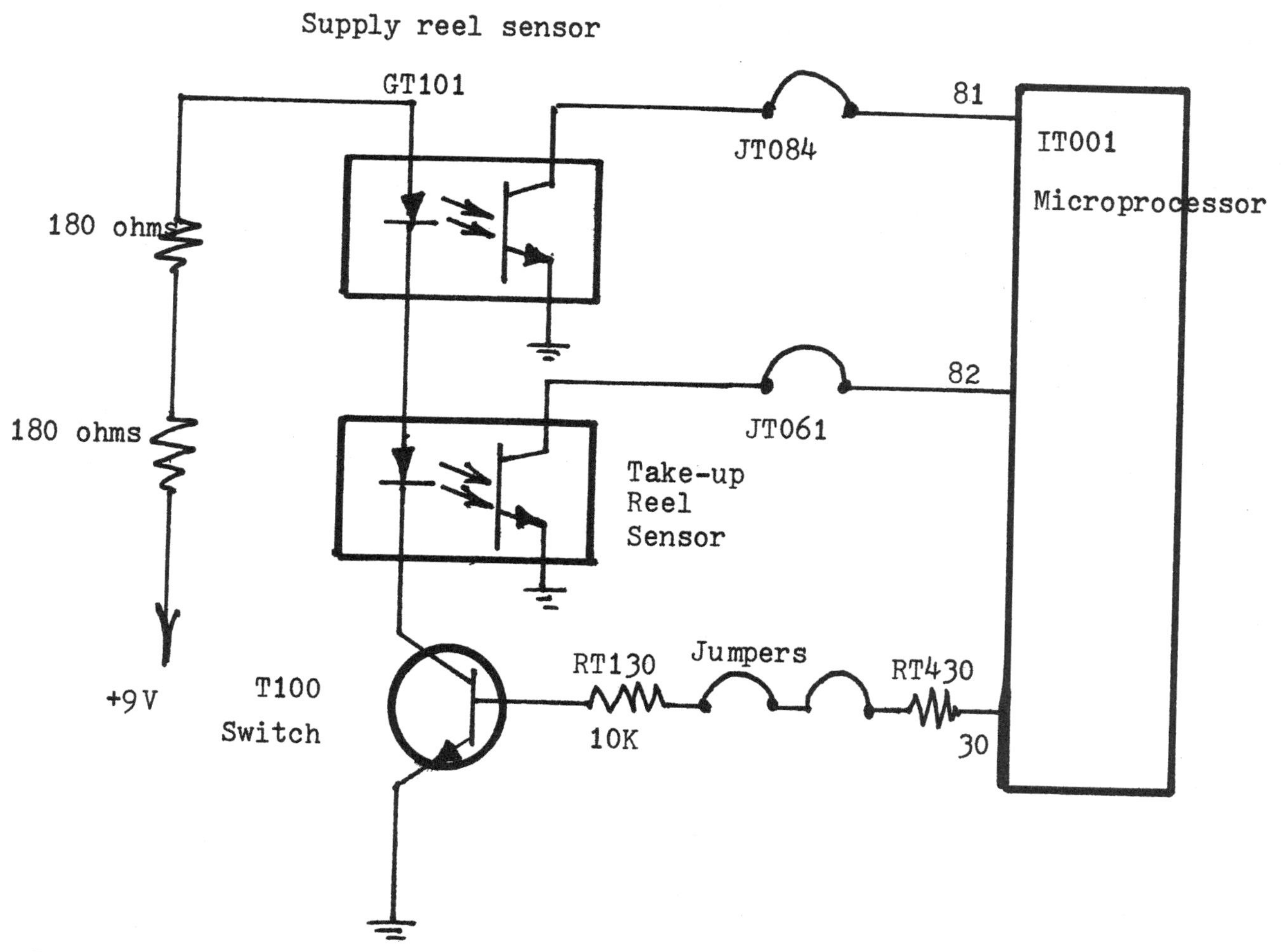

Fig. 13-9. The supply reel and take-up reel sensors control the large microprocessor IT001 in the RCA 13TVR60 TV/VCR combination.

Dew Light Problems

Some of the recent TV/VCRs have a dew circuit, somewhat like the dew sensors in camcorders. When moisture is found on the tape heads, the dew sensor circuit will not let the unit operate until the moisture has evaporated. When the VCR is taken from a cold temperature to a much warmer temperature, condensation can occur with the VCR. Likewise, when the VCR is in a cold location and then moved to a warm location, moisture can collect inside the VCR mechanism.

If the moisture has evaporated and the VCR still does not operate, suspect a defective dew sensor, dew sensor circuit, or wires on the dew sensor component. Check for a shorted dew sensor with the low-ohm scale of a DMM. Inspect the dew light and notice whether the light is on. Take a peek at the metal surface of video head drum for moisture. Leave the VCR on for up to one hour before trying to play a tape. Apply heat from a hair dryer or heat lamp to remove the moisture. Troubleshoot the dew light circuits if the light is still on and moisture is no longer found in the VCR.

Dew Light On - Sears TV/VCR

The dew light stayed on as long as the VCR was turned on, indicating trouble within the dew circuits. Q213 was tested in the circuit and was normal. A voltage measurement at the collector terminal of the transistors measured around 5 volts. The VCR was shutdown and the AL+5-volt line was discharged to ground before connecting the ESR meter. An ESR test between dew sensor connector CN202 to base of Q213 was normal. No measurement was found from the collector of Q213 to pin 7 of IC201 (*Fig. 13-11*). Pin 7 and 8 were resoldered on the servo/control IC201, which solved the dew problem.

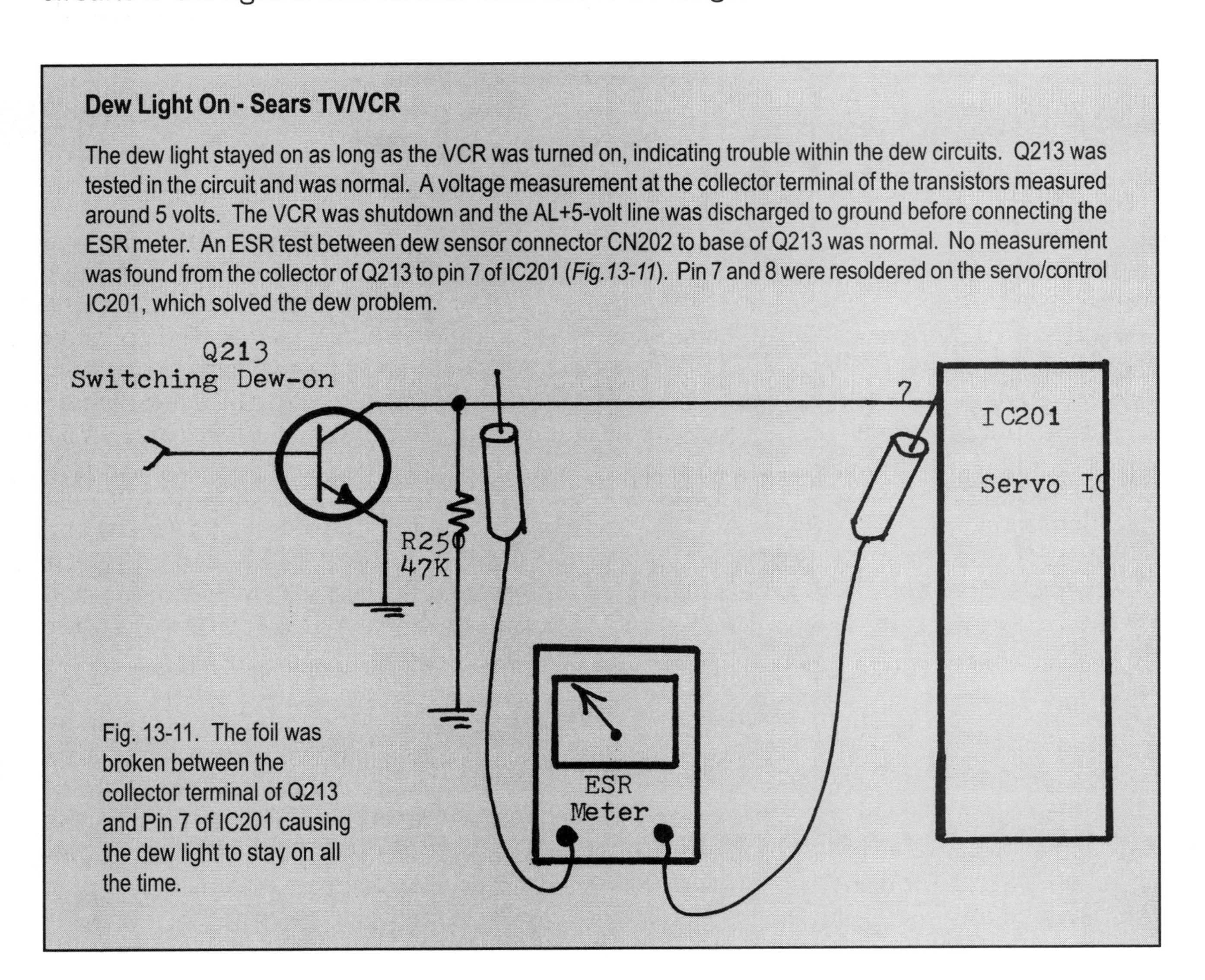

Fig. 13-11. The foil was broken between the collector terminal of Q213 and Pin 7 of IC201 causing the dew light to stay on all the time.

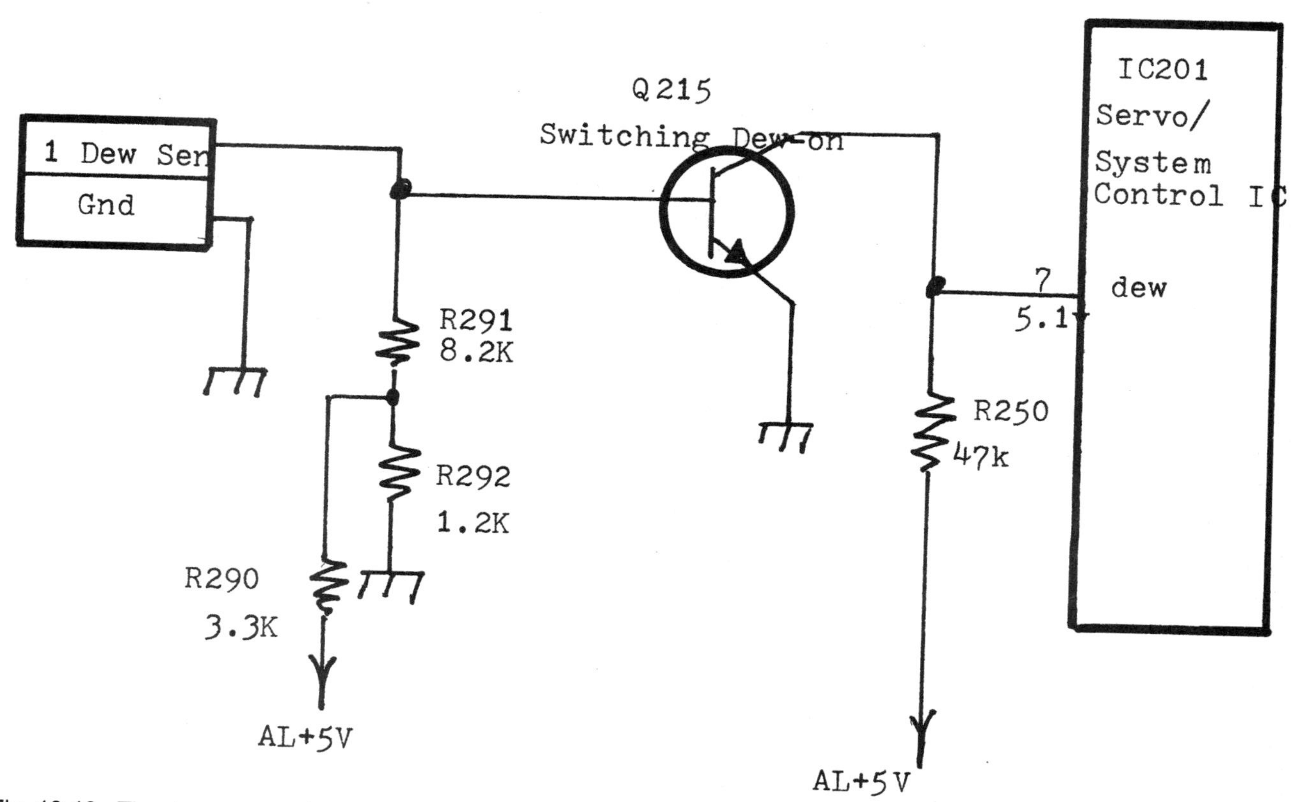

Fig. 13-10. The dew sensor circuit keeps the VCR shutdown until all moisture leaves the metal tape head drum.

Extreme moisture in the VCR can also cause the tape to cling to the metal drum and tear or eat the tape. The dew sensor circuit is designed to keep the VCR shutoff until the moisture is no longer present in the VCR. Within a Sears 9-inch TV/VCR, the dew sensor is fed directly to the base terminal of a switching-on dew transistor. The collector terminal of the dew transistor (Q215) is connected directly to Pin 7 of the servo/system Control (IC201). The dew sensor and switching-on dew transistor keep the servo system shut down until the moisture is removed from the VCR tape drum or cylinder (*Fig. 13-10*).

When the dew or stop light is on all the time, check for badly soldered joints from the dew light circuits to the servo/control IC. A break in the foil between R291 (8.2k) and R292 (1.2k), and from the switching-on dew transistor to pin 7 of IC201 can cause the dew light to stay on all the time. Replace the sensor lamp if the dew light is on and there is no VCR operation. Resolder pin terminals on the servo/control IC when the dew light stays on all the time and there is no VCR operation.

Check for open connections of the output dew circuits to the Servo/Control IC terminal. Use the ESR meter and check between dew sensor and base of switching-on dew transistor, and from the collector terminal to Pin 7 of IC201. Suspect a loose wire or pin connector plug from sensor, when dew light appears intermittent. Replace a defective dew sensor when the dew light flashes and there is no VCR operation. A flashing dew light can also result from an intermittently functioning dew switch-on transistor within the dew circuit.

Chapter 14: Servicing VCR Fast-Forward and Rewind Problems

The capstan motor rotates the tape from left to right in fast-forward and play modes then reverses the procedure in rewind operation. In larger VCRs, a separate motor might rotate the tape in play and fast-forward, while another motor rotates the tape from right to left in rewind mode. An idler wheel is found in many VCRs to change the different modes of operation. When one motor is used to rotate the forward motion, the idler wheel can then be switched to turn the tape in the opposite direction. The idler is rotated in fast-forward and in rewind modes.

The idler wheel and tire assembly swing back to the different reels for fast-forward or rewind modes. Replace the idler assembly if the idler is rough, worn, cracked, and glazed over. First try to clean the idler tire with alcohol and cloth. When the idler is worn so bad it causes erratic speed problems, replace the idler assembly. If the clutch assembly is too tight, it can cause excessive wear of the idler tire.

The idler slip clutch can become weak and cause slow or erratic speeds. An idler clutch assembly, weak from extensive wear, can cause intermittent, erratic or no rewind or fast-forward. Most idler wheel assemblies are replaced at the same time as worn or dirty motor drive belts.

Replace all stretched, loose, worn drive belts that will not stay on the pulleys. Dirty or deteriorated rubber belts can cause speed, fast-forward, and rewind problems. A worn or dirty old idler tire and belt can cause the tape to spill or be eaten and slow speed. A defective idler and tire can cause the take-up reel to not rotate.

When the tape is moving in fast-forward mode, the clear leader at the end of the videotape strikes a light-sensitive sensor and shuts down the tape movement. Likewise in rewind mode, the clear leader of the video tape strikes a different sensor and the tape stops.

Fast–Forward and Rewind Problems

A variety of problems can occur that will cause fast-forward and rewind modes to either function incorrectly or stop functioning completely. Check for bad take-up reel sensors when the rewind mode goes to stop in 7 to 10 seconds and then stops completely. A bad ratchet lever assembly can cause fast-forward to go to the stop mode; the take-up reel stops and the supply reel spills out tape (*Fig. 14-1*).

Check for a change in resistance on the take-up sensor line when the tape is inserted, the VCR rewinds, fast-forwards, and then ejects. A defective left tape end sensor can cause the tape to tear and does not stop at the end of rewind mode. Check for an open resistor in the reel PCB when play, rewind, and fast-forward go to stop and then the tape retracts back into the cassette. Suspect a SMD reel pulse amp transistor on the foil side when play mode goes to stop in a few seconds then rewinds or fast-forwards and stops with no eject of cassette. A bad reel motor assembly can cause the VCR to rewind or fast-forward for two to five seconds then stop; rewind search is okay. Check for bad solder joints on the capstan driver IC when play goes to stop after a few seconds, there is no fast-forward, and the supply post fully loads.

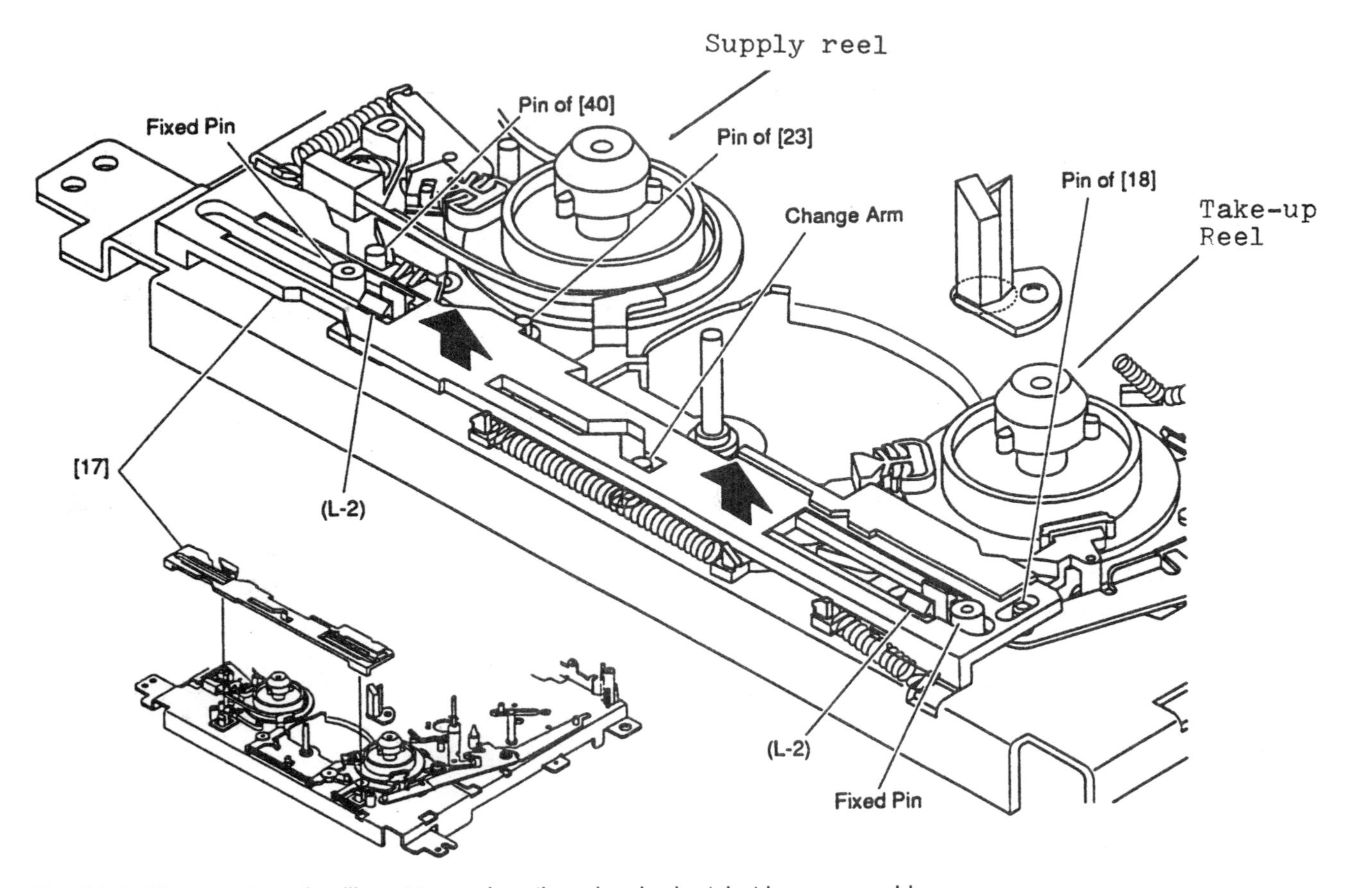

Fig. 14-1. The supply reel spills out tape when there is a bad ratchet lever assembly.

Slow, Erratic, or No Tape Movement

- ✔ If fast-forward and rewind are not working, it could be because of a bad capstan motor. Also check for a defective tape, a dirty or worn roller, and a dirty or bad switch when this occurs. A worn tire can prevent the reels from spinning in fast-forward or rewind modes as well. Replace and clean up the tire surface if this is the problem (*Fig. 14-2*).

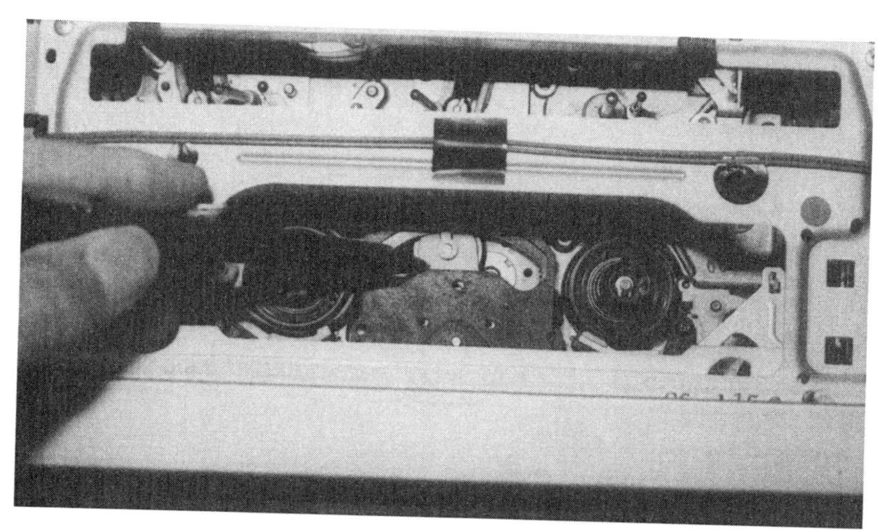

Fig. 14-2. Replacing the motor drive belts and the idler tire can cure many service problems affecting fast-forward and rewind operations.

- ✔ Check for loose or worn motor drive belts. Most technicians replace motor drive belts and rubber drive tires when there is slow, erratic or no movement of the tape.
- ✔ A bad tape can be the cause of poor, sluggish or no fast-forward or rewind operations; try a different tape to test if this is the problem or not. Discard the tape if there is an area of crinkling or tearing of the tape. A badly crinkled or wrinkled tape sometimes can damage the video heads.
- ✔ Check for loose material that might be lodged and binding the rewind/forward movement of the tape. Notice if the idler wheel goes to either rewind or fast-forward position. Also inspect the tape mechanism for jammed or dry spots. Small objects can become lodged within the tape loading area.
- ✔ Do not overlook the possibility of a bad capstan motor or switch assembly. A quick continuity ohmmeter check can determine if either component is open or closed circuit. Suspect improper or lack of supply voltage at the motor terminals when the motor does not rotate.

Slow or No Rewind

Possible Causes:

- ✔ Defective capstan motor assembly
- ✔ Bad plastic idler or idler assembly
- ✔ Worn or defective belts and drive tires

- ✔ Bad gum gear or gear idler assembly
- ✔ Bad rubber damper part
- ✔ Bad clutch gear
- ✔ Bad reel motor
- ✔ Damaged teeth in the cam gear or driving gears
- ✔ Capstan motor drive IC
- ✔ Defective servo IC (*Fig. 14-3*)
- ✔ Defective line voltage IC regulator
- ✔ Shorted take-up sensor (goes to stop from play to reverse)
- ✔ Leaky transistor on the tape LED line (tape might break)
- ✔ Bad sensor lamp (tape ejects, no other functions)

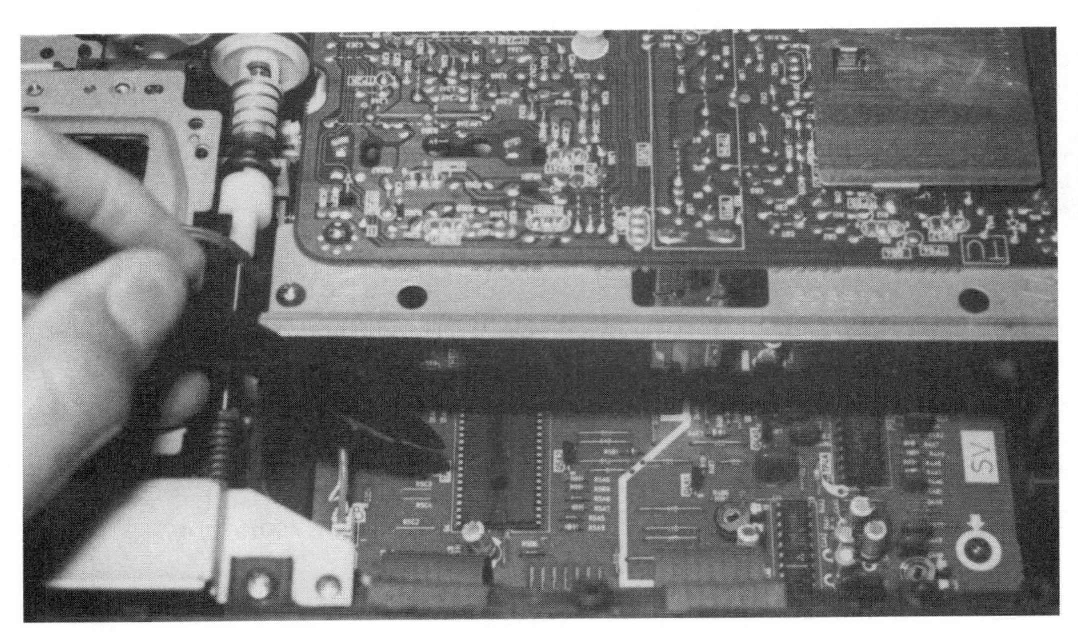

Fig. 14-3. The servo/control IC is usually the largest IC having over 50 pin terminals.

Intermittent Rewind

- ✔ Check for badly soldered connections on the capstan driver IC terminals if the VCR starts and stops, then rewinds on its own.
- ✔ A defective line voltage regulator can cause no rewind operation.
- ✔ Suspect a defective tape end sensor when the VCR rewinds when the tape is inserted.
- ✔ If the VCR will not accept tape and goes to rewind after three to five seconds of operation, it may be the result of a bad IP leaf switch; clean the dirty contacts or replace it if it's worn.

- ✔ When the VCR eats tape, the rewind function becomes intermittent, and the idler gear rides high on the supply reel, replace the idler assembly and check for a bent idler gear bracket.
- ✔ Replace the left tape end sensor when the VCR tears the tape and does not stop at the end of the rewind mode.
- ✔ Check the tape end sensors when the VCR intermittently shuts down, goes to stop, and then to rewind (*Fig. 14-4*).
- ✔ Replace the loading mechanism if there is intermittent rewind and the VCR goes to rewind when any function buttons are pressed.
- ✔ No rewind or intermittent rewind, the VCR eats tape, and the idler gear rides on the supply reel can be caused by a defective idler unit.
- ✔ Replace a bad slide switch on the side of the carriage assembly for intermittent on rewind and all other functions.

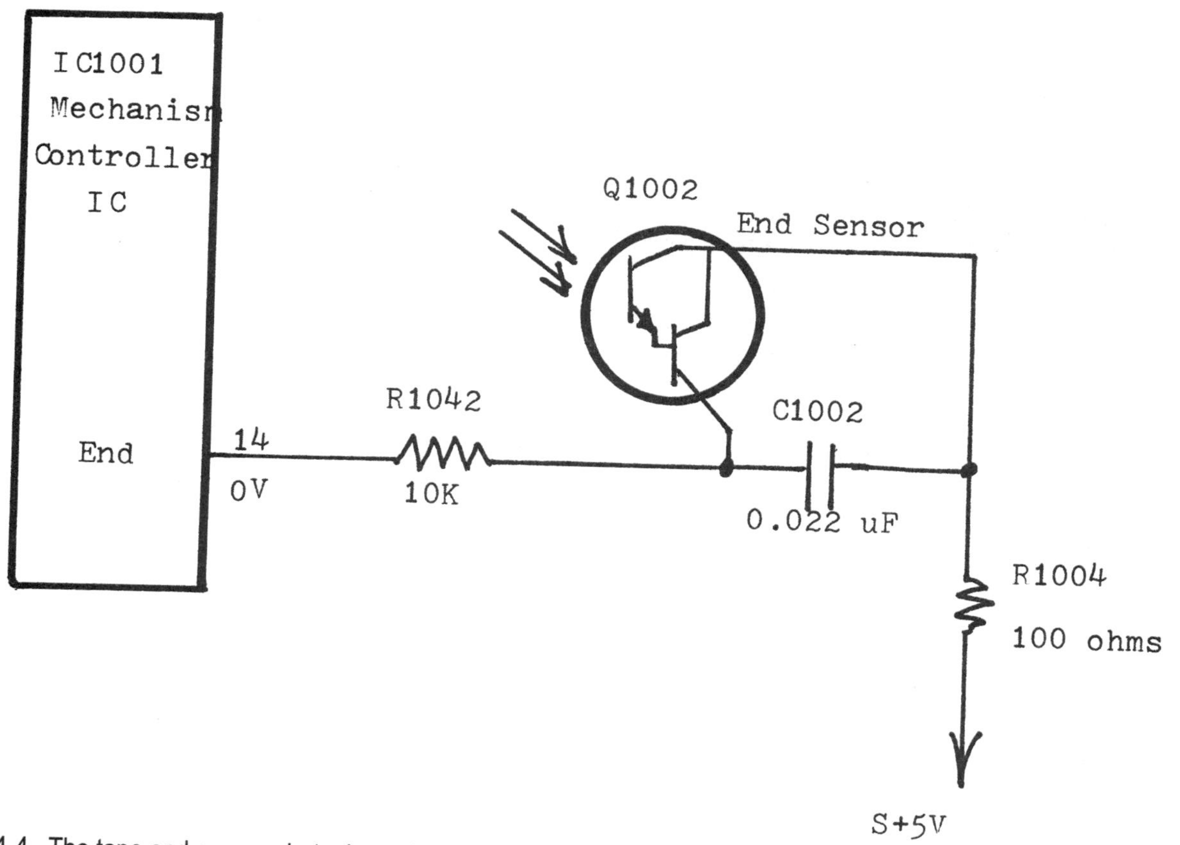

Fig. 14-4. The tape end sensor shuts down the VCR or switches the VCR into stop mode when the tape has reached the end of the supply reel.

Switches to Other Modes - Rewind

- ✔ Check the mode switch, clean and reburnish it if, when pressing the play button, the VCR goes to rewind or ejects.
- ✔ Replace the IC voltage regulator if the VCR will not load or eject and goes to rewind.
- ✔ Check the tape end sensor when play mode goes into rewind.
- ✔ Suspect leaky capacitors and diodes in the servo control circuit if when pressing the play button, the VCR goes to rewind.
- ✔ Suspect a defective syscon IC when power up switches rewind and then shuts off.
- ✔ Check for an open resistor on the sensor supply line if play mode goes into rewind; it might seem like a bad tape end sensor problem.
- ✔ Check for a bad contact on the supply sensor PCB if rewind goes to stop and the tape unloads.
- ✔ A bad cassette housing tape end sensor can cause the VCR to stop or go to rewind or fast-forward by itself.
- ✔ Replace a bad tape end sensor when the VCR tears tape and does not stop at the end of the rewind.
- ✔ Change both reel sensors when rewind stops 7 to 10 seconds after starting.
- ✔ Suspect open resistors in the supply sensor line when play mode goes into rewind and appears to be an end sensor problem.
- ✔ Check for open resistors on the reel PCB when rewind, play, or fast-forward switches to stop mode.
- ✔ Replace the reel motor assembly when the VCR rewinds and fast-forwards for two or three seconds and then stops.
- ✔ Check for an open transistor that supplies +5 volts to the take-up sensor if the VCR quits in rewind, play, or fast-forward, and then goes to stop mode after 10 to 20 seconds.
- ✔ Replace the IC voltage regulator when play mode switches to rewind and will not load or eject.

Fast-Forward Problems

- ✔ Suspect a bad AY gear assembly if the VCR will not fast-forward.
- ✔ A bad spot on the capstan motor armature can cause no fast-forward and intermittent pulling out of tape.
- ✔ Check for a bad switching regulator for no play, fast-forward, or rewind modes.
- ✔ Clean and realign the mode switch when the VCR loads, goes to fast-forward and will not eject tape.
- ✔ If the tape is inserted and then rewind and fast-forward stop working and the VCR ejects by itself, the cause may be a defective start sensor (*Fig. 14-5*).

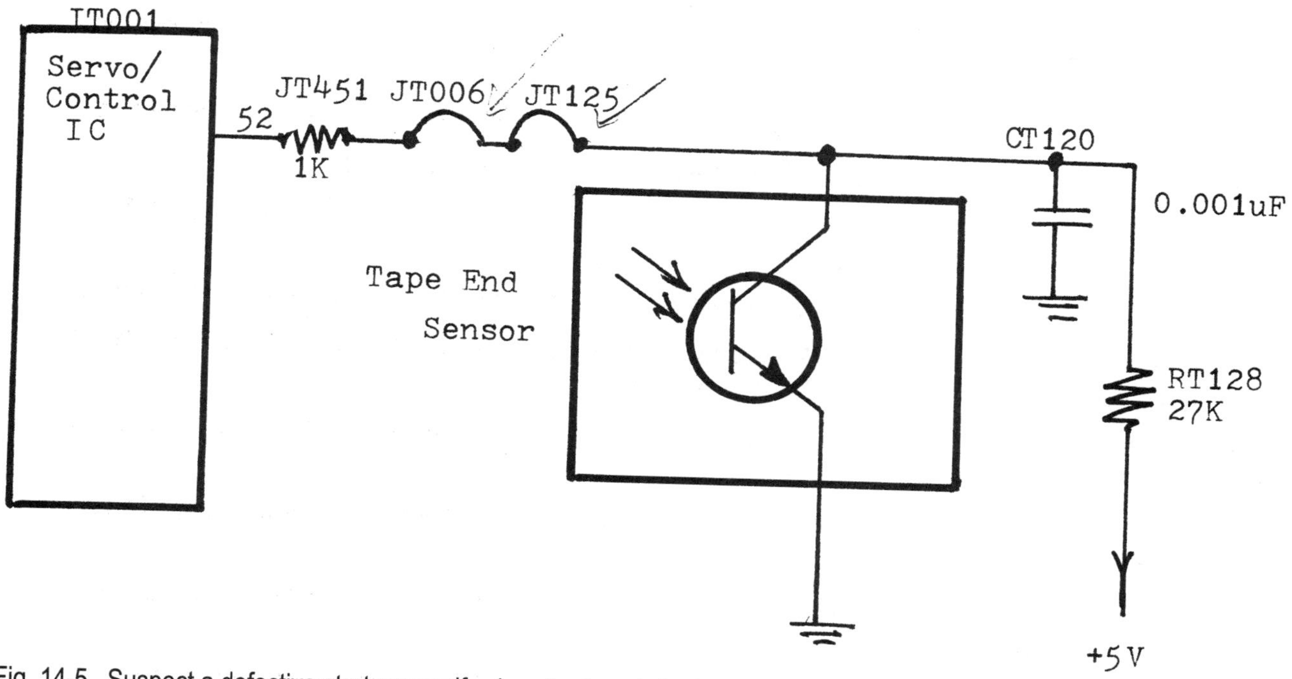

Fig. 14-5. Suspect a defective start sensor if, when the tape is inserted, the VCR ejects it and shuts down.

- ✔ A bad reel motor IC can cause the VCR to eat tape, the take-up reel will not rotate, rewind operates, but there is no fast-forward motion.
- ✔ Suspect a timer/display IC if there is no fast-forward-play mode, but the VCR accepts tape.
- ✔ Replace the servo IC when fast-forward was at a high speed and sometimes appears to work intermittently.
- ✔ Check that the carriage assembly is not out of alignment when fast-forward is sluggish and slow.
- ✔ Suspect a bad brake solenoid when the tape comes off of the reel during fast-forward operation.
- ✔ Check for a bad ratchet lever assembly when fast-forward and take-up reel stop and the supply reel spills out tape.
- ✔ A bad mode switch can make the audio sound like a chipmunk when play mode is in fast-forward.
- ✔ Replace a bad clutch idler assembly when there is no play and sluggish fast-forward (*Fig. 14-6*).

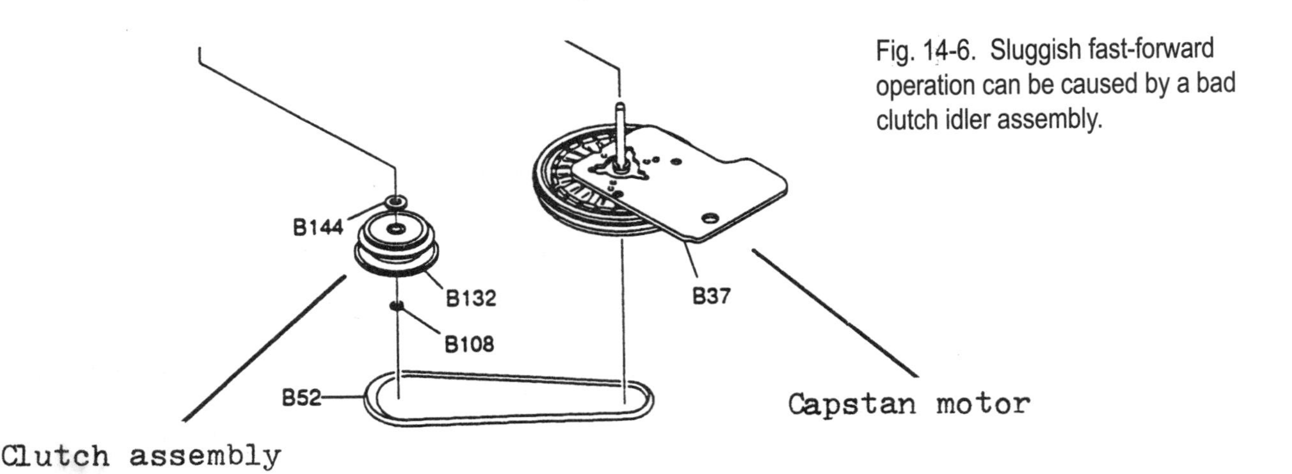

Fig. 14-6. Sluggish fast-forward operation can be caused by a bad clutch idler assembly.

No Rewind and Fast-Forward

This problem comes in many forms. When the unit rewinds and fast-forwards for two or three seconds and then stops, it can be the result of a bad reel motor assembly or a driver IC. The unit may need to have a voltage regulator IC replaced if it will not switch from TV to VCR and there is no fast-forward or rewind. No fast-forward or rewind can be caused by a leaky diode off of the servo IC, which prevents servo data from reaching the syscon IC. Check for a bad solder joint on the regulator transistor within the supply source of the capstan motor when the capstan will not rotate and there is no rewind and fast-forward operation. A leaky diode in the power supply +5 volt line will keep the VCR from fast-forward or rewind operations and cause the VCR to eat tape. Other things to look for is there is no fast-forward or rewind functions include:

- ✔ Defective AY-gear assembly
- ✔ Broken spring from the loading motor assembly
- ✔ Bad switching regulator
- ✔ Dirty idler gear assembly (clean and lubricate)
- ✔ Soft brake (does not engage and tighten the spring on the soft brake assembly)
- ✔ Bad lever holder and main cam assembly
- ✔ Improper brakes and a bad solenoid assembly (may appear to be a bad mode switch)
- ✔ Worn rubber damper on the top right mechanism
- ✔ Broken mode switch
- ✔ Defective photo interrupter under the take-up reel
- ✔ Dirty cam switch
- ✔ Bad plate assembly

- ✔ Defective main lever latch under the main cam gear (brake assembly rubs against the reel assembly)
- ✔ Broken arm gear on the underside of mechanism
- ✔ Open 1-amp fuse
- ✔ Defective regulator voltage IC

Speed Problems - Fast-Forward and Rewind

- ✔ Suspect a bad supply reel sensor with slow fast-forward and rewind modes.
- ✔ The brakes might stay on in fast-forward and rewind if these operations are too slow, which is caused by a bad plate assembly.
- ✔ Replace the servo IC when fast-forward runs at a high rate of speed and sometimes operates intermittently.
- ✔ A bad supply reel sensor can cause the fast-forward and rewind mode to run slow.
- ✔ Check the plate assembly when rewind and fast-forward operate extremely slow.
- ✔ Check for a few broken plastic clips on top of the cassette housing when the tape speed drags in rewind and fast-forward.
- ✔ Clean the mode switch and adjust the center position on the mode switch when fast-forward and rewind operation are slow.

No Play, Fast-Forward, or Rewind Operation

- ✔ Replace a F/R bracket with a broken shifter if there is no rewind, fast-forward or play modes, and unit eats tape in eject mode (*Fig. 14-7*).
- ✔ Check the loading motor assembly if there is no play, rewind, and fast-forward.
- ✔ A bad 1-amp fuse or a bad fuse holder can cause no rewind, fast-forward or play operations.
- ✔ Check for an unhooked or missing tension spring when there is no tape action in play, rewind, or fast-forward.
- ✔ Test each 10-µF, 16-volt electrolytic on the drum FG line off of the servo IC for uncontrollable drum speed with no play, fast-forward, or rewind functions and loading functions fine.
- ✔ When the VCR eats tape and will not play, fast-forward, or rewind, check for a defective diode in the power supply +5-volt line.
- ✔ A bad voltage regulator IC can cause no play, fast-forward, or rewind modes.
- ✔ A bad switching regulator will prevent the VCR from play, fast-forward, or rewind operations.
- ✔ Suspect a defective photo interrupter under the take-up reel for no counter advance, play, fast-forward, or rewind modes.

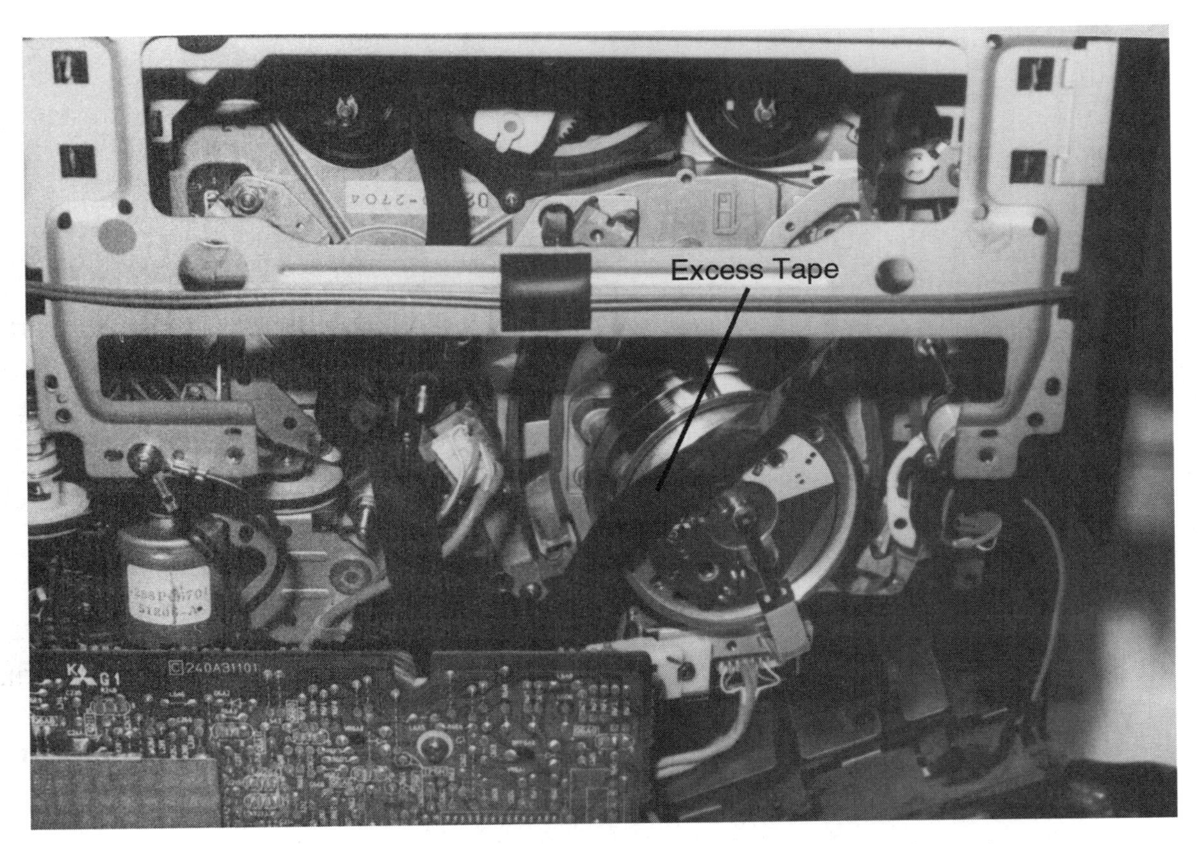

Fig. 14-7. Excess tape within the VCR can be caused by a stopped take-up reel or bad F/R bracket.

Intermittent Fast-Forward, Rewind and Play Operations

- ✔ Check for a bad connector or connection for intermittent play, fast-forward and rewind modes.
- ✔ An intermittent voltage IC regulator can cause intermittent play, fast-forward and rewind operations.
- ✔ An intermittent capstan motor drive IC can cause intermittent play, fast-forward and rewind modes.
- ✔ Intermittent shutoff in fast-forward and rewind when the unit powers up can result from a bad syscon IC.
- ✔ A defective capstan motor can intermittently blow the fuse in play, rewind, and fast-forward rotation.
- ✔ A defective servo IC can cause intermittent high speed in fast-forward mode.
- ✔ Replace the idler unit when rewind works only intermittently, the unit eats tape, and the idler gear rides high on the supply reel.
- ✔ Clean the mode switch, replace the idler and lubricate all when rewind and fast-forward modes are intermittent.

- ✔ Clean and lubricate the idler gear shaft and spring post assembly when there are intermittent fast-forward and rewind functions.
- ✔ Resolder all pins of the capstan drive IC when there is intermittent play, record, fast-forward, and rewind operations.
- ✔ A bad cam switch can cause intermittent rewind and fast-forward modes.

Noisy Rewind and Fast-Forward Functions

- ✔ Check for a bad capstan bearing holder if you hear a vibrating noise in fast-forward and rewind modes.
- ✔ A loud grinding and scraping noise in rewind and fast-forward can be caused by a bad drive clutch reel.
- ✔ Replace a broken drive motor bracket if a grinding noise is heard.
- ✔ A bad cassette holder assembly can cause a vibrating noise in fast-forward and rewind modes.
- ✔ Suspect a bad mode switch when the VCR loads and squeals then shuts down.
- ✔ Check for broken plastic clips on top of the cassette housing causing a rubbing noise and the to tape drag in rewind and fast-forward.
- ✔ Suspect a broken bracket when the VCR squeals in play and rewind.
- ✔ Clean and lubricate a bad capstan bearing if a loud squealing noise in play, fast-forward and rewind occurs (*Fig. 14-8*).
- ✔ Clean the brake pad in the take-up reel for a squealing noise.
- ✔ A bad loading motor belt can cause a high-pitched squeaking noise.
- ✔ Replace a bad idler assembly for a thumping noise.
- ✔ A screeching from the loading motor belt during load or unload can result from a bad bracket assembly.

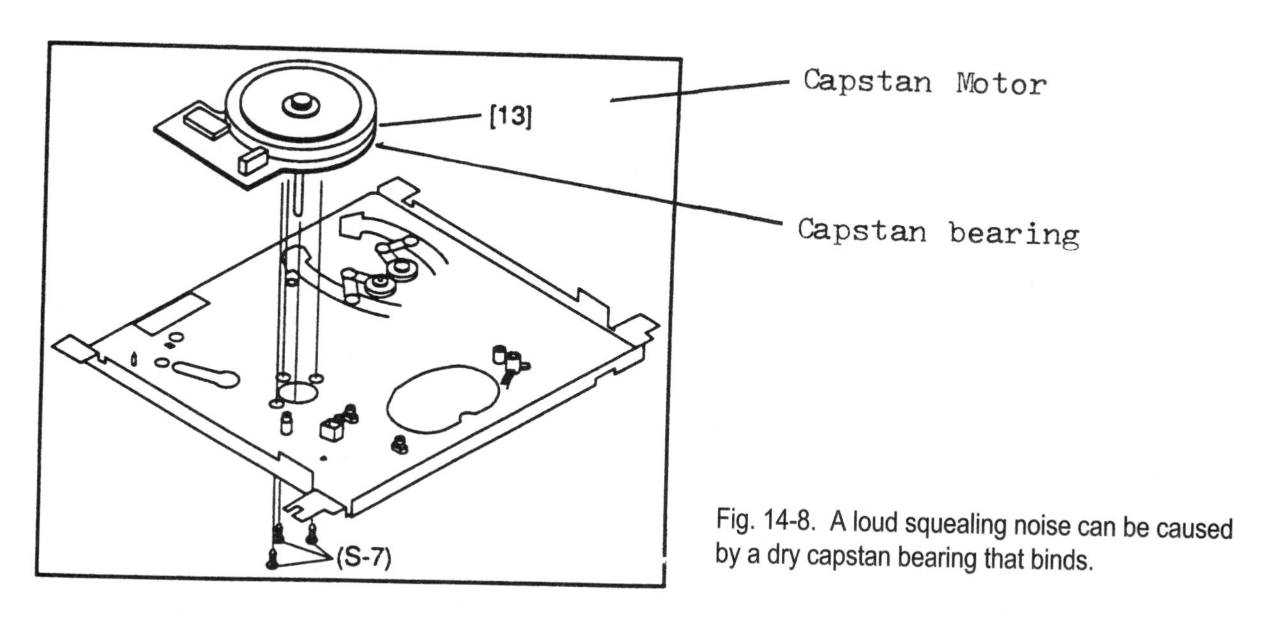

Fig. 14-8. A loud squealing noise can be caused by a dry capstan bearing that binds.

Take–up Reel Problems

The take-up spool operates at a constant rate of speed to take-up or wind excess tape around the reel and to prevent spill-out or eating of tape. A motor belt, shifting idler, or gear drives the take-up reel (*Fig. 14-9*). If the take-up reel becomes sluggish, slows down, or stops, the excess tape spills out and clogs up the pinch roller and capstan with excess tape. The tape spills out when the take-up reel stops. The excess tape wrapped around the capstan can cause a higher drive speed making the audio sound like a chipmunk. A bad or worn tire belt can cause the take-up reel to eat tapes.

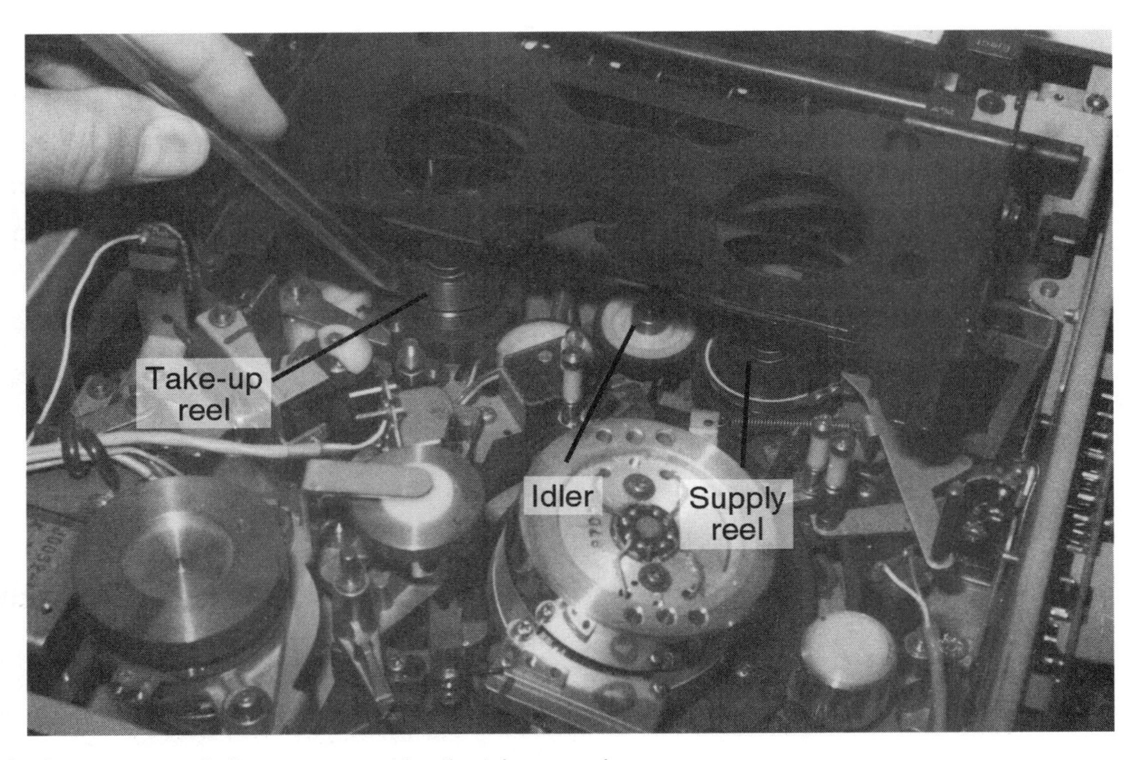

Fig. 14-9. A motor, motor belt, or gear can drive the take-up reel.

The take-up reel has a reflected surface on the bottom that reflects light to a take-up sensor component. Dirt, dust, or smoke on the reflector can cause intermittent stopping of the take-up reel and shutdown of the VCR. Clean off all dirt and residue from the reflection surface if the VCR stops after a few seconds.

A bad idler tire can cause slow and sluggish rewind or fast-forward rotation, the tape to be eaten, and the unit to stop. First clean the idler tire and see if the speed returns to normal, if not, replace the idler tire.

The take-up sensor is located underneath the take-up reel assembly. The take-up sensor sends a signal to the system control microprocessor to stop the machine, if correct speed is not maintained at the take-up reel, tape will spilling out. A defective take-up sensor might play or record the tape for several seconds and then the VCR shuts down. The take-up reel sensor might be directly connected to the system control/servo IC and an isolation resistor is found between the sensor and the servo IC (*Fig. 14-10*).

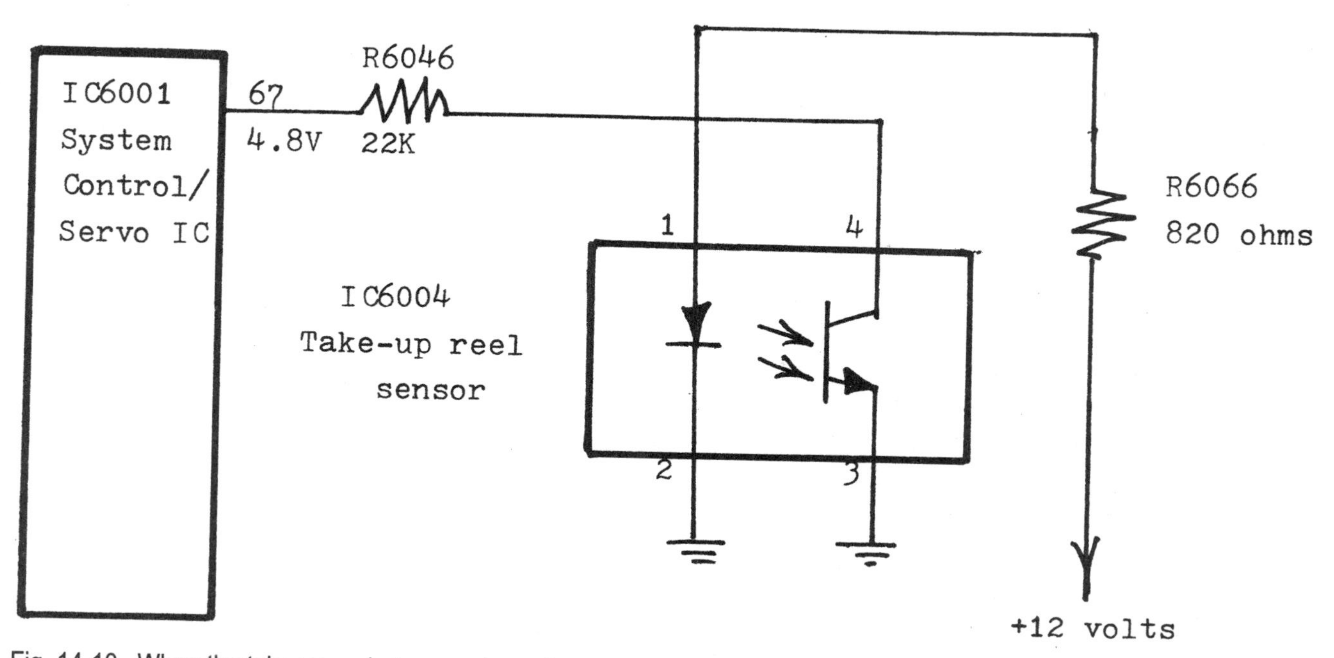

Fig. 14-10. When the take-up reel stops or slows down, the take-up reel sensor shuts down the VCR via IC6001.

Take-up Reel Does Not Work

- ✔ The leaky capstan driver IC becomes hot with no playback and take-up reel action.
- ✔ A bad take-up motor drive IC can cause no take-up, the VCR loads and the capstan rotates.
- ✔ A bad reel motor IC can cause no fast-forward rotation, the take-up reel will not move, unit eats tape, and rewind does rotate.
- ✔ A bad take-up reel sensor can cause the rewind mode to stop in 7 to 10 seconds and go to stop mode.
- ✔ A bad ratchet lever assembly can cause the fast-forward to go to stop mode, the take-up reel stops, and the supply reel spills out tape.
- ✔ A defective clutch assembly can cause the take-up reel to not rotate.
- ✔ Check for a leaky transistor off of the control circuits when the take-up reel stops in play after 6 to 10 seconds, guides retract, and rewind is okay.
- ✔ If the take-up reel magnet falls, the play mode shuts down after a few seconds and there is no take-up reel pulse.

- ✔ A dragging take-up reel brake can cause erratic take-up and quivering audio (*Fig. 14-11*).
- ✔ Suspect a tight clutch assembly or worn idler if there is wow and flutter, the audio drags, and the take-up reel quivers.
- ✔ Replace the idler plate assembly if there is no eject, it stops in play/record mode, and the take-up reel does not rotate.
- ✔ Check for a bad loading belt when the take-up reel turns as power is turned on.
- ✔ If the VCR stops for anywhere from a few minutes up to a couple of hours, it may be a bad take-up reel sensor.
- ✔ Replace the take-up reel sensor when the VCR plays for two to four seconds and then shuts down.
- ✔ Replace a defective take-up reel sensor if the VCR shuts down during play, rewind, and fast-forward.
- ✔ Suspect a take-up reel sensor when the tape guides go into loading the VCR by themselves.
- ✔ Replace the limiter post assembly when the take-up reel will not rotate.
- ✔ Suspect a bad take-up reel photo sensor if the VCR shuts down right away.
- ✔ A broken cam track gear can cause no take-up reel action.
- ✔ A badly soldered connection on the take-up reel sensor can cause the VCR to stop or shut down.
- ✔ A dirty or worn mode switch can cause the take-up reel to stop.

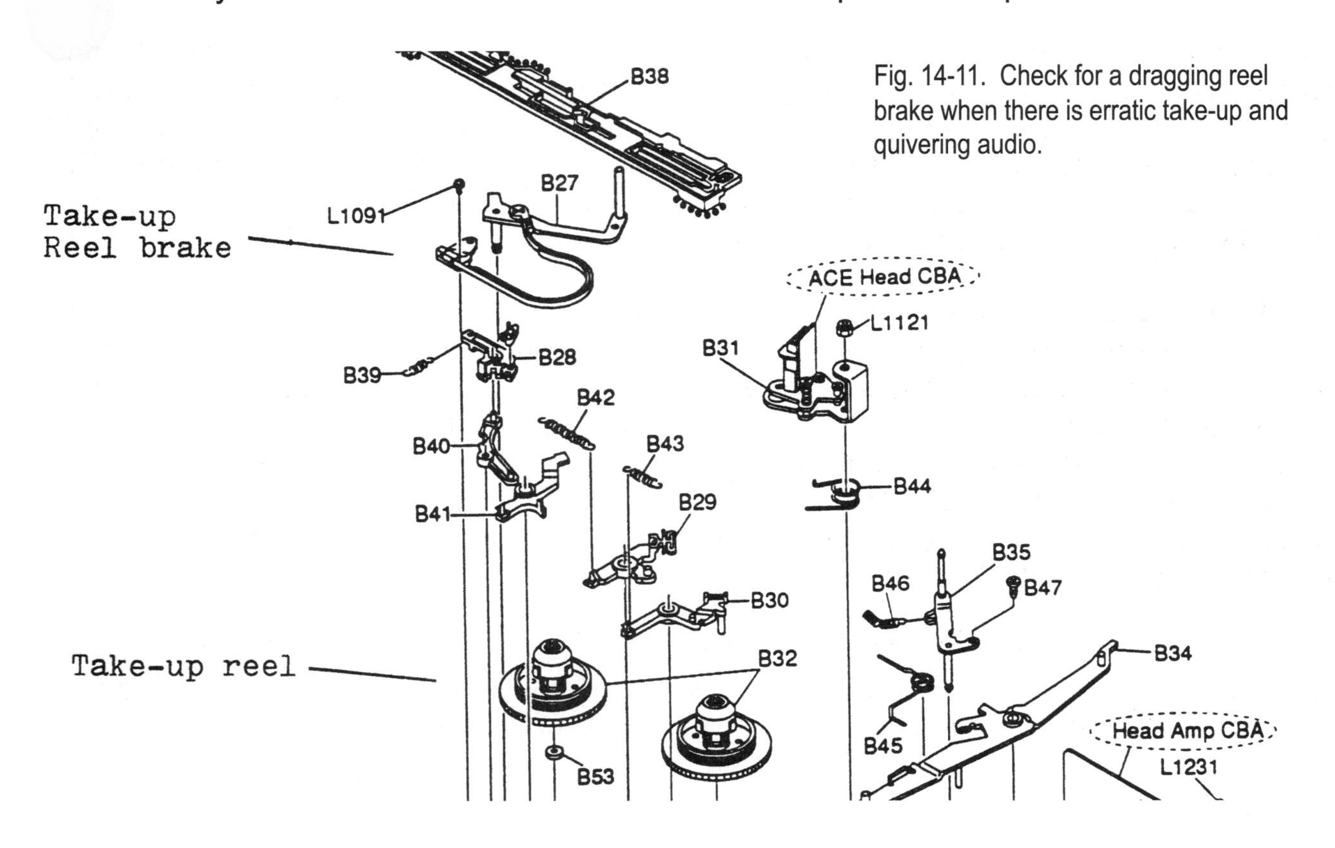

Fig. 14-11. Check for a dragging reel brake when there is erratic take-up and quivering audio.

Chapter 15: Servicing VCR Audio Problems

The VCR sound circuits consist of recording and playback of the audio. Although the audio playback and record circuits are quite common to most TV/VCR combo circuits, several different audio circuits are employed to provide the same results. The audio playback from the TV or VCR circuits is common to the audio output and speaker circuits and also to the audio ACE head. The various audio playback and record circuits are selected through an input select IC component.

The audio record circuits are selected by the input IC select circuit, which consists of audio injected from either an external jack or the TV audio circuits. The input select IC switching is controlled by the servo/system control IC. The audio record signal is then fed to the audio process IC. The audio playback-record circuits in a Panasonic PV-M2021 VCR include an input select (IC7101), audio record/playback process IC (IC4001), 15.75 kHz trap (FL4001), and the audio output IC (IC4151).

The TV–Audio Record Circuits

The audio output circuits of the TV chassis are inserted into the input select IC and switched directly to the TV audio output IC. The audio output IC is common to both TV and VCR output sound circuits. The TV audio signal, to be recorded on the tape head circuits, is selected by the input select IC and fed into an audio process IC. In most TV/VCR audio circuits, the input select IC switches the TV sound, TV signal to be recorded, the VCR tape playback (playback) and recording circuits.

In a 9 inch Sears TV/VCR combo, the recorded TV audio signal, is switched into the audio circuits with the input select IC701. The correct controlled switch, of the input select IC, is controlled by the servo/system control IC. The TV audio signal is then passed on to the input terminal 57, of the audio process IC401. The TV audio signal goes through an ALC stage, line amp, mute circuits, in and out of Pin 59 of IC401. This TV audio signal is injected at Pin 60 again through a record-on switch, to a recording amp, out of Pin 64, to an auto bias circuit, then out of Pin 64 to the connector CN851, and all of this is controlled inside IC401 (*Fig. 15-1*).

The TV-record audio signal from Pin 1 of IC401 is found at connector CN851 and fed onto connector CN287 then to the audio track head winding. The recorded TV signal is then recorded on the cassette tape. Both the TV audio signal and the VCR recording

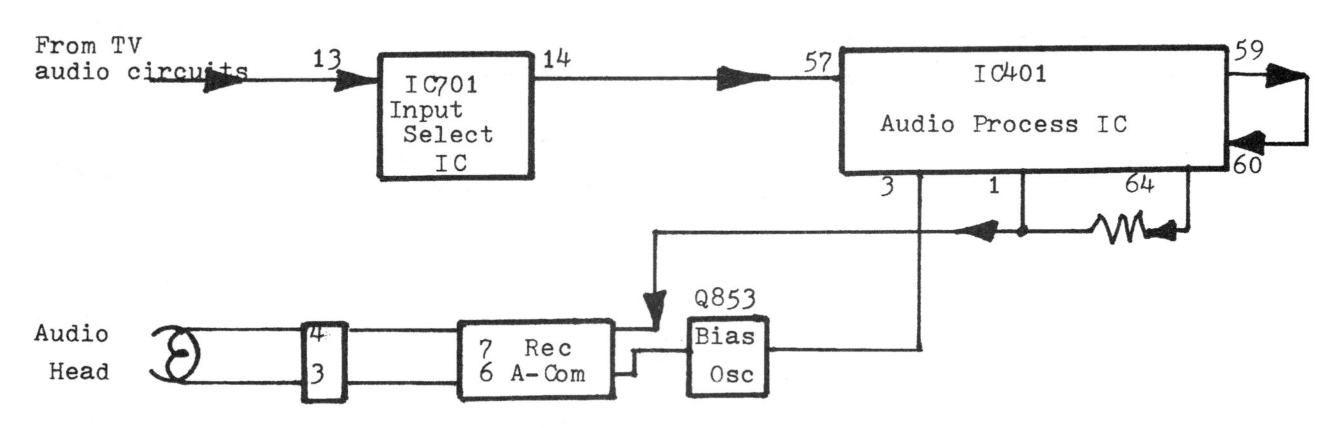

Fig. 15-1. The TV audio signal path to be recorded by the input select and audio process IC to the audio head in a 9-inch TV/VCR.

signals follow the same path. Usually when service problems exist in the audio playback and recording circuits, suspect a defective input select and audio process IC because both the playback and record audio signals have the same signal paths.

In a Sears TVSA1320 TV/VCR combination the TV audio signal is injected at the input select IC, goes through an internal limiter and FM detector stage switched inside the input select IC (IC301), and follows the same path to the audio process IC (IC401). The input select IC of the Sears 9-inch TV/VCR is different in that the TV audio signal comes from the video/chroma IF IC701, which does not include the FM circuits or attenuation and AF amp circuits as found in the 13 inch Zenith input select IC (IC301) (*Fig. 15-2*).

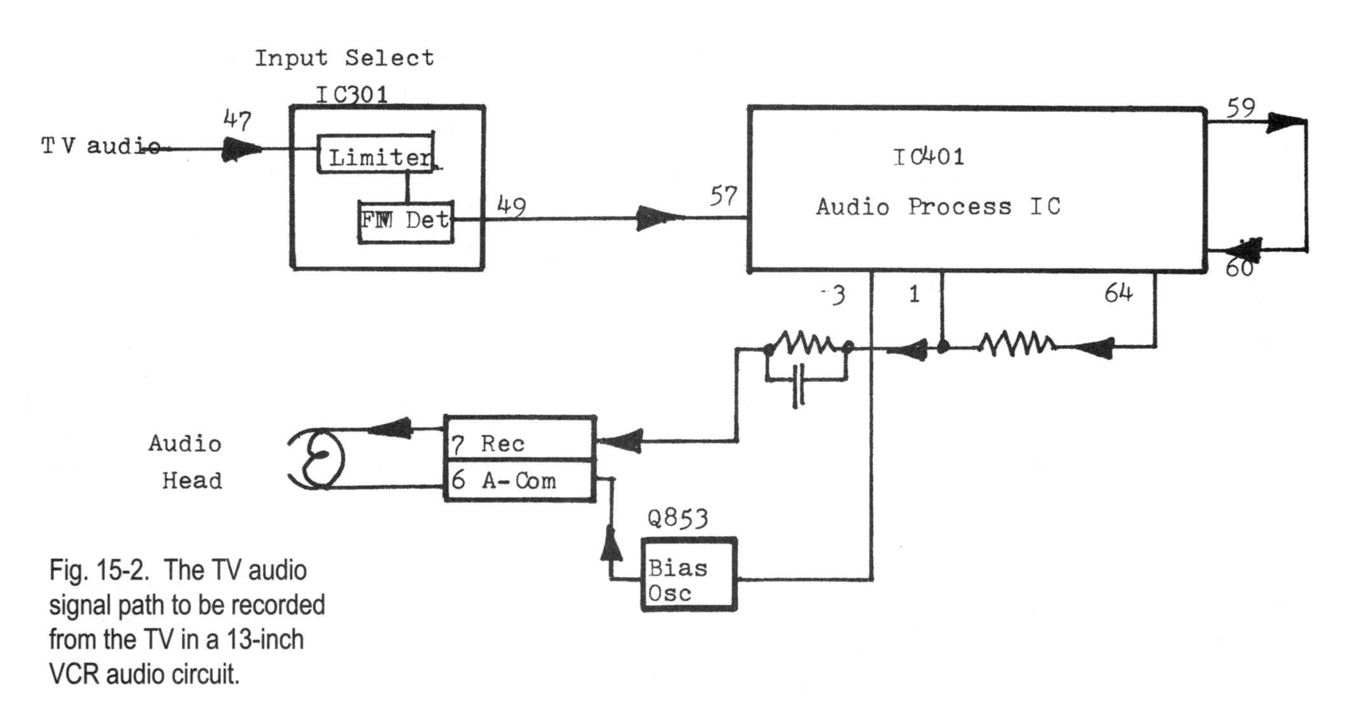

Fig. 15-2. The TV audio signal path to be recorded from the TV in a 13-inch VCR audio circuit.

The Audio Playback Circuits

The TV audio and audio playback signals are selected in the Video, chroma/IF IC (IC301) in a 13-inch Sears TV/VCR combo. The servo/system control IC provides correct switching inside IC301. An audio process IC (IC401) has internal playback-on switching, SP/LP-on, and an equalizer amp with the signal out of Pin 8 to Pin 9 of IC401. The playback signal from Pin 9 goes to a switching circuit, line amp, and a mute circuit inside IC401 to the output terminal (59).

The audio playback signal is fed to Pin 50 of IC301 where the signal is boosted with an AF amp and attenuator circuit before being passed on to Pin 51 (*Fig. 15-3*). The playback signal from Pin 51 is coupled to the input terminal (1) of the audio output IC801. Here the audio signal is controlled by a volume control and amplified out of Pin 7 to the PM speaker and headphone jack. Notice the audio signal is fed directly to the audio output IC801 like the playback audio from the recorded cassette or tape. Also, the playback and record audio signal use the same circuit path throughout most TV/VCR circuits.

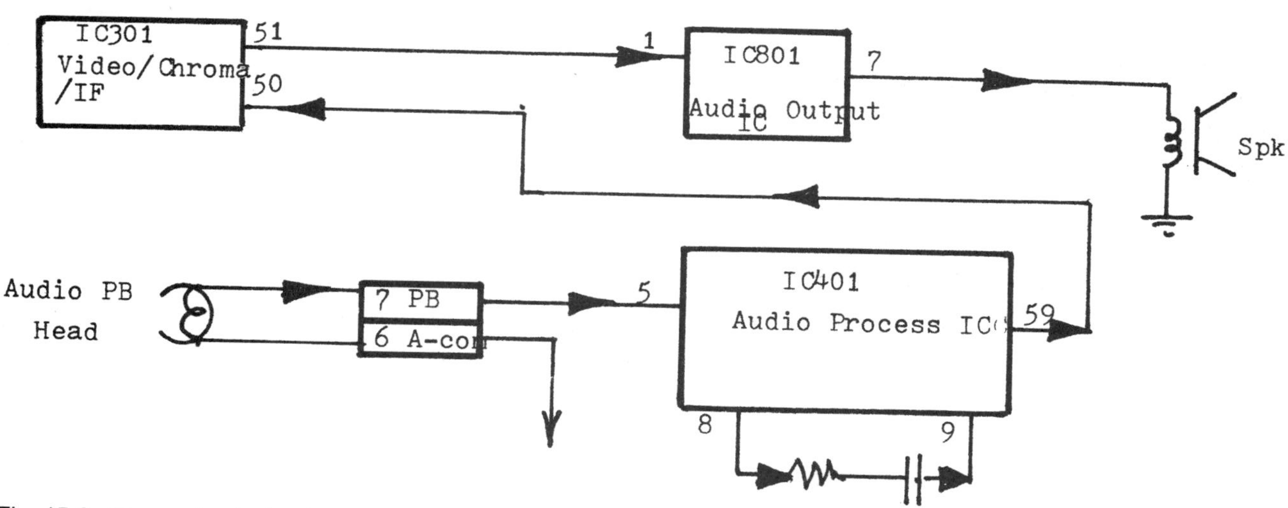

Fig. 15-3. The audio playback path from audio head to audio output IC in a 13-inch TV/VCR.

The TV audio signal to be recorded from a TV channel broadcast is fed into Pin 18 of IS001 in an RCA 13TVR60 TV/VCR. IS001 is an audio head switch IC component that switches the audio input, contains playback, ALC, record, amplifier, and bias circuits. The TV audio input signal at Pin 18 is fed internally to a TV-line switch and a playback-record switch; it then passes through an amplifier circuit, is switched to a record circuit, and out Pin 21 of IS001. The TV signal is capacity coupled to Pin 4 of the connector BS031. Here the TV audio signal is applied to the VCR audio head and onto the revolving tape of a cassette (*Fig. 15-4*).

The audio and playback circuit of the 13-inch RCA TV/VCR contains the audio head, audio erase head, and control head in one component. A separate full erase head and the audio erase head are exited by a bias oscillator erase circuit. The playback-record CTL and record bias is fed to the control head from the system control circuits.

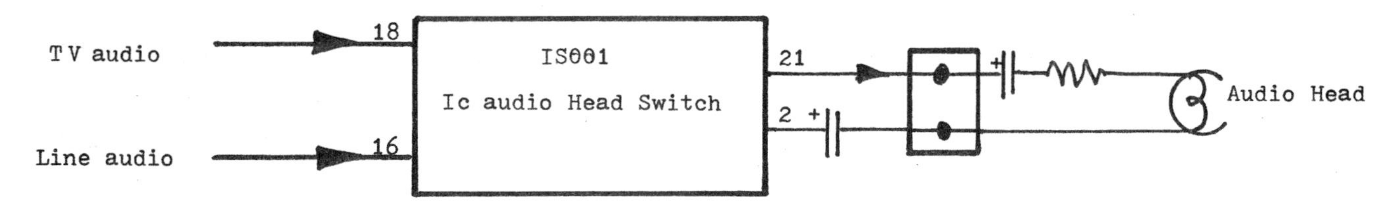

Fig. 15-4. Block diagram of the TV audio signal to be recorded in an RCA 13TVR60 TV/VCR.

Intermittent Audio Recording – Sears TVSA1320

In a Sears TVSA1320 TV/VCR the audio signal was normal on the TV audio and was intermittent when a tape was played. Since the audio process (IC401) provides both playback and record audio signal, an intermittent waveform was found at Pin 59 of IC401 (*Fig. 15-5*). At first IC401 was suspected of being intermittent and all pin terminals were resoldered, but this did not correct the playback audio signal.

The audio head (ACE) was cleaned, but to no avail. The VCR was shutdown and the low scale of the DMM was clipped to Pins 4 and 3 of the audio tape head. By prodding and moving the audio head terminals, the DMM moved up and down. Replacing the intermittent ACE head solved the audio playback and record-audio signal.

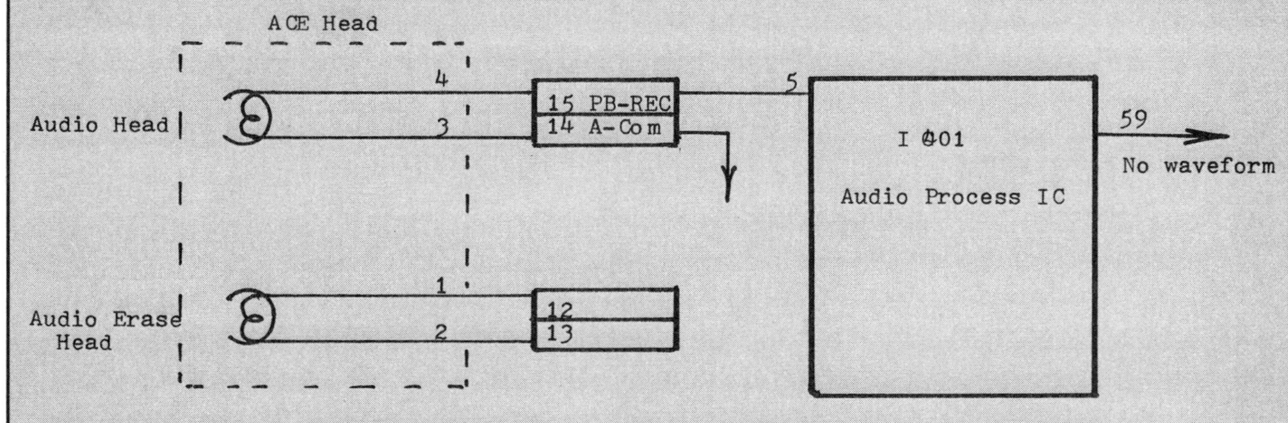

Fig. 15-5. A bad audio (ACE) head caused intermittent audio recording in a Sears 13-inch TV/VCR.

The Bias Oscillator Circuits

Like the recording circuits in a cassette player, a simple bias oscillator circuit is found tied to the audio erase head and full erase head. The bias oscillator provides an oscillator frequency to erase a previously recorded cassette and bias the audio system for linear recording. A separate transistor is used as a bias oscillator circuit in a 9-inch Sears

TV/VCR. The automatic bias circuit is found within the audio process IC401 and fed out of Pin 3 and 4 to the bias oscillator transistor (Q853). In the RCA 13-inch TV/VCR the bias oscillator circuit contains two separate transistors.

The bias oscillator frequency is fed to the audio and full erase head. The bias oscillator also sends a signal to the A-Comm terminal on the audio erase head. The switching transistors Q855, Q854, Q852, and Q851 provide correct switching of voltage to the bias oscillator circuits and switches in the bias oscillator frequency to the control heads. Notice the bias oscillator signal is fed to the audio erase head and is in series with the full erase head (*Fig. 15-6*). The bias operates only when the VCR is in record mode and the DC supply voltage is applied to the bias oscillator circuits with switching transistor (Q854).

Transistor (Q853) provides the bias oscillator frequency to the audio erase and full erase head that are connected in series. The center tap of the oscillator transformer (T851) is also applied to the bottom half of the audio erase head for linear recording. Switching transistor (Q855) provides a direct switching path to ground at the emitter terminal of the bias oscillator (*Fig. 15-7*). Transistor (Q854) provides switching the record-off in record mode. A defective bias oscillator circuit can cause no or poor recording and erasing of a previous recording.

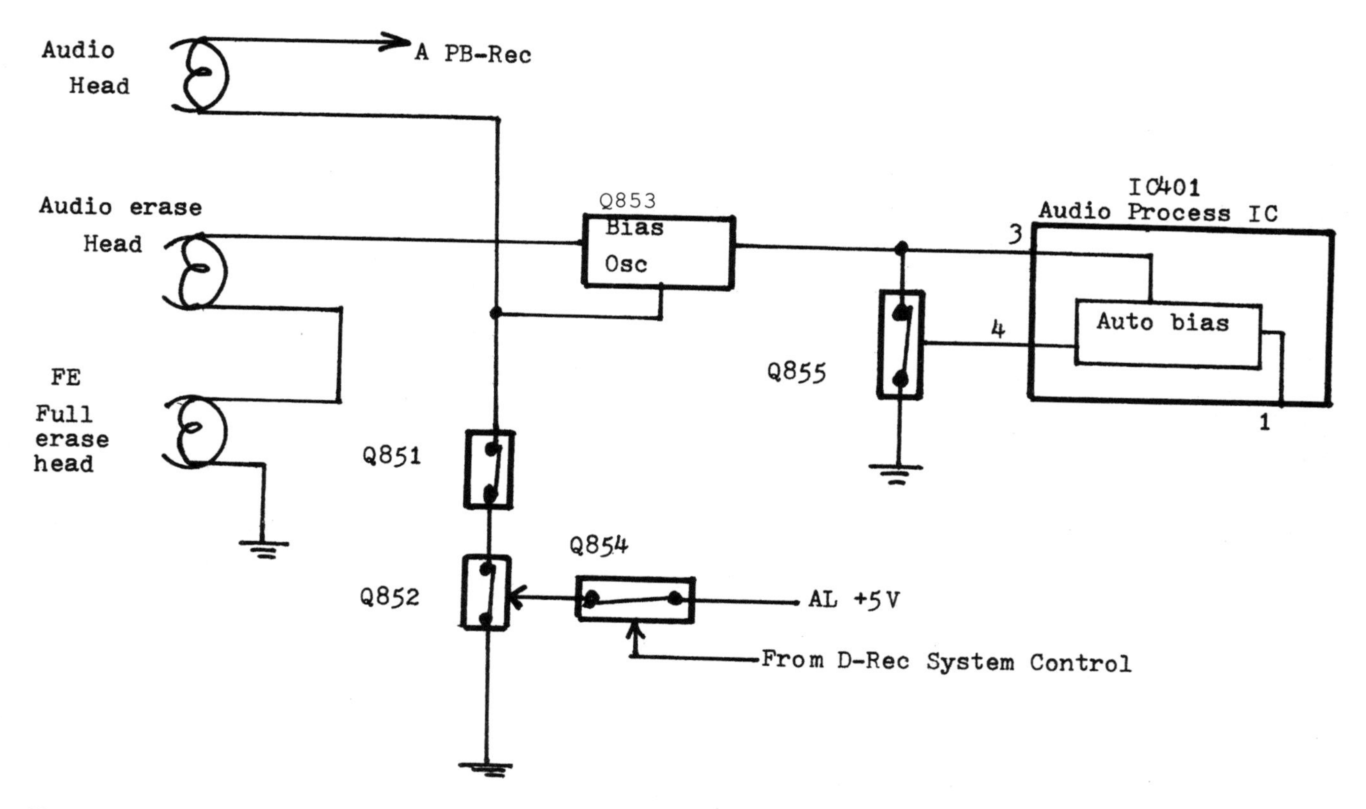

Fig. 15-6. A block diagram of a Sears 9-inch TV/VCR bias oscillator circuit.

Take a scope waveform off of the ungrounded side of the audio erase or full erase head to determine if the bias oscillator circuits are functioning in record mode. Suspect a defective bias oscillator transistor (Q853) improper supply voltage (4.9V), on the collector terminal of Pin 61 of IC401. Check each switching transistor with in-circuit tests when no erase signal is found on the erase heads. Intermittent recording can result from poor transformer terminal connections and an open primary winding of T851. Do not overlook an open erase head for poor recordings or for not erasing the previous recording.

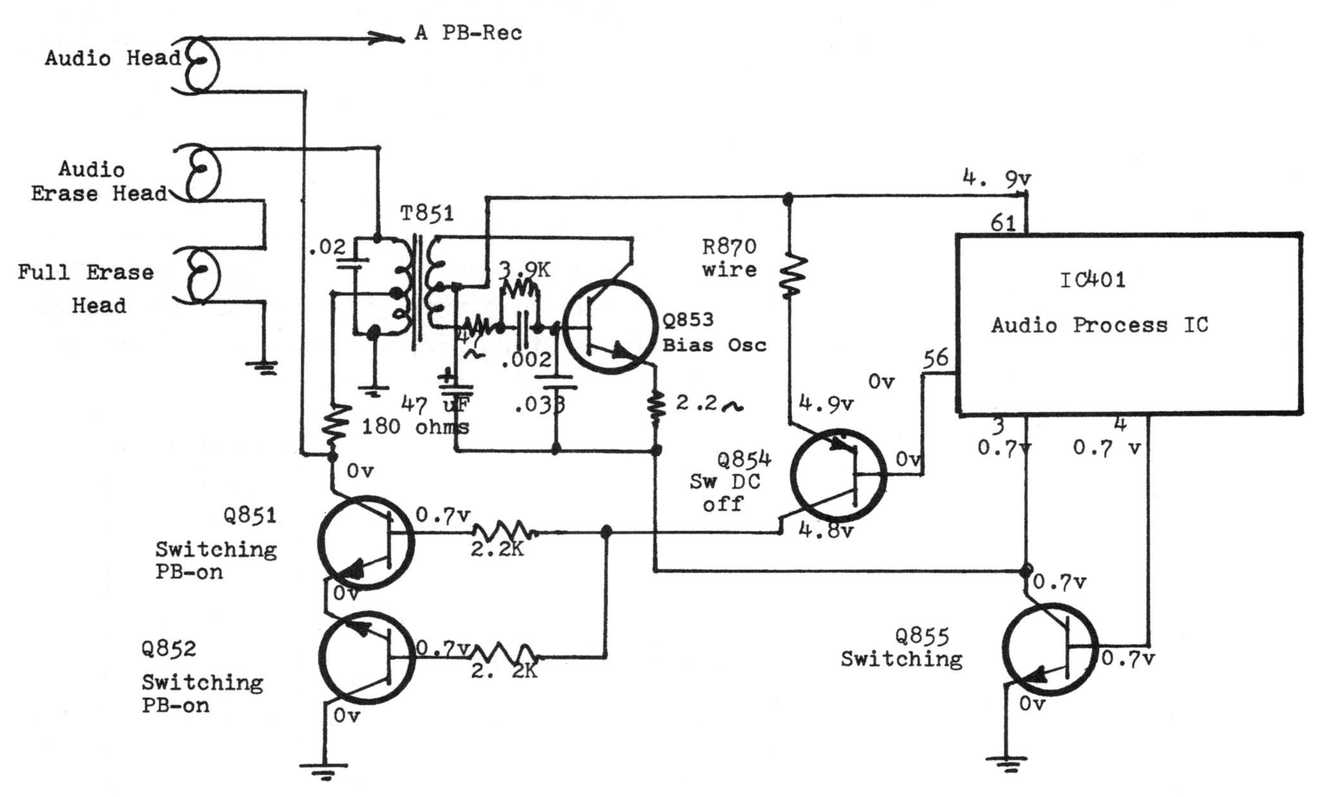

Fig. 15-7. Switching transistor Q855 provides a ground path for the bias oscillator to turn on the oscillator circuits.

The Erase Heads

Most TV/VCR combinations have a full erase head and a stationary audio erase head. The full erase head erases the entire tape and is only on during recording. The erase head takes off any previous recording found on the tape. An audio erase head erases only the audio track found upon the tape. Both erase heads are excited by the bias oscillator circuit. Both erase heads are also found mounted along side the tape path and erase the tape as it passes over the magnetic field of the erase heads.

The recorded sound signal is applied to the edge of the sound track as the tape moves past the magnetic head. In playback, the audio head picks up the recorded sound from the tape as it passes. The ground side of the audio head can also be excited by the bias oscillator circuit, providing linear fidelity to the audio sound tracks.

The control head winding is also included in the same component as the audio heads. The ACE head might include the audio, erase, and control head windings. During recording, the control head winding records a bunch of 30-Hz pulses. These pulses synchronize the video heads during playback so that they pass over the previously recorded video. The control head provides a steady picture on the TV screen. The audio control head should be cleaned at the same time as the video heads to ensure properly recorded picture and audio.

The ACE head unit is connected directly to the servo/system control IC. In a 9-inch Sears TV/VCR combination the control head is mounted inside the audio head and is connected to connectors CN287 and CN851. The control signal CTL is wired to Pins 91 and 92 of IC201 (*Fig. 15-8*).

A misaligned, open, or defective erase head can result in the previous recording not being completely erased when recording a new program, often called cross talk. The video or picture breakup can be caused by the old control track not being completely erased. Noisy sound in the background or a loss of sound indicates the tape was not fully erased. A poor adjustment of the control audio head screw can indicate poor tracking and the picture might break up into several lines.

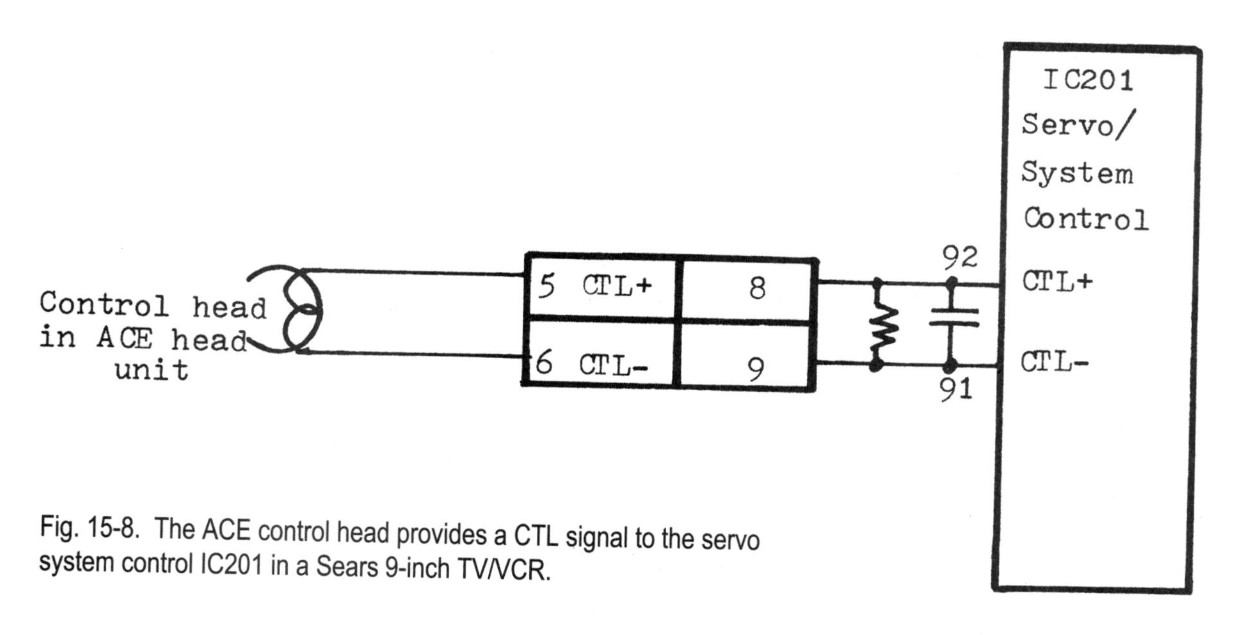

Fig. 15-8. The ACE control head provides a CTL signal to the servo system control IC201 in a Sears 9-inch TV/VCR.

Troubleshooting Audio Playback and Recording Circuits

Troubleshooting the audio playback circuits using a scope and DMM can locate defective components within the sound recording and audio playback circuits. Since the same record and playback circuits are used for both modes, insert a cassette and try to locate the defective component in playback mode. A quick waveform test at Pin 13 of the input select IC (IC701) can indicate if the audio sound is found at the input terminal. Check the waveform at Pin 59 of the audio process IC (IC401) for both playback-audio signal and record-audio signal (*Fig. 15-9*). A quick waveform test with the scope at Pin 64 of IC401 indicates the record-audio signal is being fed to the audio head winding.

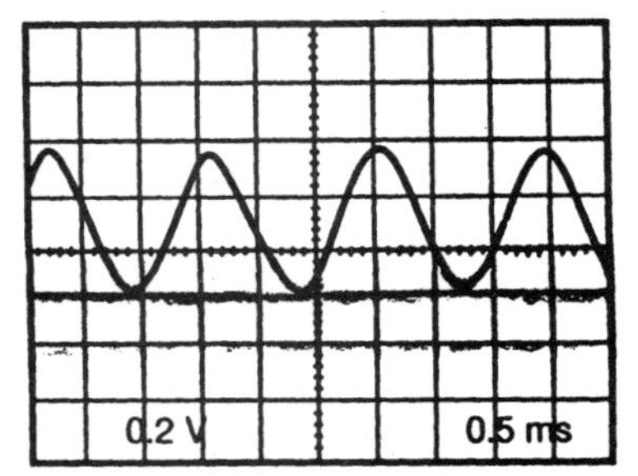

WF9 MAIN 2/3 SCHEMATIC DIAGRAM
IC701 PIN 13

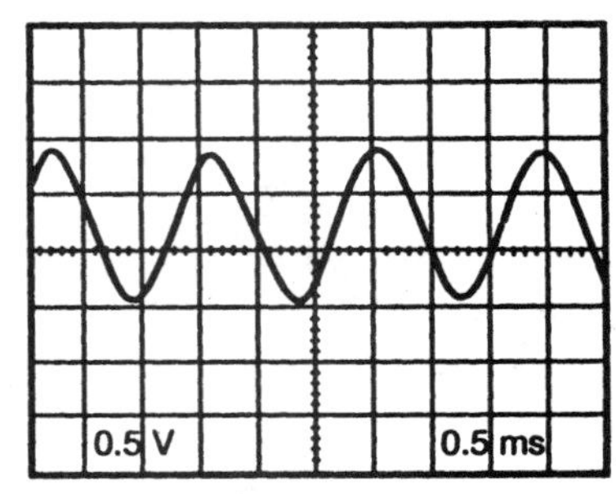

WF10 MAIN 2/3 SCHEMATIC DIAGRAM
IC401 PIN 59

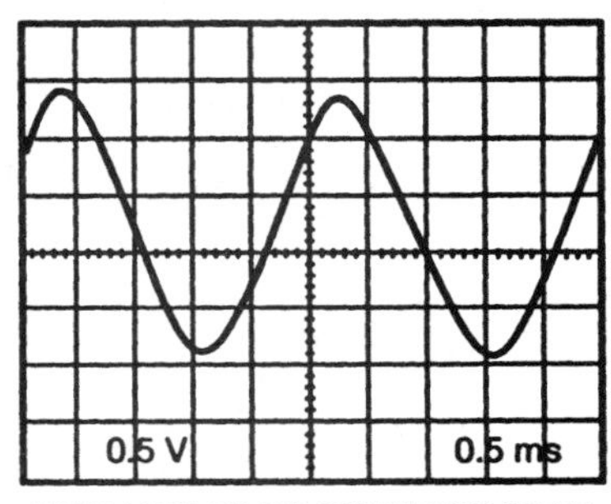

WF11 MAIN 2/3 SCHEMATIC DIAGRAM
IC401 PIN 64

Fig. 15-9. By checking the waveforms in the audio circuits, you can tell where the signal stops.

When the audio-record tape path is normal with no audio recording suspect problems within the bias oscillator circuits or the recording heads. Improper erasing on the audio track can result from an open or defective audio erase or full erase head. A portion of the picture bad and intermittent sound can result from a defective control head winding, inside the audio ACE head.

A low-ohm continuity winding test can indicate if the audio head winding is open. A quick scope test on the audio erase or full erase head can indicate if the bias oscillator circuits are functioning. A poor audio head winding connection or poor connectors and plug-in sockets can cause intermittent audio. Do not overlook a defective switching transistor in the bias oscillator circuits when there is no bias operation and erasing of the cassette.

Audio Power Output Circuits

In most TV/VCR combinations the power output IC is used in both TV sound and playback audio operations. A single IC component is found in most audio output circuits of the TV/VCR units. In a Sears 9-inch TV/VCR the sound input is found at Pin 1 of IC801 through a coupling electrolytic (C802). The amplified audio output is found at Pin 7. The earphone jack is provided for silent listening and when the headphone jack is removed the audio appears over a PM loud speaker. A mute transistor is connected to Pin 3 of IC801 to shut off the unwanted sound.

The audio amp IC (IA01) in an RCA 13-inch TV supply voltage is found at Pin 3. The +13.8 supply voltage is taken from the UA supply source. The TV audio and VCR playback signal is found at the input terminal (8) of the power output IC (IA01). The amplified sound is found on Pin 2 and capacity coupled by a 470 µF electrolytic to the earphone and speaker circuits (*Fig. 15-10*).

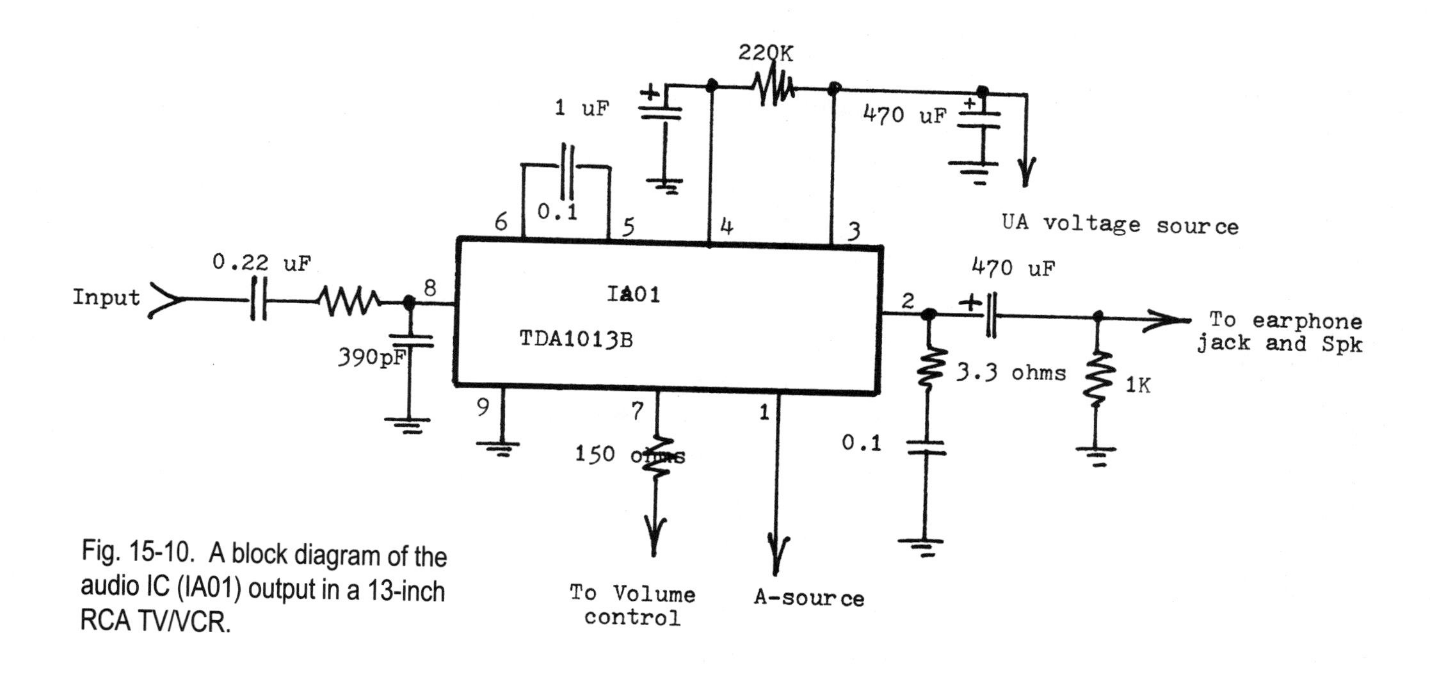

Fig. 15-10. A block diagram of the audio IC (IA01) output in a 13-inch RCA TV/VCR.

Headphone Circuits

Most of the low-priced TV/VCR combinations use a very simple headphone circuit between audio output IC and PM speaker. The audio signal from Pin 7 is capacity coupled by C801 to a self-shorting headphone jack. When the headphone plug is pulled out of the earphone jack (JK801), the audio is switched to the PM speaker. The PM speaker is switched out of the circuit when the earphones are plugged into JK801 (*Fig. 15-11*).

In the RCA 13TVCR60 the headphone jack receives the power output audio through a plug-connector BE01. When the headphones are plugged into jack (BE02), the speaker is disconnected from the sound output circuits. A 180-ohm resistor is inserted into the headphone circuits lowering the sound output. Notice that each headphone winding has a 180-ohm resistor wired in series with the audio output signal (*Fig. 15-12*).

Intermittent headphone operation can be caused by poor switching contacts inside the earphone jack. Spraying cleaning fluid inside the entire jack area and plugging the headphone in and out several times can clean the shorting terminals inside the headphone jack area. A dirty shorting switch can cause the dead sound problem in either the headphone jack or PM speaker. When the earphone jack appears to cause an intermittent audio symptom and will not clean up or has broken internal leaf springs, replace the headphone jack. Sometimes cleaning around the ground or outside terminal of the headphone jack (on the front side) can cure intermittent headphone operations.

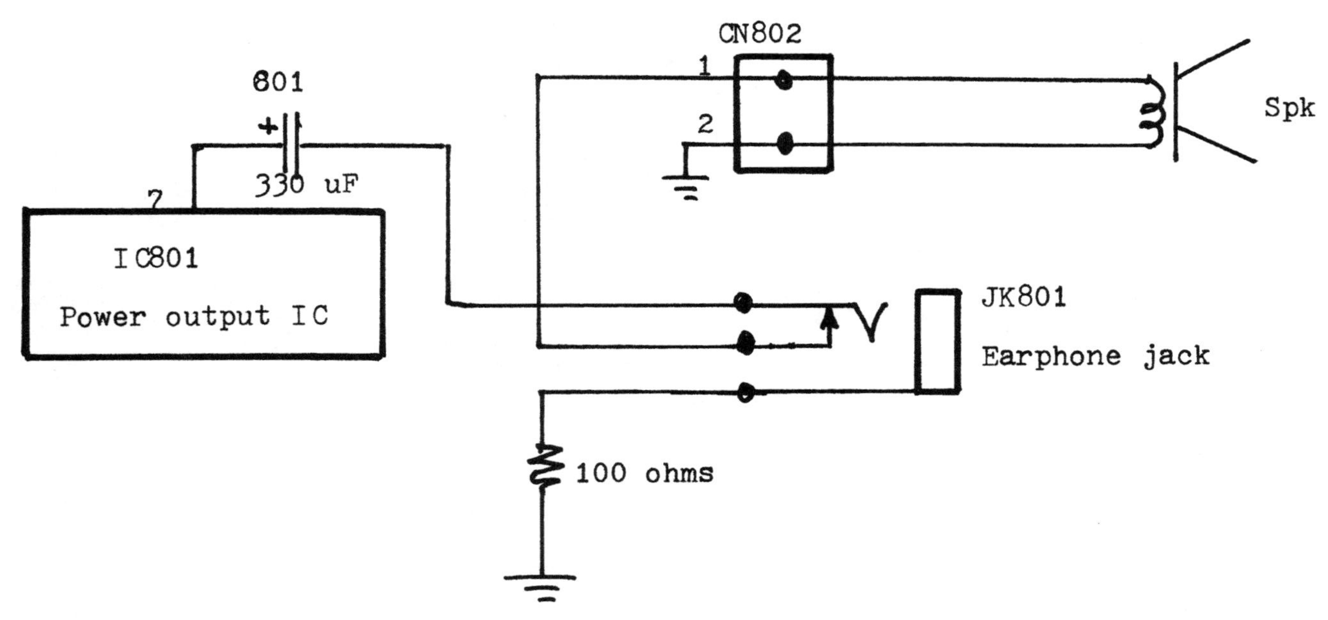

Fig. 15-11. The speaker circuits are switched into the audio output with earphone jack JK801 in a Sears 9-inch TV/VCR.

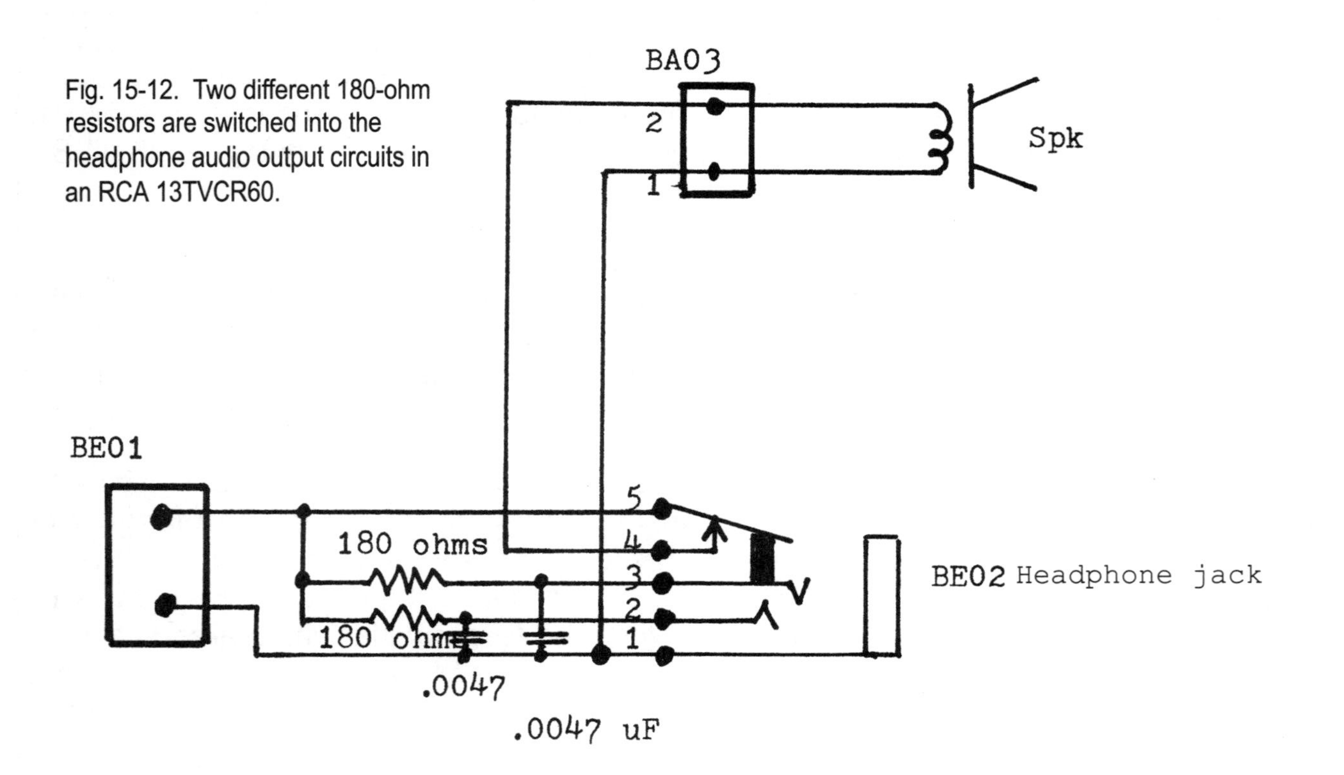

Fig. 15-12. Two different 180-ohm resistors are switched into the headphone audio output circuits in an RCA 13TVCR60.

Troubleshooting the Power Output Audio Circuits

The audio output circuits can be signal traced with the scope or an external audio amplifier. In fact, the playback audio signal can be traced from the audio tape head back to the power output IC using an external audio amplifier. Insert a recorded cassette for audio tests. The audio playback signal is very weak, but can be heard from the audio tape head in the external audio amplifier.

Also, the audio signal from the audio generator can signal trace the audio circuit throughout the audio circuits. Simply inject the audio signal at the audio head and signal trace the audio signal at each audio stage. When the audio stops, you have located the defective circuit or stage. Now take critical voltage tests on audio transistors or IC components.

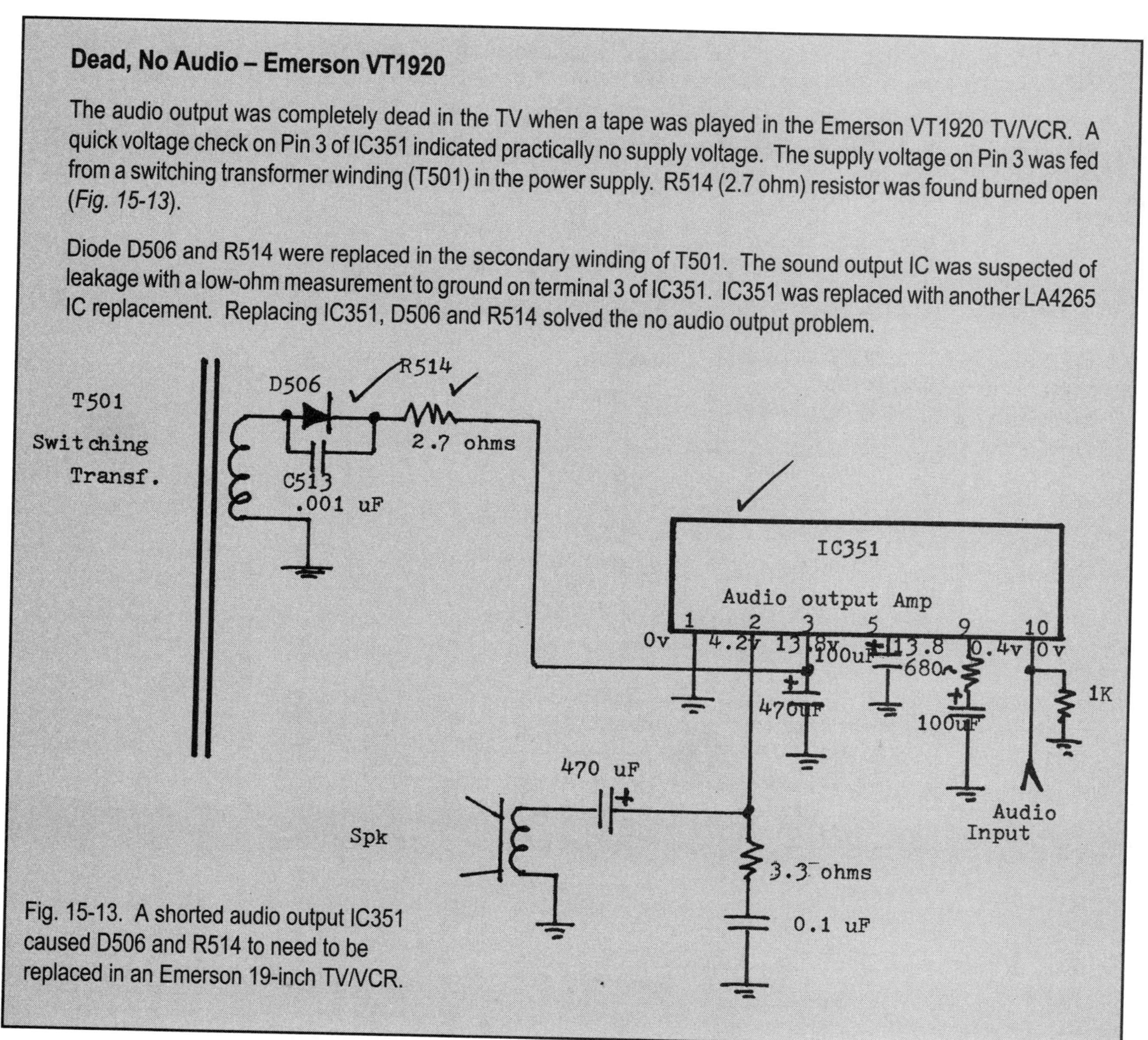

Dead, No Audio – Emerson VT1920

The audio output was completely dead in the TV when a tape was played in the Emerson VT1920 TV/VCR. A quick voltage check on Pin 3 of IC351 indicated practically no supply voltage. The supply voltage on Pin 3 was fed from a switching transformer winding (T501) in the power supply. R514 (2.7 ohm) resistor was found burned open (*Fig. 15-13*).

Diode D506 and R514 were replaced in the secondary winding of T501. The sound output IC was suspected of leakage with a low-ohm measurement to ground on terminal 3 of IC351. IC351 was replaced with another LA4265 IC replacement. Replacing IC351, D506 and R514 solved the no audio output problem.

Fig. 15-13. A shorted audio output IC351 caused D506 and R514 to need to be replaced in an Emerson 19-inch TV/VCR.

Signal trace the audio to Pin 5 of the audio process IC401. Recheck the audio playback signal at Pin 59 of IC401 and onto the AF amp (Pin 50) of the input select IC301. Again, signal trace the playback audio signal to the input terminal of the audio output IC801 and to the speaker.

The defective audio output circuit can cause garbled audio, distorted audio, no audio, weak audio, and intermittent audio at the PM speaker. Suspect a leaky audio output IC for weak and distorted recording. A leaky output IC can lower the supply voltage (Vcc), at the supply pin of the audio output IC. Often the leaky or shorted output IC and a bad coupling electrolytic can cause weak and intermittent sound in the speaker. Check the mute transistor or switching of the mute circuits when there is no sound at the speaker.

No Audio – Panasonic PVM2021

In a Panasonic PVM2021, when the volume control was accidentally turned up very high, the sound stopped. Initially, the speaker was suspected of an open cone winding, but it tested okay. The audio was signal traced to the input terminal 2 of IC4151 (*Fig. 15-14*). No sound was found at the output pin (8). The supply voltage was measured low at Pin 9. IC4151 (AN5265) was replaced with a NTE1663 universal sound output IC and the sound returned to normal operation.

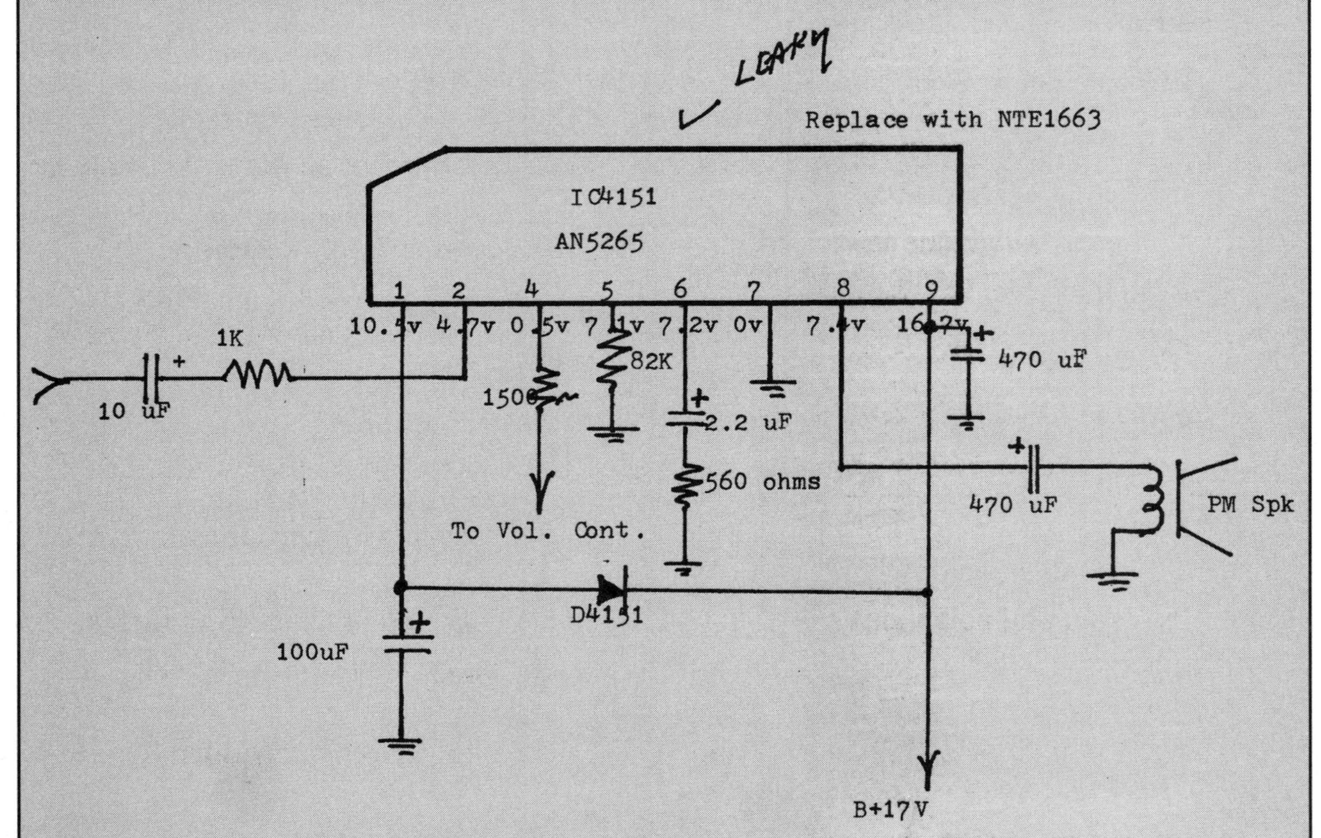

Fig. 15-14. A leaky IC4151 output IC caused a no audio symptom in a Panasonic PVM2021 TV/VCR. It was replaced with a universal NTE1663 replacement part.

Intermittent or No Audio and Video Playback

- ✔ Check for a shorted 47-μF, 6.3-volt electrolytic and burned resistor when there is no picture or sound in playback mode.
- ✔ A shorted 220-μF, 6.3-volt electrolytic on the video head amp board can cause no audio or video in playback mode.
- ✔ For no audio or video playback resolder the loose connections on the fuse and chassis connectors.
- ✔ Check for open or leaky transistors in the +12-volt regulator on the main PCB of the power supply section for no video and audio in playback and tuner.
- ✔ Distorted audio and video in playback can result from a defective 330-μf, 18-volt electrolytic on the UNSN 5-volt line in the power supply.
- ✔ Do not overlook an open fusible resistor with no audio and playback operation.
- ✔ Suspect a defective servo IC for no audio or video in the playback mode.
- ✔ Poor audio and video in playback can result from a defective 1000-μF, 16-volt capacitor in the power supply.
- ✔ For no audio and video in playback mode, check for a defective 12-volt transistor regulator.
- ✔ Check the auto process IC for no audio in playback mode.
- ✔ A bad input IC switch can cause no video and sound in playback mode.
- ✔ Suspect a defective SMD playback switch transistor on the video head assembly for no video and audio in playback mode.
- ✔ Intermittent video and audio in playback can be caused by badly soldered connections on the input video IC.
- ✔ Check the 10-μF, 16-volt electrolytics on CTL pulses for poor video and audio at all speeds when there are lines in the video.
- ✔ Resolder loose line connectors for no audio or video in the VCR.

Mechanical Audio and Video Playback Problems

- ✔ Suspect a defective transistor on the playback 5-volt switch for no video and audio in the playback mode.
- ✔ A no video and audio problem in the VCR can result from a bad fuse connection.
- ✔ No audio or video in playback and eating tape can be caused by a bad pinch roller bracket.
- ✔ A broken or loose supply tension band can cause a no video and distorted audio symptom.
- ✔ Intermittent audio and video and no rewind at tape end can be caused by a defective master cam gear.

- ✔ A defective pinch roller bracket can cause no audio or video in playback mode.
- ✔ Check for a bad cam and mode switch when audio and video are intermittently unstable.
- ✔ Replace a bad brake assembly if SP picture goes from normal to snow and the audio growls then returns to normal again.

Intermittent, Bad, or No Audio in Playback

Possible Causes:

- ✔ Open coil in the audio process circuits
- ✔ Bad connector (resolder all pins)
- ✔ Open ground on audio head
- ✔ Defective or bad control head of the audio/CTL or full erase head
- ✔ Intermittent audio in playback can be caused by a defective wire on the audio/CTL control head.
- ✔ Broken wire on the audio heads.
- ✔ Defective electrolytic capacitors
- ✔ Open 10-μF electrolytic in the audio process circuits can cause no audio in playback.
- ✔ Open 1-μF, 50-volt capacitor on the audio main PCB
- ✔ Defective 4.7-μF, 5-volt electrolytic on the main PCB
- ✔ Open 1.5-ohm fusible resistor in the audio 12-volt line
- ✔ Open or dried-up electrolytics on the 5-volt line
- ✔ Defective transistor regulator of audio 9-volt line
- ✔ Defective transistors in the audio section on the main PCB
- ✔ Defective 12-volt regulator transistor
- ✔ Flat pack IC on the capstan phase error line
- ✔ Defective audio processor IC
- ✔ Audio process IC (audio fades out after playing for a short while)
- ✔ Dirty or worn switching transistor
- ✔ Dirty or worn audio relay switch terminal

Audio Recording Problems

- ✔ Check the audio processor IC for no audio record mode.
- ✔ Suspect a SMD audio defeat transistor on the foil side for no audio record with normal tuner video.
- ✔ No audio record and no full erase can be caused by a leaky IC on the record 9-volt line to the bias circuits.

- ✔ Check for a leaky zener diode on the 9-volt main PC, for no audio record with normal video playback.
- ✔ A defective bias oscillator transistor can cause no audio record mode; take a scope waveform on the audio heads to check it.
- ✔ No audio record can be caused by a defective audio process IC or a bad input select IC.
- ✔ The VCR will not record audio with a leaky 0.47-μF capacitor on the 50-volt line.
- ✔ Suspect a bad A/C head assembly for no audio in record mode.
- ✔ No audio record or playback can occur with an open 3.3- to 10-μF, 50-volt coupling capacitor in the audio head circuit.
- ✔ Check for a leaky or shorted 0.1-μF, 50-volt capacitor off of the audio process IC when there is no audio record or play back.
- ✔ A defective 3.3-μF, 50-volt electrolytic on the audio record playback module can cause no record or playback.
- ✔ A poor audio recording can be caused by an open 470-μF, 16-volt electrolytic in the power supply secondary winding.
- ✔ Replace the erase head for no audio recording.
- ✔ No audio record can be caused by a bad full erase head.
- ✔ Check for a bad plug on the full erase head for intermittent audio recording.
- ✔ Poor audio recording can result from an open bias oscillator transformer.
- ✔ Intermittent audio recording can be caused by poor connections on the bias oscillator transformer.
- ✔ Suspect a bad record safety tab switch when audio cuts in and out during recording.
- ✔ A bad 4.5-MHz ceramic filter can cause poor audio recording or no recording of audio.

Audio Erase Problems

- ✔ When the audio erase or full erase heads will not erase the previous recordings, check the bias oscillator circuits and for defective erase heads.
- ✔ A shorted 220-μF, 10-volt electrolytic on the audio/video board can cause a no audio erase symptom.
- ✔ An open bias oscillator transistor can cause no erase of the previous recording.
- ✔ Replace the open audio erase head for no erase in record mode.
- ✔ A bad erase head harness can cause no erase of the previous recording.
- ✔ A bad soldered connection at the full erase head can cause no erase of the previous recording.
- ✔ Poor bias oscillator winding connections can cause intermittent audio record symptom.

- ✔ Check for improper voltage at the bias oscillator transistor caused by a defective regulator IC in the record 9-volt line for no audio record and no full erase of the ACE head.
- ✔ Check for open low-ohm resistors in the bias oscillator supply voltage for no audio erase.

Weak or Low Audio

- ✔ Suspect open or loss of capacity in electrolytic coupling capacitors for no or weak audio.
- ✔ Check for very low open coupling capacitors (0.47 μF, 25 volts) for very low audio in the speaker.
- ✔ Test each filter capacitor with the ESR meter for ESR problems and weak audio.
- ✔ Low or weak audio can result from a defective input select or audio output IC.
- ✔ Sub another large coupling capacitor across the speaker electrolytic for weak or no audio.
- ✔ A defective full erase head can cause low or no audio in the speaker.
- ✔ A change of resistance in resistors in the audio output circuits can cause low or weak audio symptom.
- ✔ Low and muffled audio can result from a badly soldered connection on the jack-switch at the rear panel.
- ✔ Low or intermittent audio can result from a bad R/P switch.

Intermittent Audio

- ✔ A leaky audio transistor can become extremely hot and shutoff the sound after 15 minutes of operating.
- ✔ Check for a defective audio output IC that is overheating and causing the sound to cut in and out.
- ✔ A defective mute transistor can cause intermittent sound in the speaker.
- ✔ Suspect a defective connector when intermittent sound occurs.
- ✔ The speaker cuts in and out when there is a defective 10-μF, 50-volt electrolytic coupling capacitor.
- ✔ A defective audio relay can cause an intermittent sound symptom.
- ✔ Clean the audio switching relay contacts if audio fades in and out.
- ✔ Resolder all jumper wires for intermittent audio in the audio circuits.
- ✔ Do not overlook a bad speaker voice coil or poor speaker connections for intermittent audio.
- ✔ Dirty headphone jack switching points can cause intermittent sound in the speaker.

Intermittent Sound – Symphonic TVCR901

In a Symphonic TVCR901 the TV/VCR would play for 15 or 20 minutes and then the audio would become intermittent. Sometimes when this occurred, it would shut off, but not shutdown. Q1012 (2SA1318) was found to be leaky when the transistor got hot, but checked okay out of the circuit.

Distorted Audio

Most of the audio distortion problems are related to the audio output stages. A leaky audio output IC can cause weak and distorted sound. Check for a defective output IC for frying, popping noises, and distortion in the speaker. A leaky input select IC can cause no audio or distorted audio. An open or leaky electrolytic coupling capacitor from IC to speaker can cause distorted audio.

Check for a change of resistance of low-ohm resistors within the audio output circuits if the sound is distorted. Leaky or shorted electrolytics in the power audio output IC circuits can also cause distorted audio. A speaker with a dropped or frozen cone can cause a distorted sound (*Fig. 15-15*).

Suspect a leaky line voltage IC regulator for distorted audio after warm up. Low supply voltage to the audio output IC can be caused by open or large electrolytics in the power supply.

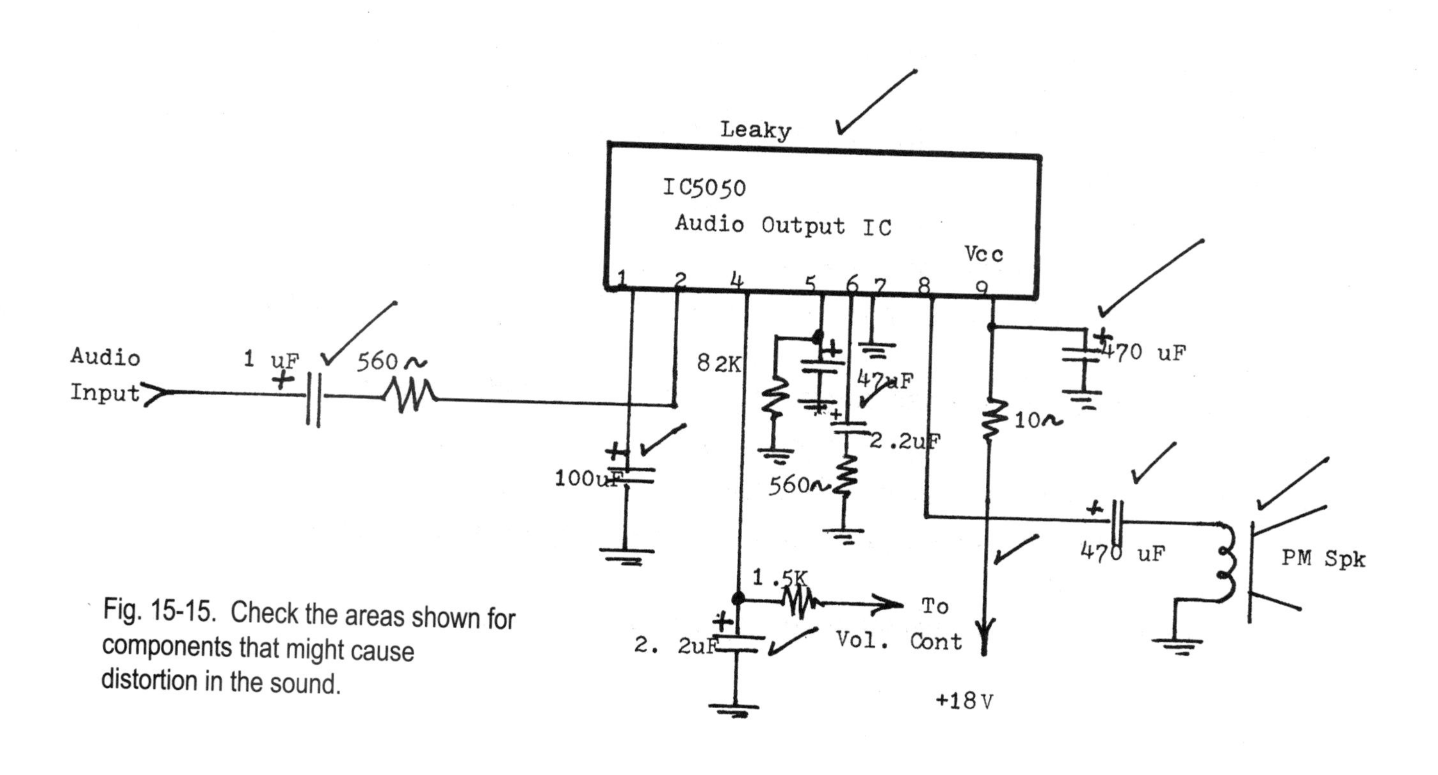

Fig. 15-15. Check the areas shown for components that might cause distortion in the sound.

Garbled, Motorboat-Type, and Squealing Sounds – Zenith SMV-1341S

A loud burping noise occurred when the set was plugged in a Zenith SMV-1341S TV/VCR. Sometimes the chassis would start up and at other times it was dead. Replacing the 6-volt regulator IC1003 (AN7806F) cured the intermittent start-up problem. IC1004 and IC1502 were replaced to cure the burping sound problem.

A defective 33-µF, 6.3-volt electrolytic in the servo circuits can cause erratic and garbled audio. A bad servo IC can cause a garbled sound in EP mode. A 33-µF electrolytic in the servo circuit phase error line can cause erratic and garbled audio in playback.

Although a very loud squealing sound came from inside another Zenith SMV-1341S TV/VCR, the noise was caused by a horizontal driver transformer and not actually in the audio section. The TV and VCR sections both were fine except for the squealing noise. Replacing the horizontal driver transformer (T9001) cured the loud squealing noise.

Muffled audio can result from a glazed pressure roller in the VCR. Low and muffled audio can be caused by a poor jack-switch on the rear panel. Check for small electrolytic capacitors (10 µF, 16 volts) in the audio section for distorted audio hum and hum in the playback mode. Sound can be distorted and garbled when the automatic head cleaner assembly leaves a residue on the heads.

Noisy Audio

- ✔ Poorly soldered connections on a coil in the audio circuits can create a static noise.
- ✔ Bad relay contacts can cause a cracking or static sound in the speaker.
- ✔ Erratic and noisy audio can result from a bad switching relay; clean up the relay points or replace it.
- ✔ Poor coil board connections can cause static even when the unit is in mute mode.
- ✔ Noise in the audio can be caused by a 2200-µF electrolytic in the motor 12-volt regulated supply or if there is a bad capstan motor.
- ✔ Internal arcing of bypass capacitors can cause a loud static noise in the audio.
- ✔ Resolder poor connectors for a loud buzzing noise in the sound.
- ✔ Install star washers on the ground screens and shields when there is a buzzing noise in the audio.
- ✔ Retighten metal screws on the preamp shields for noise in the audio.
- ✔ When a noise is generated as the capstan motor rotates, try to shunt the motor terminals with a 100-µF, 35-volt capacitor to ground to help eliminate the noise from the motor or replace the capstan motor.
- ✔ A clutch assembly that is too tight can cause audio drag and take-up reel quivers.
- ✔ A bad audio/CTL head can cause noisy audio in the playback mode.
- ✔ A loud static noise can be caused by a defective audio process IC.
- ✔ Weak audio with a popping noise can be caused by a hot audio output IC.

- ✔ The rushing noise in the audio can result from a defective audio ceramic filter.
- ✔ An intermittent buzzing noise can be caused by an intermittently functioning transistor in the sound circuits.
- ✔ Check for a defective or dirty relay points when the audio squeals in playback.
- ✔ Check for badly soldered jumper wire connections when there is static in the audio.
- ✔ A bad audio play switch can cause hissing or a buzz in play mode.

Humming Noise

- ✔ Replace large filter capacitors (2200 to 4700 μF, 35-volt) within the power supply circuits when hum bars appear in the picture and a there is a buzz in the audio.
- ✔ Check for a badly soldered joint or connections on large filter capacitors for a loud hum; replace all electrolytics in the low-voltage power supply that have ESR problems.
- ✔ Check small electrolytics (10 to 47 μF) in the audio circuits for low hum occurring in the speakers.
- ✔ Poor grounds and shields can cause a low hum in the front end of the audio circuits.
- ✔ A defective audio output IC can cause a squealing and hum noise.
- ✔ The defective audio record/playback process IC can cause hum and a buzz in the audio playback mode.
- ✔ Poor switching in record and audio playback can cause a humming noise in the speaker.
- ✔ Check the defective servo IC for wow and flutter in the sound because wow and flutter in the audio can be caused by a defective servo IC. Snow bands moving through the picture are also common with this problem.
- ✔ Check for open capacitors in the servo IC circuit for wow and flutter problems.

Poor Audio Muting

The audio mute system in the Sears 13-inch TV/VCR originates from Pin 3 of the audio output IC. The actual muting of the sound output is controlled by the system control microprocessor (IC201). Q801 takes the mute control signal and shuts off the audio inside IC801 (*Fig. 15-16*). The mute button on the front of the TV/VCR or on the remote can turn off or mute the sound before it goes to the speaker.

A defective sound output IC can cause no muting of the sound in the TV/VCR section. Test the audio mute transistor with in-circuit tests for open or leaky conditions. A shorted or leaky mute transistor can cause no sound in the speaker. Suspect a defective system control IC if the mute signal is not reaching the muting components. Check the remote control when the remote will not turn the speaker audio off. Sometimes a weak battery in the remote control transmitter will not activate the remote control mute button.

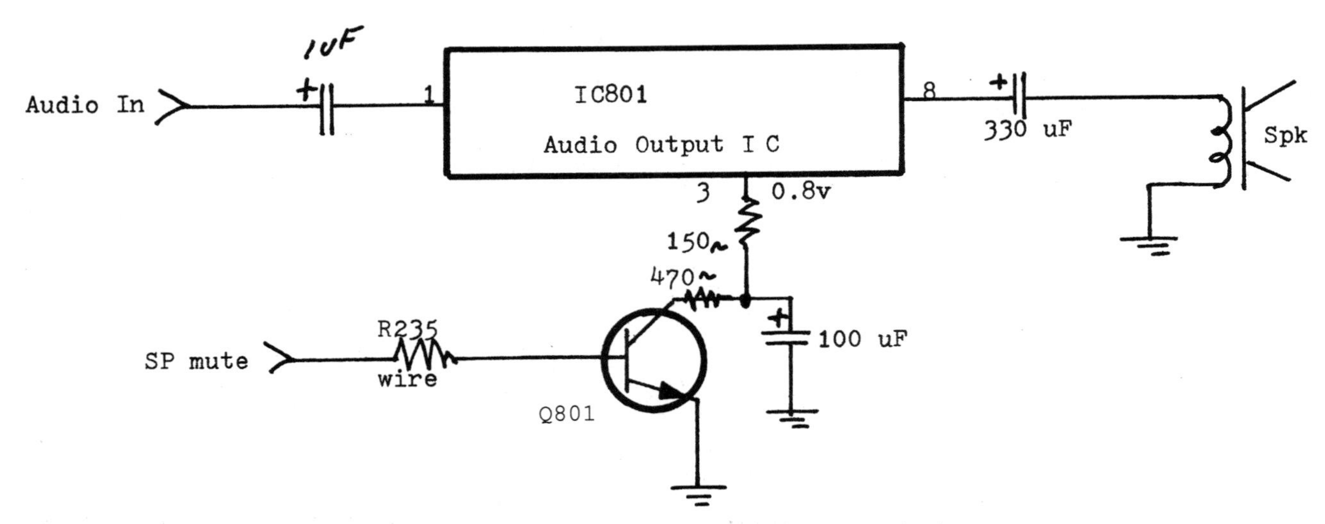

Fig. 15-16. Q801 controls the audio mute circuits in a Sears 13-inch audio output circuit.

Intermittent audio muting can be caused by a defective main microprocessor in the VCR circuits. Check for a defective volume control transistor located in the mute transistor circuit when there is no muting of the sound. Bad foil connections within the audio output circuits can cause the muting to shut off the sound at times. Intermittent muting of the sound can be caused by bad soldered connections in the mute circuits.

Switching Transistor Problems

There are many switching transistors and diodes found in the present day TV/VCR combination's circuits. For example, the Sears 13-inch TV/VCR part list of the video/ audio signal process circuits is loaded with switching diodes. Switching transistors can be located in many TV/VCR circuits. Often switching transistors provide the turn-on of the voltage source, ground returns, or switching of the different circuits.

Switching transistor (Q851) switches the B+ voltage to the bias oscillator circuit and to the erase heads in record mode. Transistors Q854 and Q855 switch the playback circuits on. Q852 provides turning the record and the +5 volts off of the bias recording circuits. Switching transistor (Q801) shorts out the audio to ground in the muting circuit. Switching transistors Q852 and Q802 are transistors with built in resistors and are shown with a square box in the schematic.

Suspect a defective switching diode or transistor when the circuit does not function properly. Most switching diodes and transistors can be checked with an in-circuit test using the DMM or on a transistor or diode tester. Intermittent operations caused by a switching diode can be checked by removing one terminal from the circuit and checking the diode out of the circuit. The intermittent diode can be located with several coats of freeze spray. Simply replace the intermittent diode if in doubt. Replace all critical switching diodes and transistors with the original part number.

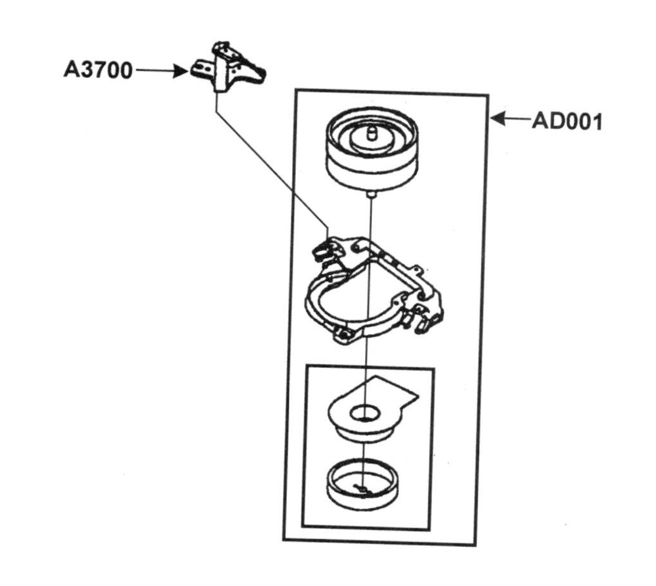

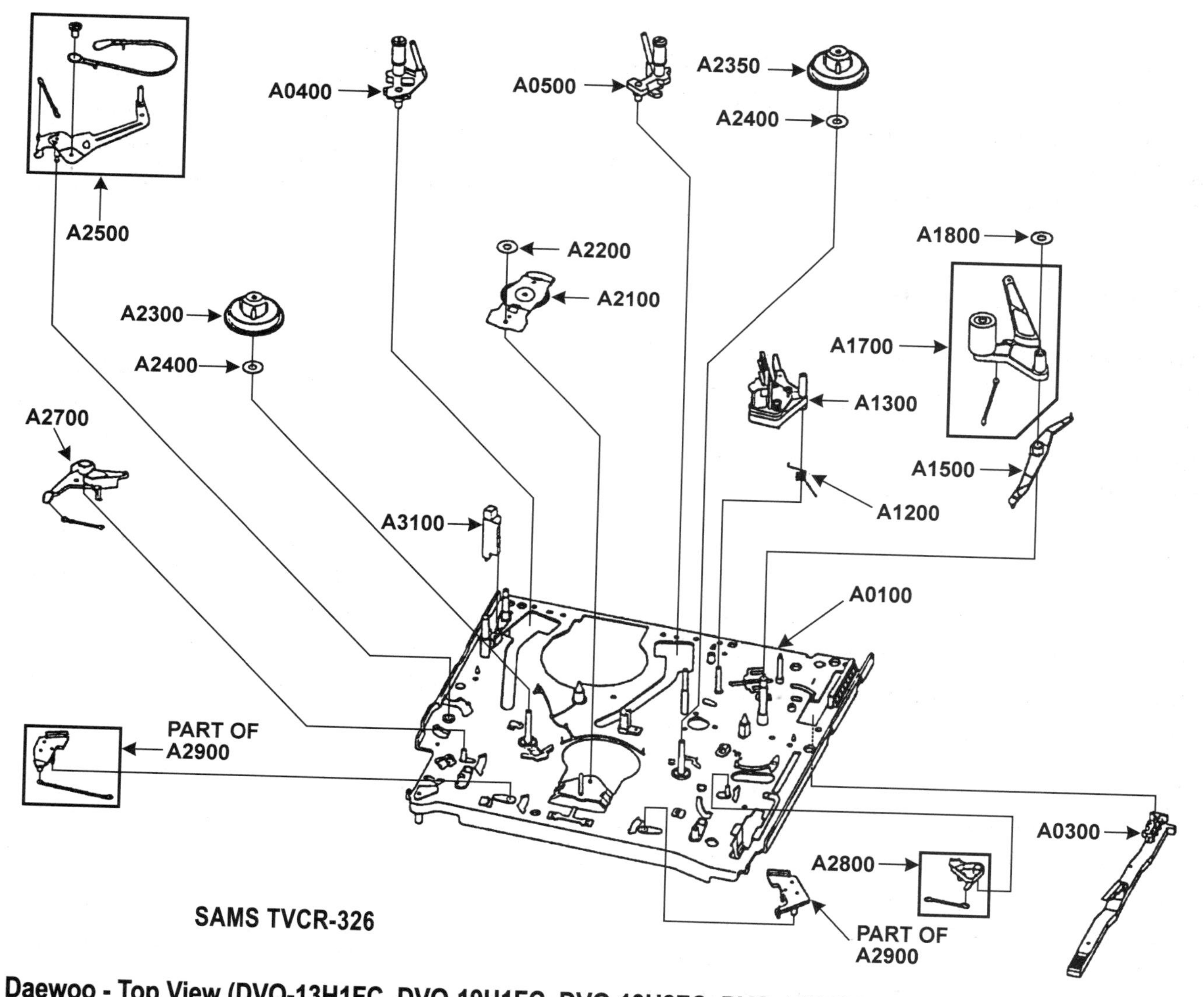

SAMS TVCR-326

Daewoo - Top View (DVQ-13H1FC, DVQ-19H1FC, DVQ-13H2FC, DVQ-19H2FC)

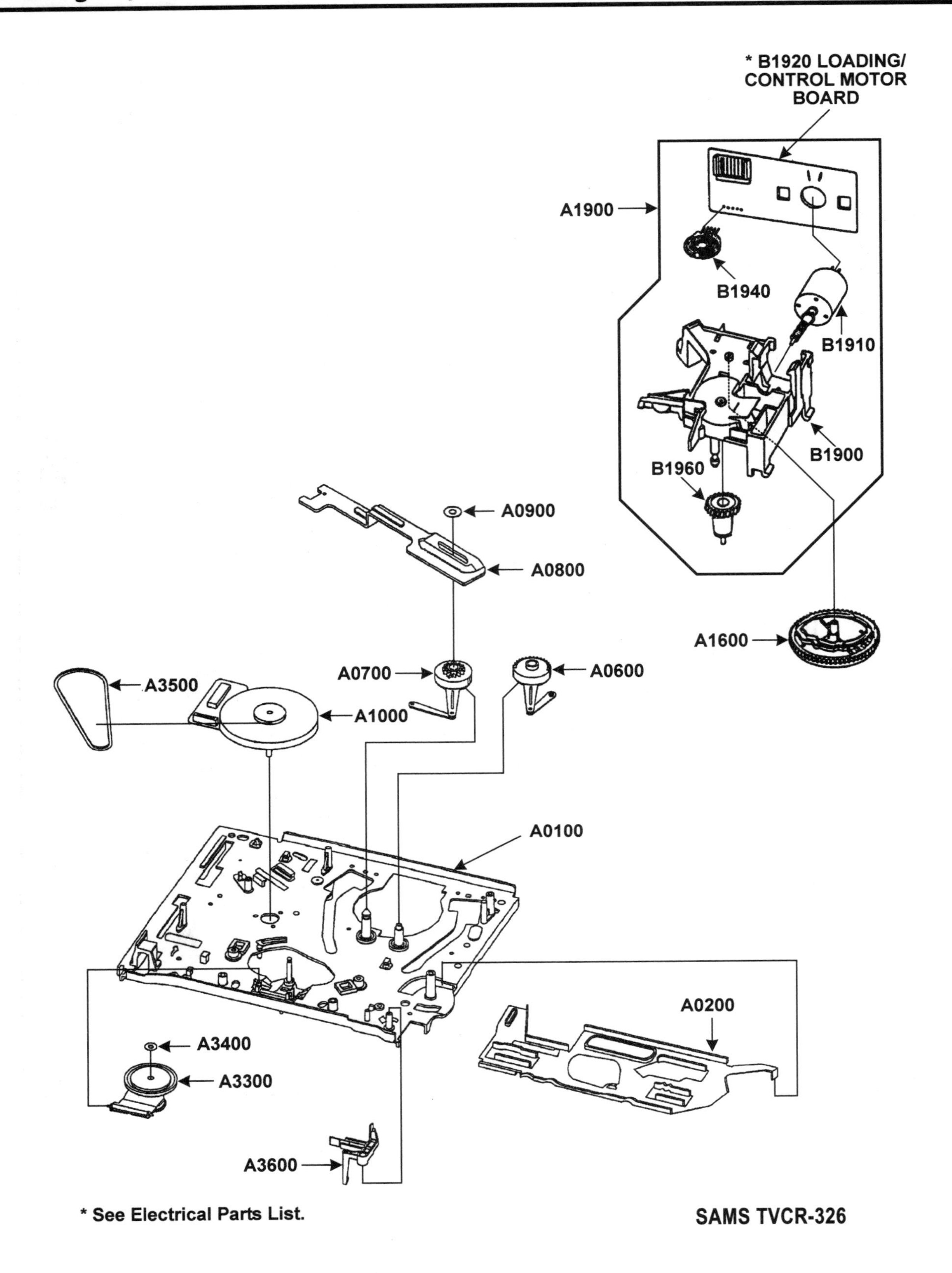

Daewoo - Bottom View (DVQ-13H1FC, DVQ-19H1FC, DVQ-13H2FC, DVQ-19H2FC)

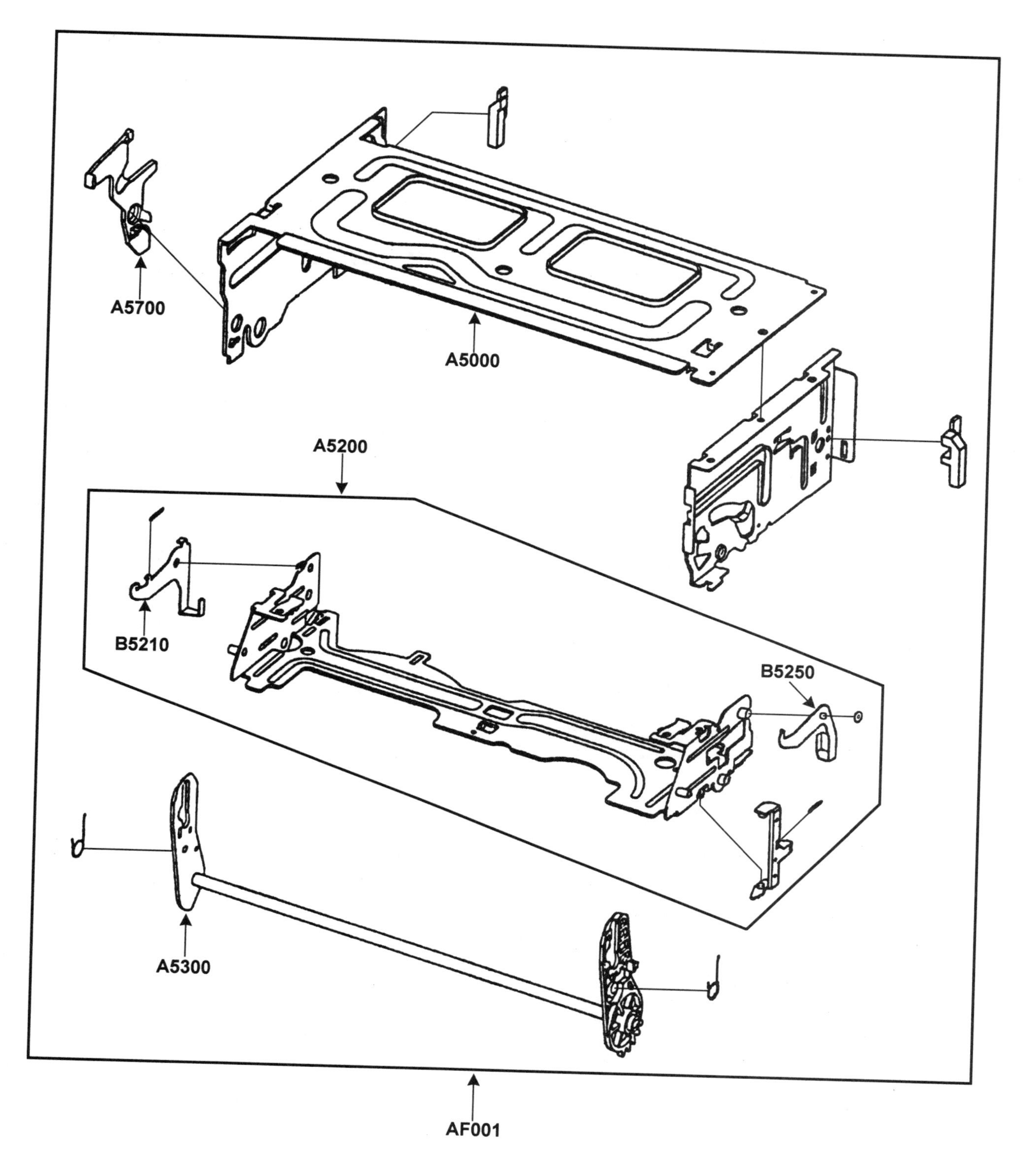

SAMS TVCR-326

Daewoo - Tape Loading Mechanism (DVQ-13H1FC, DVQ-19H1FC, DVQ-13H2FC, DVQ-19H2FC)

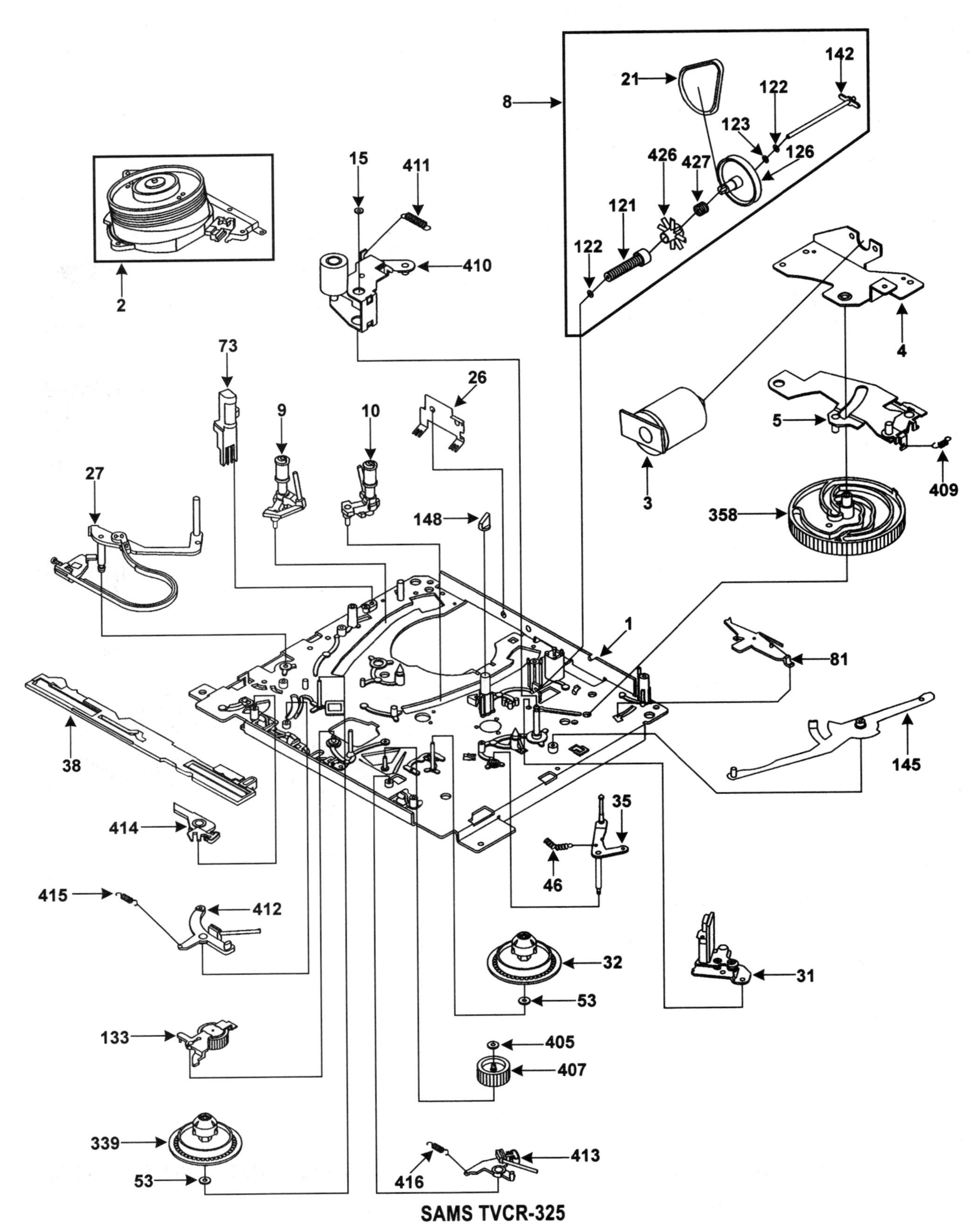

SAMS TVCR-325

Philips Magnavox - Top View (CCZ190AT31, CCZ191ZT31, CCZ192ZT31, CCZ194ZT31, CCZ196ZT31)

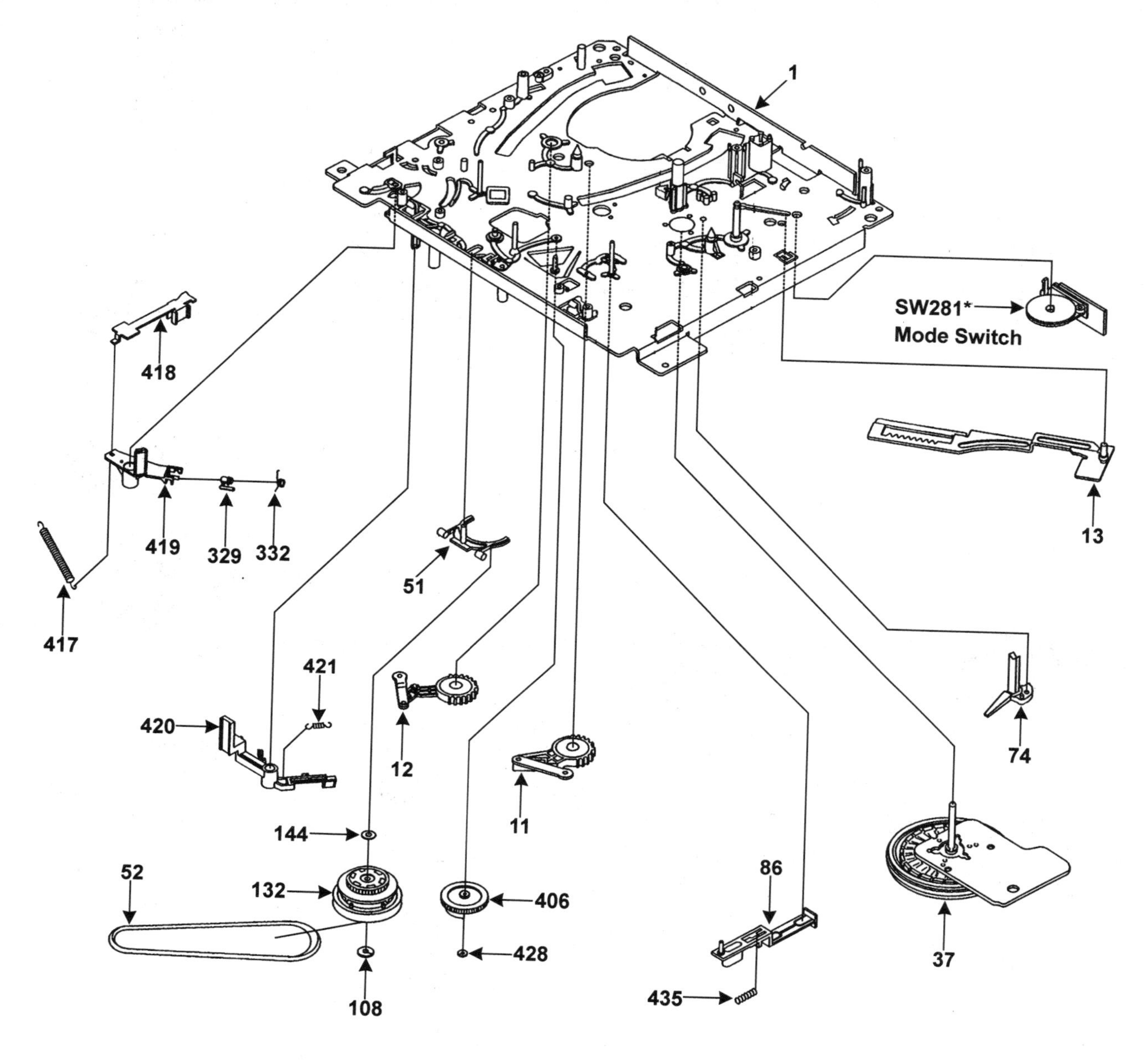

*See Electrical Parts List.

SAMS TVCR-325

Philips Magnavox - Bottom View (CCZ190AT31, CCZ191ZT31, CCZ192ZT31, CCZ194ZT31, CCZ196ZT31)

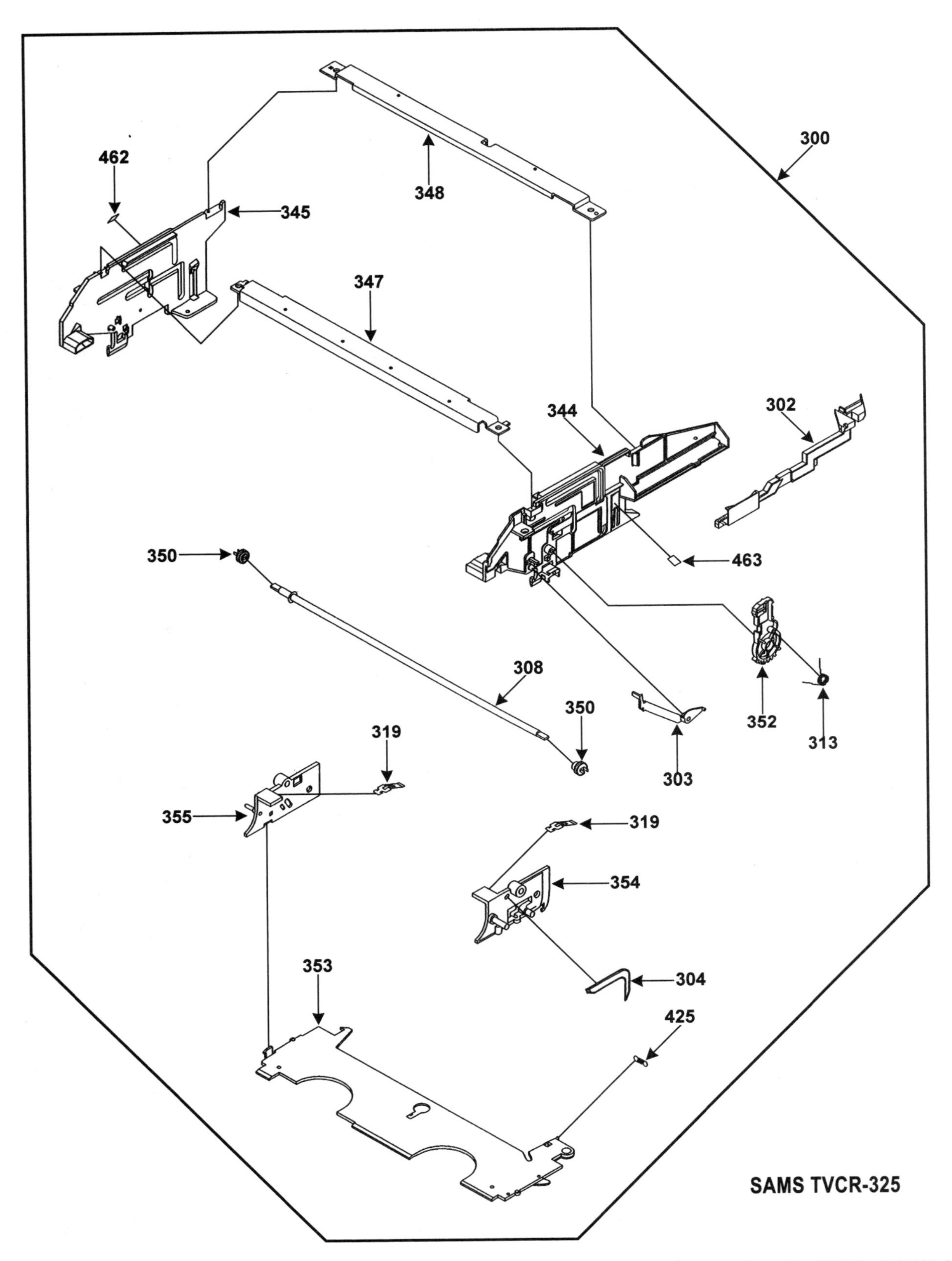

Philips Magnavox - Tape Loading Mechanism(CCZ190AT31, CCZ191ZT31, CCZ192ZT31, CCZ194ZT31, CCZ196ZT31)

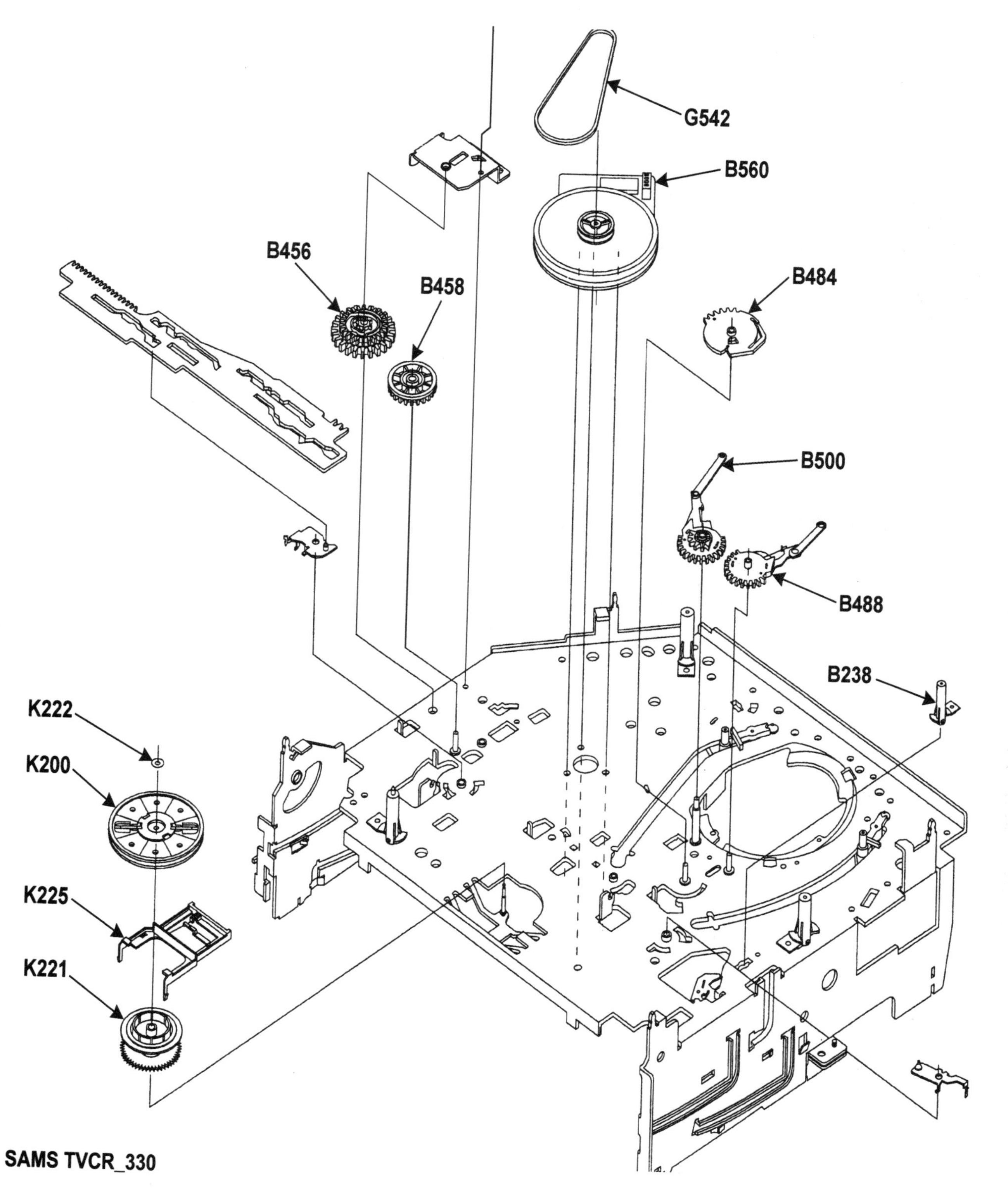

SAMS TVCR_330

Samsung - Top View (CXJ1352, CXJ1353, CXJ1331, CXJ1931)

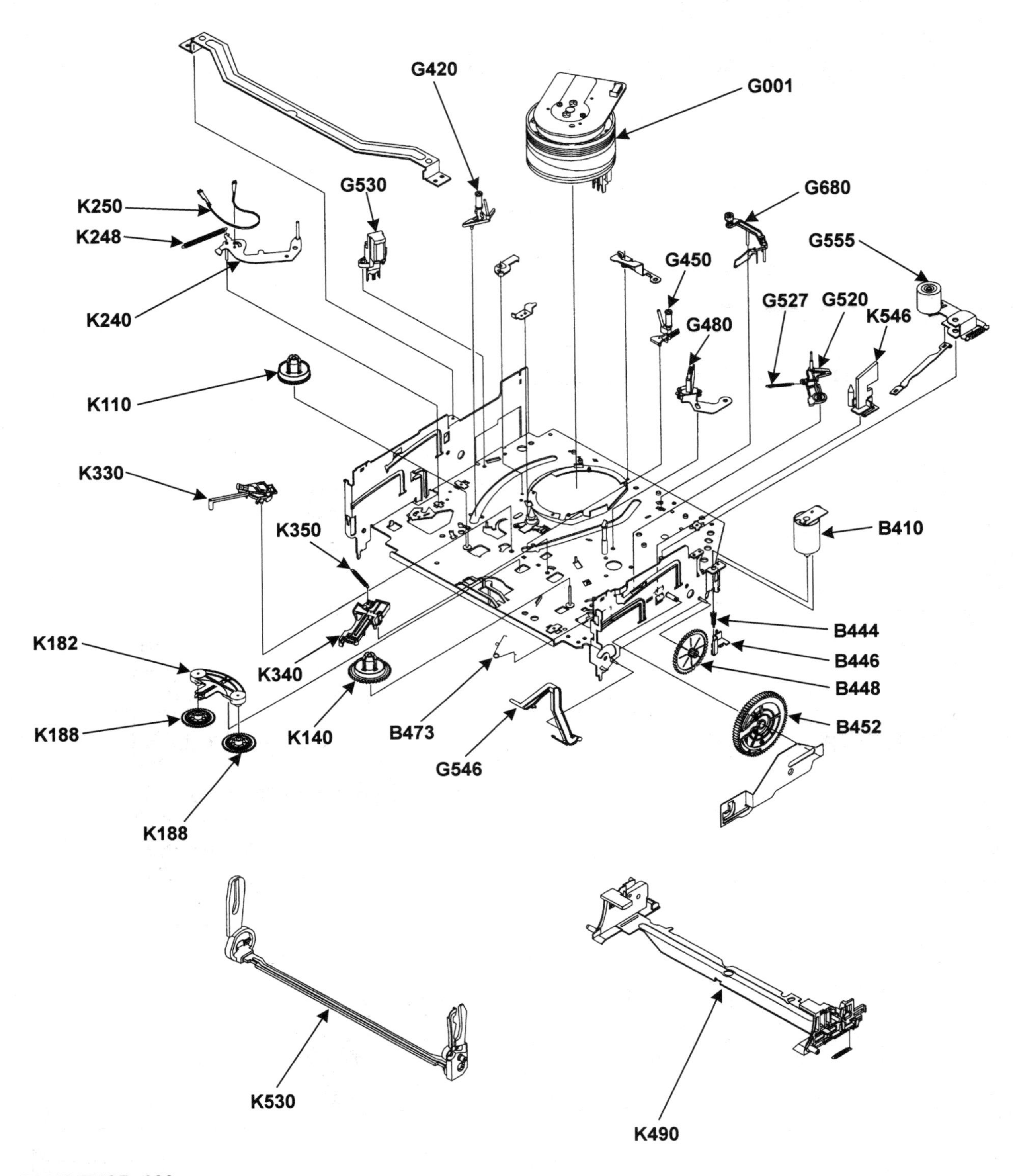

SAMS TVCR_330

Samsung - Bottom View, Casette Loading Assembly Included (CXJ1352, CXJ1353, CXJ1331, CXJ1931)

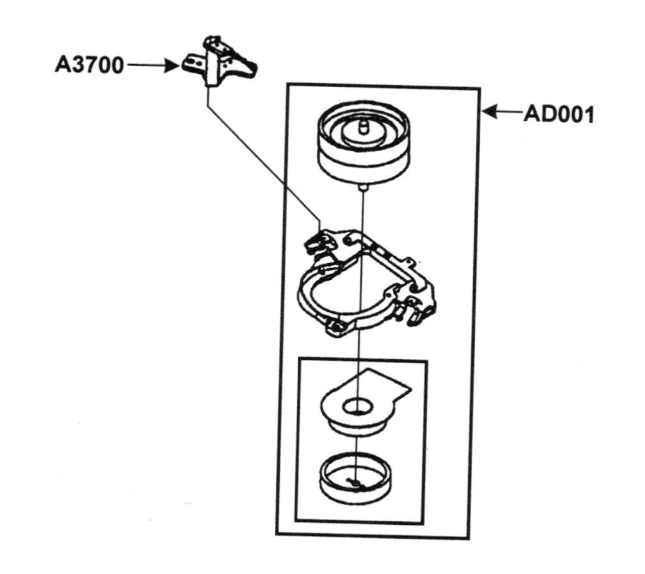

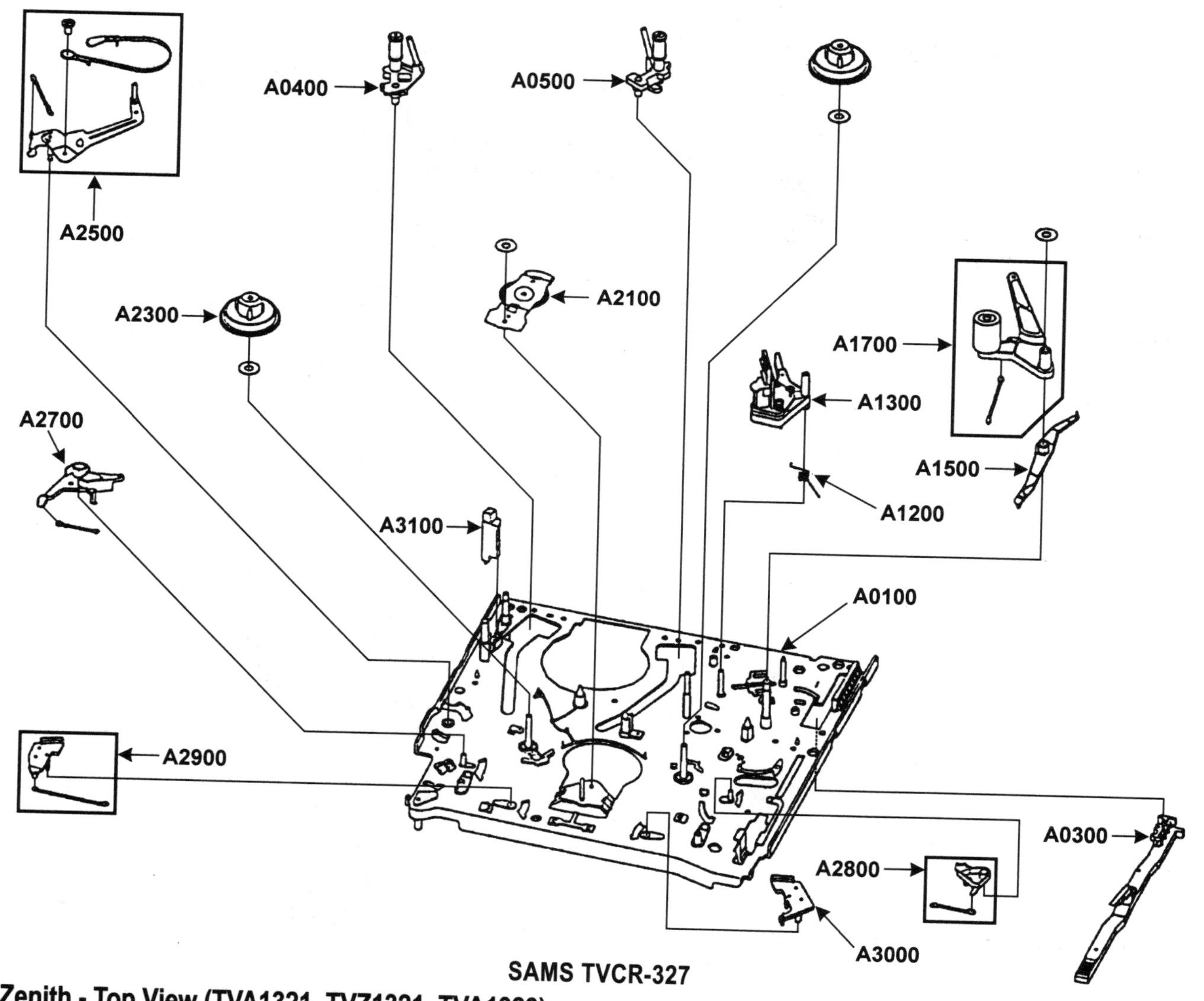

SAMS TVCR-327

Zenith - Top View (TVA1321, TVZ1321, TVA1923)

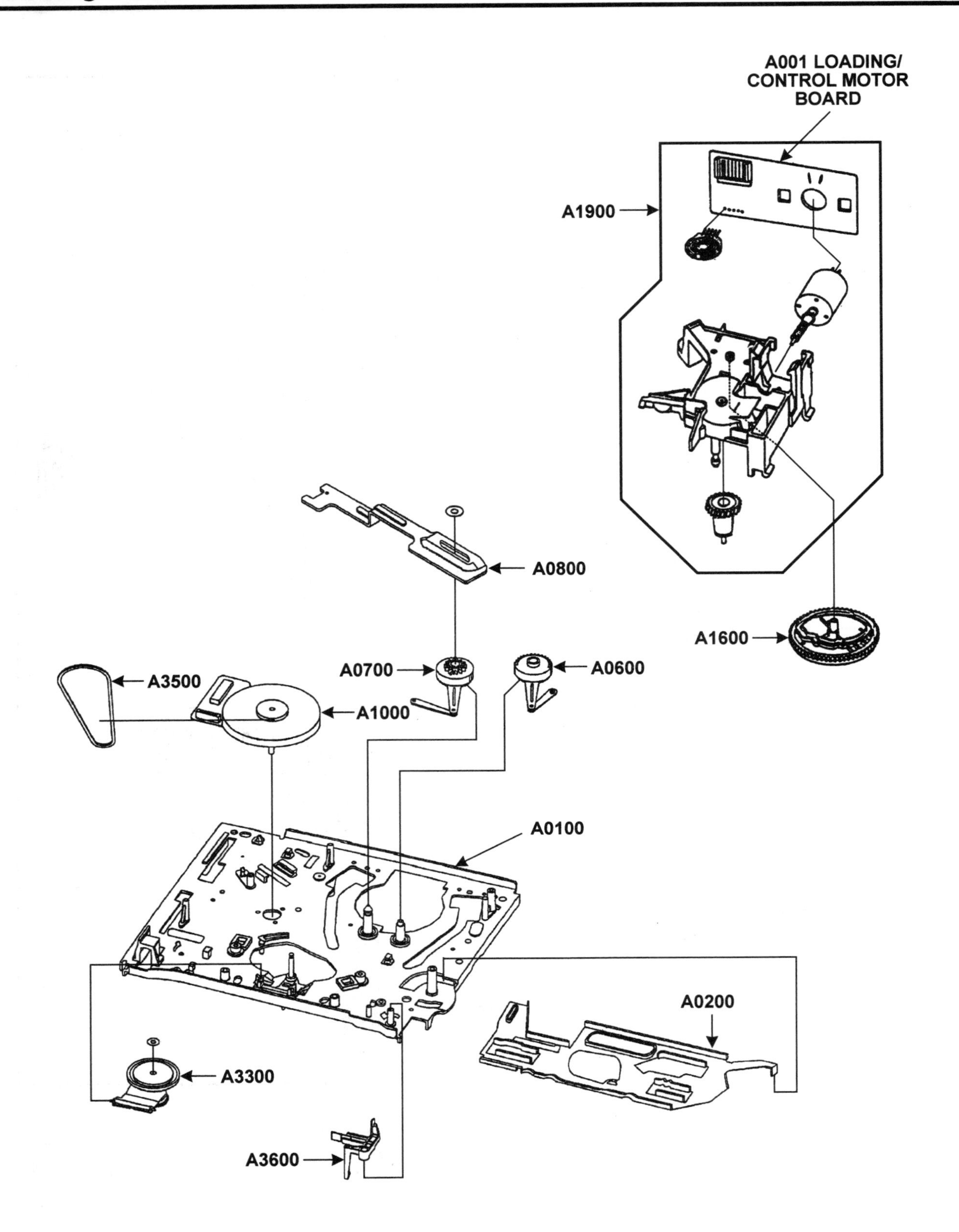

SAMS TVCR-327

Zenith - Bottom View (TVA1321, TVZ1321, TVA1923)

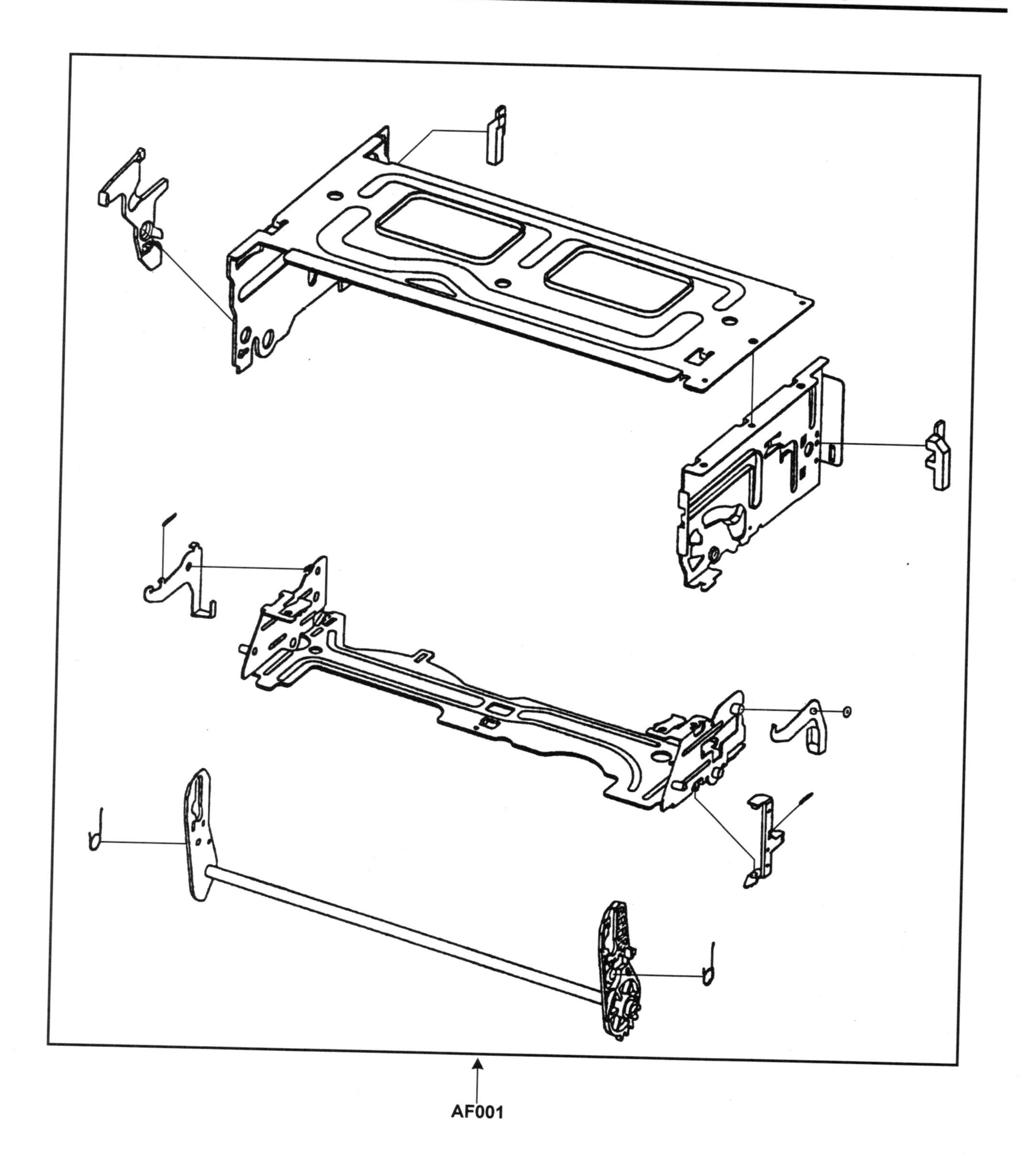

SAMS TVCR-327

Zenith - Tape Loading Mechanism (TVA1321, TVZ1321, TVA1923)

Glossary of Abbreviated Terms

A

AC Alternating Current
A/C Auto Control
ACC Automatic Color Control
ACK Automatic Color Killer
AE Audio Erase
AE Head Audio Erase Head
AFC Automatic Frequency Control
AFC DET Automatic Frequency Control Detector
AFT Automatic Fine Tuning
AFT DEF Automatic Fine Tuning Defeat
AGC Automatic Gain Control
AGC DET Automatic Gain Control Detector
ALC Automatic Level Control
APC Automatic Phase Control
ASB Assemble Mode
ASSY Assembly
AUX Auxiliary
A. Head Audio Head
A. Mute Audio Mute

B

B. EMPH Burst Emphasis
BGP Burst Gate Pulse
BIAS ADJ Bias Adjustment
BOT Beginning of Tape
BPF Band Pass Filter
BU Back Up
BUF Buffer

C

CASE Cassette
CAP Capstan
CAP F/R Capstan Forward/Reverse
CAP HI Capstan High
CAP MID Capstan Medium
CAP VCC Capstan Motor Supply Voltage
CARR Carrier
CCD Charged Coupled Device
CH Channel
CHROMA Color
CLK Clock
CLOCK Clock (Syscon to Servo) (Sy-SE) CONV Converter
CPM Capstan Motor
CTL Control
CY Cylinder
CYL-M Cylinder Motor
CYL SENS Cylinder Sensor

D

DC Direct Current
DD UNIT Direct Drive Motor Unit
DEM De-emphasis
DET Detector
DEV Deviation
D GND Digital Ground
DN Down
DPG Drum Pulse Generator
DPU Drum Pick-up
DRUM F/F Drum Flip/Flop

E

E-E Electronic-to-Electronic / Electric-to-Electric
EF Emitter Follower
EOT End of Tape
EP Extended Play
EQ Equalizer
EXT External

F

FBC Feedback Clamp

FG Frequency Generator

FGB AMP Amplifier Frequency Generator

FL SW Front Loading Switch

FM Frequency Modulation

FM DLM FM Demodulation

FM MOD FM Modulator

FSC Frequency Sub Carrier

G

GAIN EQ Gain Equalizer

GCA Gain Control Amplifier

GEN Generator

GND Ground

H

HD Horizontal Deflection

HP Headphone

HPF High Pass Filter

HSS Horizontal Sync Separator

I

ID Indent Pulse

IF Intermediate Frequency

INST Insert Mode

INT Interrupt or Internal

IR Infrared Rays

INV Inverter

L

LIM Limiter

LM Loading Motor

LNC Line Noise canceller

LOAD Loading

LP Long Play

LPF Low Pass Filter

LUMI or LUMA Luminance or brightness

LUMA Color

M

MIX Mixer

MOD Modulator

MPX Multiplexer

M.BRAKE Main Brake

MS SW Mechanism State Switch

N

NC No Connection

NC MIX Noise Cancel Mixer

NOR Normal

NR Noise Reduction

O

OSC Oscillator

OREQ OR Equivalent

OSD On-Screen Display

OSP On-Screen Programming

P

PB AMP Playback Amplifier

PB CTL Playback Control

P.CON Power Control

PD Phase Detection

PG Pulse Generator

P/F Frequency Picture Intermediate

PLL Phase Lock Loop

PNR Peak Noise Reduction

PREAMP Preamplifier

PRE-PB Pre-Playback Voltage

PS Power Supply

PWM Pulse Width Modulation

PWM TV Pulse Width Modulated

R

RCH	Right Channel
RCR	Remain Chroma Reduction
RC/PC	Record Control/Playback Control Signal
REC-C	Recording-Chrominance
REC-Y	Recording-Luminance
RECT	Rectifier
REC-VR	Record Volume Control
REEL BRK	Reel Brake
REEL S	Reel Sensor
REG	Regulator or Regulated
REJ	Reject
RF	Radio Frequency
R.INH	Record Inhibit

S

SAW	Surface Acoustic Wave
S.CLK	Serial clock
S/K CLK	Servo-to-Keyboard Clock
SCL	Serial Clock Input
S.COM	Sensor Common
SDA	Serial Data
SEC/SEP	Separator
SEG	Segment
SER	Search Mode
SI	Serial Input
SIF Frequency	Sound Intermediate
SLP	Super Long Play
SO	Serial Output
SP	Standard Play
SUB LPF	Subcarrier Low Pass Filter
SW	Switch
SWNC	Switch Noise Compensator
S/N	Signal-to-Noise Ratio
SYNC SEP	Sync Separator

T

TP	Test Point
TR	Transistor

V

VCC	Voltage Supply
VCO	Voltage Controlled Oscillator
VD	Vertical Drive
VDD	Voltage Supply
VIF	Video Intermediate Frequency
VP	Vertical Pulse
VR	Variable Resistor
V-REF	Voltage Reference
Vs	Voltage Ground
VSS	Voltage Sync Separator
Vss	Voltage Super Source
V-SYNC	Vertical Sync
VT	Voltage Tuning or Tuning Voltage

Y

Y/C	Luminance/Chrominance

Acknowledgements

Special thanks goes to Sams Technical Publishing's **TVCRfacts** and to the many electronic technicians whose TV and VCR problems, symptoms, and repairs appear throughout these pages: Tom Krough, Tom Rich, Richard Martin, and David Held.

A great deal of thanks also goes to Michael B. Danish of Mikes Repair Service at PO Box 217, Aberdeen Proving Ground, MD 21005, for the availability of *TV Case Histories* and *TV/VCR Repair Problems*.

Without the help of these outstanding electronics technicians, this book could not have been written.